Introduction to Numerical Analysis

International Mathematics Series

Consulting Editor: A Jeffrey, University of Newcastle upon Tyne

Titles in the series

Discrete Mathematics: Numbers and Beyond
S Barnett

Complex Variables and their Applications
A Osborne

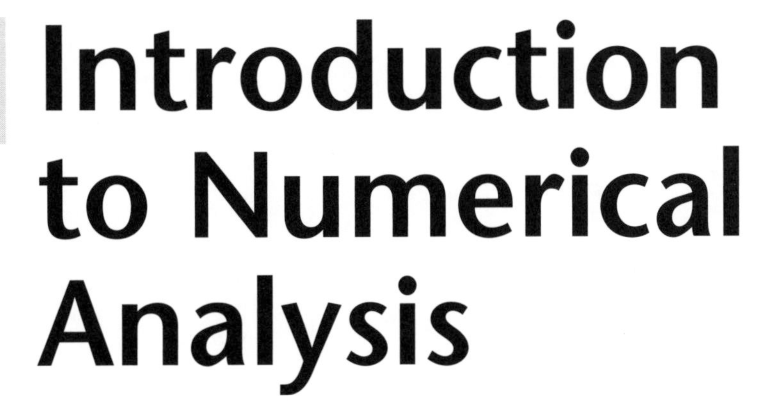

Introduction to Numerical Analysis

ALASTAIR WOOD

School of Computing and Mathematics
University of Bradford

ADDISON-WESLEY

HARLOW, ENGLAND ■ READING, MASSACHUSETTS ■ MENLO PARK, CALIFORNIA
NEW YORK ■ DON MILLS, ONTARIO ■ AMSTERDAM ■ BONN ■ SYDNEY ■ SINGAPORE
TOKYO ■ MADRID ■ SAN JUAN ■ MILAN ■ MEXICO CITY ■ SEOUL ■ TAIPEI

Addison Wesley Longman Limited
Edinburgh Gate
Harlow
Essex CM20 2JE
England

and Associated Companies throughout the World.

Cover designed by OdB Design and Communication, Reading, UK.

Typeset by 59

Produced by Addison Wesley Longman Singapore (Pte) Ltd., Printed in Singapore

First printed 1999

ISBN 0-201-34291-X

British Library Cataloguing-in-Publication Data
A catalogue record for this book is available from the British Library

Contents

Preface

Numerical Analysis, in all its disguises (computational mathematics, numerical methods, quantitative techniques, and so on), is probably one of the more challenging mathematical subjects put before A-level and university students.

The concept of numerical approximation is not an easy one to grasp. To follow the process of replacing problem A (the *exact* problem with solution A*) by problem B (the *model* problem with solution B*) and then of obtaining an approximate solution, say C*, to B, is difficult enough. What is then required is to relate solution C* to the desired (but unknown) solution A*.

The intention of this text is to base the material on a diet of numerical methods that might reasonably be digested by a typical first-year university undergraduate student who is taking a full-year introductory course on numerical methods, with the exception of spline interpolation and the numerical solution of ODEs. This description includes students of mathematics, of engineering and of other subjects having a significant numerical content. Such a reader will utilise most of the text, as opposed to 'the odd chapter'. The more knowledgeable reader will spot the material we have omitted in order to maintain this focus – such as non-linear systems, BVPs, 'best' approximation (with the exception of discrete least-squares analysis), matrix eigenvalues and eigenvectors, and so on. For similar reasons, the subjects covered are taken to the level required to develop realistic working algorithms, but not far beyond. For example, numerical quadrature stops at Newton-Côtes rules and Romberg integration (Gaussian rules, adaptive rules, multi-dimensional integration, and so on are omitted), and the numerical solution of ODEs treats only Runge-Kutta methods and their analysis. References are given to more in-depth descriptions of specific topics and to areas not covered in this text.

Perhaps the greatest help given to the subject in recent years has come from the rapid advances made both in computer hardware and in software applications. Generic packages such as *DERIVE*, *MatLab*, *Maple* and *Mathematica*, to name but a few, provide excellent interactive environments for implementing algorithms, producing graphical output, and executing the symbolic operations associated with mathematics but, until recently, not normally associated with computers. It is tempting to let the software dictate the subject and design material that, ultimately, ends up selling features of the software. This book attempts instead to use software where appropriate to enhance the understanding of a mathematical concept or to implement a particular numerical algorithm.

It is for this purpose that the reader will encounter a number of *DERIVE Experiments* in the text[1]. The functions described in the text are available from the publishers, Addison Wesley Longman, via their website `http://www.awl-he.com/`

The direction of this text is towards the practical, and it is hoped that the *DERIVE Experiments* will stimulate the reader to experiment further. The facility to 'suck it and see' is a powerful tool for discovering the limitations of numerical techniques, and a range of them are developed and implemented here. Nonetheless, the text is designed to retain a classical feel in that the supporting analysis to describe the behaviour of most of the numerical methods is presented with respect to the numerical output produced. Theorems are kept to a minimum (there is just one formal proof), and a basic working knowledge of calculus and matrix algebra is sufficient for most of the analysis and algorithm development.

A few more advanced topics are included to appeal to the drop-in reader looking simply for a solution. These topics, which may be skipped without prejudice, include rational approximation, spline interpolation and the numerical solution of ordinary differential equations.

To put a face to a subject that naturally underpins a wide range of disciplines, biographic extracts are included for those mathematicians and scientists whose work contributed specifically to the methods described in the text.

Acknowledgements

To write a text on aspects of numerical analyis is to acknowledge the plethora of books and research articles that have been published previously.

For this endeavour in particular, appreciation is expressed to the developers of *DERIVE* for providing a flexible tool for numerical experiment, and the award-winning *MacTutor History of Mathematics Archive* at the University of St Andrews is acknowledged for stocking the meat and grist that went into the biographies [16]. Many thanks are due to the publishers for their patience and the elasticity afforded to deadlines. And finally, thanks to the many students on whom numerical methods have been 'implemented' and teaching mechanisms tried.

A S Wood
School of Computing and Mathematics
University of Bradford, September 1998

[1] In the author's experience, *DERIVE* is the algebra package most easily used *ab initio* and it has minimal hardware requirements (being implemented on anything from a monochrome DOS-based 386 PC to the latest SVGA Pentium/Windows platform).

0 Beginnings

What is a *numerical method*? Why are numerical methods required? What is meant by the phrase *numerical analysis*? What is an *algorithm*? Computers don't make errors, do they?

In this chapter, some motivation for the development and use of numerical methods is given. It is clearly desirable to believe in the need for the area of mathematics that forms the basis of this text. The *DERIVE* package is introduced as a suitable platform for implementation and analysis.

0.1 Why numerical methods?

Most of us are familiar with the analytical tools required for solving a wide range of mathematical problems, such as the determination of the values of x that satisfy the equation

$$3x^2 + x - 2 = 0, \tag{0.1}$$

or the evaluation of the definite integral

$$\int_0^2 (\mathrm{e}^x - x)\,\mathrm{d}x, \tag{0.2}$$

or the development of the solution of the first-order differential equation

$$\frac{\mathrm{d}y}{\mathrm{d}x} = -y + x \tag{0.3}$$

subject to $y(0) = 1$. In each case solutions (numerical values in the first two cases and a function of x in the third) may be obtained by applying the rules and laws governing algebraic and calculus processes.

Example 0.1 Using the formula $x = (-b \pm \sqrt{b^2 - 4ac})/2a$, the roots of eqn (0.1) are

$$x = \frac{-1 \pm \sqrt{1 + 24}}{6} = \left\{-1, \frac{2}{3}\right\}.$$

The primitive of $e^x - x$ is $e^x - \frac{x^2}{2}$, and the integral (0.2) has the value

$$I = \int_0^2 (e^x - x)\,dx = \left[e^x - \frac{x^2}{2}\right]_0^2 = e^2 - 3.$$

The differential equation (0.3) is linear, $y' + p(x)y = q(x)$, and may be solved with an integrating factor $\mu = e^{\int p(x)\,dx}$ to give

$$y = \frac{1}{\mu}\left(\int \mu q\,dx + c\right)$$

where c is a constant of integration (determined from the condition $y(x_0) = s$). For eqn (0.3), $p(x) = 1$, $q(x) = x$, $\mu = e^{\int dx} = e^x$, and

$$y = e^{-x}\left(\int e^x x\,dx + c\right) = x - 1 + ce^{-x}.$$

At $x = 0$, $y = 1$, giving $c = 2$ and $y(x) = x - 1 + 2e^{-x}$.

For each problem in Example 0.1, an exact (analytic) solution is obtained by applying *analytical methods*. Suppose now that each problem is modified:

$$3x^5 + x - 2 = 0; \qquad \int_0^2 e^{-x^2}\,dx; \qquad y' = -y\ln(y+1) + x.$$

In the first case, no general closed-form expression exists for the solution of quintic polynomial equations. However, a graph of $y = 3x^5 + x - 2$ shows at least one value of x in the interval (0.5, 1) for which the expression equals zero (see Figure 0.1(a)). So, how might this value be determined?

In the second case, e^{-x^2} has no primitive in terms of standard functions, yet a graph of $y = e^{-x^2}$ shows a non-zero area under the curve, between $x = 0$ and $x = 2$ (see Figure 0.1(b)). After all, this is what integration is about – 'calculating the area under a curve'. So, how can this area be determined?

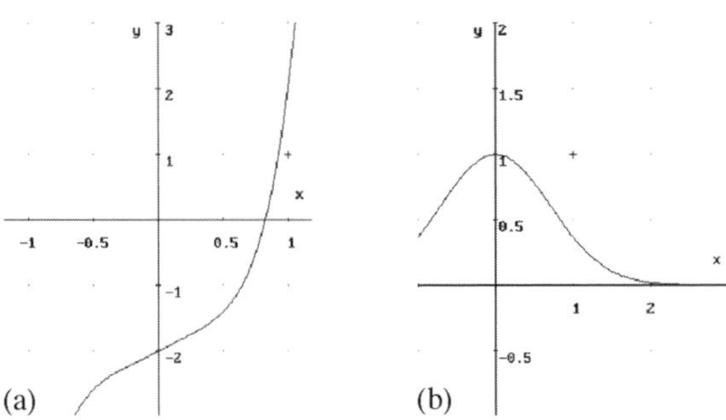

Figure 0.1 (a) A root of $3x^5 + x - 2 = 0$ lies in $0.5 < x < 1$. (b) The area under $y = e^{-x^2}$.

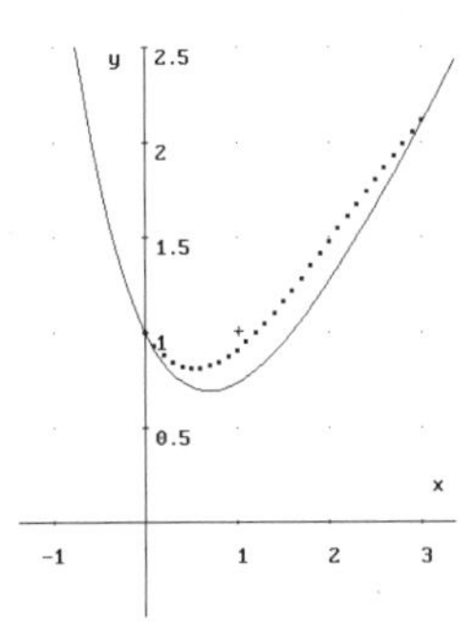

Figure 0.2 Solution curves for $y' = -y+x$ (solid line) and $y' = -y\ln(y+1)+x$ (dots), subject to $y(0) = 1$.

In the third case, the differential equation is no longer linear, yet a solution does exist (see Figure 0.2). Non-linear equations can usually only be solved analytically on a case-by-case basis. How then might a solution of the modified differential equation be obtained?

It is because of questions such as the previous three that a new approach is required, using *numerical methods*, for solving a broad range of problem descriptions. A numerical method does exactly what it says, and determines a numerical estimate of a (theoretical) analytical solution.

0.1.1 Algorithms

To make wide use of numerical methods, each method needs to be described in an unambiguous manner, usually taking the form of a set of instructions, called an *algorithm*, which may be implemented on a *computer*.

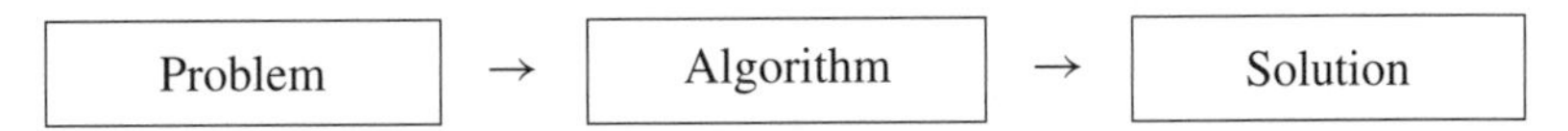

The 'input' to an algorithm is the problem description and the 'output' from an algorithm is the solution. The internal workings of the algorithm should be transparent to the problem (for example, if the algorithm solves $ax^2 + bx + c = 0$, it should handle any set of values of a, b and c).

CHARLES BABBAGE (1792–1871)
English mathematician

LIFE
Babbage graduated from Cambridge and, aged 24, was elected a Fellow of the Royal Society. He became Lucasian Professor of Mathematics at Cambridge (1827), a position he held for 12 years, although Babbage never taught.

WORK

Babbage originated the modern analytic computer. By 1834 he had invented the principle of the analytical engine, the forerunner of the modern electronic computer. He published *Reflections on the Decline of Science in England* (1830), a controversial work resulting in the formation (1831) of the British Association for the Advancement of Science. In his most influential work *On the Economy of Machinery and Manufactures* (1834) he proposed an early form of operational research.

The computation of logarithms made Babbage aware of the inaccuracy of human calculation. He became obsessed with mechanical computation and spent £60 000 in pursuit of it. A government grant of £17 000 was given, but later support was withdrawn. He felt angered by the way the Royal Society was run: 'The Council of the Royal Society is a collection of men who elect each other to office and then dine together at the expense of this society to praise each other over wine and give each other medals.' Although Babbage never built an operational, mechanical computer, his design concepts have been proved correct. Recently such a computer has been built following Babbage's own design.

Example 0.2

The problem of solving the quadratic equation $ax^2 + bx + c = 0$ where a, b and c are real constants and $a \neq 0$ has several alternatives for the roots, given by

$$x = \frac{-b \pm \sqrt{b^2 - 4ac}}{2a}.$$

If $b^2 - 4ac > 0$ the roots are real and distinct, if $b^2 - 4ac = 0$ the roots are real and equal, and if $b^2 - 4ac < 0$ the roots form a complex conjugate pair. This problem is described by the values of a, b and c, and one algorithm (recipe) to automate the solution process is given below.

Line	Comment	
1	**Problem**	`input` a, b, c
2		$d = b^2 - 4ac$
3	Real & distinct	`if` $d > 0$ `then`
4		$x_{1,2} = (-b \pm \sqrt{d})/2a$
5	Real & equal	`else if` $d = 0$ `then`
6		$x_{1,2} = -b/2a$
7	Complex	`else`
8		$x_{1,2} = (-b \pm i\sqrt{-d})/2a$
9		`end`
10	**Solution**	`output` x_1, x_2

By 'tracing your finger' from top to bottom, a set of instructions is read, one by one. If a logical `if` statement is true then the instructions immediately below it are executed, otherwise the block is 'skipped'.

For example, if $a = 2$, $b = 4$ and $c = 2$, your finger will 'read' lines 1 (input), 2, 3 (false), 5 (true), 6, 9, 10 (output).

Numerical analysis concerns the development, implementation and analysis of (numerical) algorithms. The analysis addresses questions of how accurate a numerical solution is, how efficient the algorithm is (in terms of computational effort), and how stable it is (with regard to errors in the input data and to errors appearing in the calculation process itself).

This text does not at all attempt to provide a comprehensive discussion on formal algorithms, but 'recipes' are given where it is felt that they will enhance the understanding of a particular numerical method.

As for the question of whether computers make errors or not, consider the following illustration (implemented in *MATLAB* version 4.2c).

Example 0.3 If $a = 1, b = 10^{-16}$ and $c = 10^{-16}$, the *MATLAB* expression $a+b+c$ returns the value 1 and $c + b + a$ returns the value 1.00000000000000 (using `format long`). Both results appear to equal 1. In the first expression, a is added to b and the sum is added to c. In the second expression, c is added to b and the sum is added to a.

What *is* of concern is the fact that $(c+b+a)-(a+b+c)$ in *MATLAB* 'equals' $2.220446049250313 \times 10^{-16}$ and $c+b+a-a-b-c$ 'equals' $2.204460492503131 \times 10^{-17}$. Can we ever believe in arithmetic again?

For implementation, we shall confine ourselves to using *DERIVE*.

0.2 *DERIVE* basics

DERIVE is a *computer algebra package* that can be used to implement symbolic and numerical calculations, such as those discussed in this text. In addition, *DERIVE* has a good plotting facility that is useful for understanding the results of symbolic/numerical calculations.

Throughout the text, *DERIVE Experiments* are suggested, the aim being to help the reader to implement the 'theory' with minimal fuss. Here we introduce several basic *DERIVE* commands that will be useful[1]. The *DERIVE Experiments* appearing in this text are based on version 3.01 for MS-DOS.

---

[1] This text is not a comprehensive guide to *DERIVE* and those who wish to become more familiar with the package are directed to the manual [9] and to the books by Kutzler [14, 15] which provide an introduction to the DOS and Windows versions, and give many additional references to manuals and textbooks.

You are encouraged to use this section as a hands-on tutorial session on *DERIVE* – work carefully through each of the Examples.

0.2.1 Basic commands and notation

User interaction with *DERIVE* takes place via *commands*, a valid list of which is displayed in a *command line* towards the bottom of the screen. The boot-up command line is

```
Author Build Calculus Declare Expand Factor Help Jump soLve
  Manage
Options Plot Quit Remove Simplify Transfer Unremove moVe
  Window approX
```

A command is executed by pressing the key corresponding to the upper-case letter in the command. Thus, the 'Author' command is executed by pressing the `<A>` key (there is no need to press `<Enter>`). In the text commands appear in `typewriter` font with the 'hot key' underlined; thus `Author` means 'press `<A>` to execute the 'Author' command'.

Some commands have one or more *arguments* – these are additional bits of information required by the command. For example, `Author` prompts for an expression to be input (terminated with `<Enter>`).

Some commands display a list of *sub-commands*. For example, `Calculus` displays the sub-commands

```
Differentiate Integrate Limit Product Sum Taylor Vector
```

`<Esc>` aborts the current command/sub-command, and expressions are input into *DERIVE* with the `Author` command.

Example 0.4 `Author 3x^2+x-2`, followed by `<Enter>`, enters the polynomial expression $3x^2 + x - 2$. The *DERIVE* screen shows the 'highlighted' expression with the *expression number* `#1:` to the left.

The `Plot` command graphs a 'highlighted' expression. *DERIVE* can deduce whether a 1D or 2D graph is required. If no *Graphics Window* is open, the position of the graph is specified (select `Beside` for convenience). A second `Plot` draws the graph. `Algebra` returns control to the *Algebra Window*.

Example 0.5 To plot the polynomial $3x^2 + x - 2$, enter `Plot Beside <Enter> Plot` (see Figure 0.3). `Algebra` reactivates the *Algebra Window*.

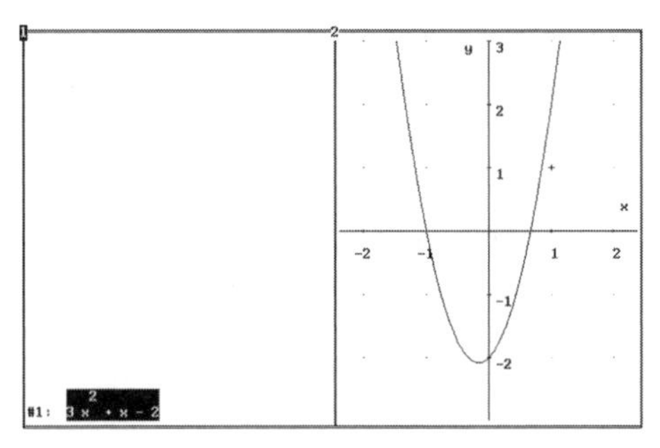

Figure 0.3 A *DERIVE* plot of the polynomial expression $3x^2 + x - 2$.

The `soLve` command prompts for an expression number and then attempts to solve the 'highlighted' expression for any unknowns, if appropriate.

Example 0.6

To solve the polynomial equation $3x^2 + x - 2 = 0$ execute `soLve` `<Enter>`. Two expressions are displayed, $x = -1$ and $x = \frac{2}{3}$, with expression numbers `#2:` and `#3:` (see Figure 0.4). `soLve` equates the expression $3x^2 + x - 2$ to zero and solves the resulting equation exactly.

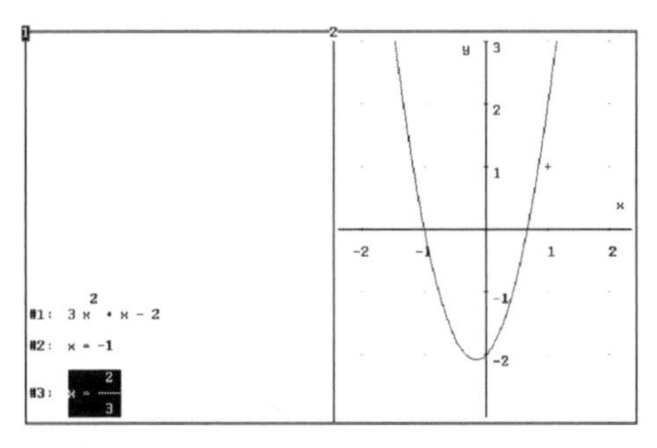

Figure 0.4 Using `soLve` to find the roots of an equation.

For expressions that reduce to a numerical value, `Simplify` can be used.

Example 0.7

`Author 34+3/7-5(9-6) Simplify` results in the fraction $\frac{136}{7}$. `Author int (expx-x,x,0,2) Simplify` gives the value $e^2 - 3$.

The approX command generates a decimal value for an exact arithmetic expression or provides an estimate to an expression that cannot be expressed in terms of basic functions (this is often the case with definite integrals).

Example 0.8 To obtain a decimal approximation to the result of the previous integral, expression #7:, execute approX <Enter> to give 4.38905.

The value of the definite integral $\int_0^2 e^{-x^2}\,dx$ is estimated by Author int(exp (-x^2),x,0,2) approX with the result 0.882081.

It should be evident that *DERIVE* uses exact arithmetic, unless told otherwise (such as with the approX command). However, this can lead to difficulties for problems without an analytic solution.

Example 0.9 Author and soLve 3x^5+x-2. The equation $3x^5 + x = 2$ is returned – there is no closed-form expression for values of x satisfying the equation. That there exists at least one solution is clear from a graph of the polynomial expression, <↑> Plot Plot Algebra (see Figure 0.1(a)).

The answer is to modify the *arithmetic precision* with Options Precision Approximate <Enter>. *DERIVE* confirms the new mode of arithmetic and now uses built-in numerical methods to solve specified problems.

Example 0.10 Highlight, <↑>, and soLve the quintic expression. A search interval is suggested, initially $[-10, 10]$, that may contain a root. The graph shows that $[0.5, 1]$ is a better interval, so type .50 <Tab> 1. <Enter>. *DERIVE* responds with $x = 0.828552$.

Exact arithmetic is recovered with Options Precision Exact.

Finally, Manage Substitute allows the evaluation of an expression containing one or more variables for specific values of the variables.

Example 0.11 To evaluate $3x^5 + x - 2$ at $x = 1$, Manage Substitute #* <Enter> 1 <Enter> Simplify (* is the appropriate expression number).

0.2.2 Functions

DERIVE supports many *intrinsic functions*, such as $\sin x$, $\cos x$, $\tan x$, e^x (see Example 0.7), $\ln x$ and $\cosh x$. A function may appear in an expression with its argument x appearing in () if required. Functions may be arguments to other functions.

Example 0.12 Author sinx and Author sin(x) both give $\sin x$, whereas Author sin2/3 and Author sin(2/3) give $\frac{\sin 2}{3}$ and $\sin\frac{2}{3}$, respectively.
Author expsinx gives $e^{\sin x}$.

DERIVE contains many *combination functions* that execute an algebraic or calculus process, such as int(f,x,a,b) which gives the definite integral $\int_a^b f(x)\,dx$ (see Example 0.7). Combination functions are introduced as needed (the DERIVE index entry lists the functions used in the text).

The power of DERIVE is the facility to allow *user-defined functions*.

Example 0.13 Author p(x):=3x^5+x-2 defines $p(x)$ as the polynomial $3x^5 + x - 2$. Author p(1.5) Simplify gives the (exact) value of $p(x)$ at $x = 1.5$ as $\frac{713}{32}$. approX provides a 6-digit decimal representation, 22.2812.

Complicated user-defined functions can be described, thus reducing the need for repeat input of long expressions. Examples are given, where appropriate, in the text.

0.2.3 Saving, clearing and loading expressions

Often you will enter several expressions to execute a sequence of calculations. To preserve these expressions, DERIVE can save the *Algebra Window* to a file (with extension .mth) – the sequence is Transfer Save Derive followed by the filename (with the DOS path, if needed) and <Enter>.

Example 0.14 Key in Transfer Clear Y (see Example 0.15) and Author

```
i(f):=int(f,x,a,b)
v(n):=vector(i(x^j),j,0,n)
v(2)
```

Make sure that you Simplify v(2). The screen will show four expressions, #1: to #4:. To store the session in the file vint.mth in the root directory c:, Transfer Save Derive c:\vint <Enter>.
Now *wait* until Example 0.16.

To finish a DERIVE session you can either (a) Quit to exit the DERIVE package or (b) use Transfer Clear to erase the working environment, but remain within DERIVE.

Example 0.15 To clear the integration session of Example 0.14, key in `Transfer Clear` – the *Algebra Window* is cleared and all user-function definitions are lost. `Author i(x)` – you do *not* get the function $i(x)$.

Now `Transfer Clear`. Since the current expressions have not been saved, the question `Abandon expressions (Y/N)?` appears. In this case, press the <Y> key.

The final 'tip' concerns loading a sequence of DERIVE expressions that is stored in a file. The command sequence is `Transfer Load Derive` followed by the name (and path, if required) of the desired `.mth` file.

Example 0.16 To recover the expressions saved in the file `vint.mth`, key in `Transfer Load Derive c:\vint` <Enter>.

Alternatively, if you simply want to access the function definitions contained in a file, in the same way that sin, cos etc. are available, use the sequence `Transfer Load Utility`. Here the file is loaded in the 'background' – you do not see the expressions on screen, although they are now in memory.

Example 0.17 Key in `Transfer Clear` and then `Transfer Load Utility c:\vint` <Enter>. Now `Author` and `Simplify i(x)` to satisfy yourself that the function definitions *are* available.

Summary

This chapter aimed to give reason for *numerical methods*, and in Section 0.1 various examples were given in which traditional *analytical methods* fail. In Section 0.1.1 the notion of an *algorithm* was introduced as a 'recipe' for implementing a numerical method, and in Section 0.2 a whistle-stop tour of DERIVE was given, this being the means used in this text to help you implement and understand some of the numerical ideas put forward.

1 Taylor series

In this chapter the technique of *Taylor series* expansions is developed for constructing a function f given limited information about the function. As you work through this text you will find that Taylor series, and the associated *Taylor's theorem*, form both the starting point and the tool for the analysis of many areas of numerical analysis. For this reason the topic deserves a chapter all to itself, right at the beginning.

BROOK TAYLOR
(1685–1731)
English mathematician

LIFE

Musicians and artists were often entertained at Taylor's house. This had a lasting influence on Taylor, who published a definitive work on the mathematical theory of perspective and obtained major mathematical results on the vibrations of strings (an unpublished work *On Musick* was to be part of a joint paper with Newton). Taylor's life was blighted by unhappiness, illness and tragedy. Both his wives died in childbirth. His first wife was not rich enough to satisfy his father and the two men argued bitterly, finally parting ways. Towards the end of his life Taylor devoted much time to religion and philosophy and, according to Taylor himself, works that bear his name were motivated by coffee-house discussions about Newton's work on planetary motion and Halley's work on the roots of polynomials. However, his writing was so terse and hard to understand that he often received no credit for his innovations.

WORK

Taylor wrote on many subjects: magnetism, capillary action, thermometers, perspective and calculus. In 1708 he produced a solution to the problem of the centre of oscillation which, since it went unpublished until 1714, resulted in a priority dispute with Johann Bernoulli. In 1712 Taylor was elected a Fellow of the Royal Society and was appointed to the committee for adjudicating the claims of Newton and of Leibniz to have invented 'calculus'. 1714–19 was Taylor's most productive period. His *Methodus Incrementorum Directa et Inversa* (1715) provided mathematics with the 'calculus of finite differences' and he invented 'integration by parts'. This work contained the celebrated *Taylor series* formula, the importance of which remained unrecognised until 1772 when Lagrange proclaimed it the basic principle of 'differential calculus'. Taylor devised the principles of perspective in *Linear Perspective* (1715) and gave the first general treatment of vanishing points. He described an experiment to discover the law of magnetic attraction (1715) and an improved method for approximating the roots of an equation by giving a new method for computing logarithms (1717).

1.1 Introduction to function approximation

Let f be a real function of one real variable x. If the *function value* $f(x_0)$ is the *only* information on f (see Figure 1.1(a)), then it is reasonable to ask

how might $f(x)$ be estimated for values of x other than $x = x_0$?

A simple solution is to set the value of $f(x)$ at $x \neq x_0$ to equal $f(x_0)$. In other words, f is approximated by a *constant* function, p_0, defined by

$$p_0(x) = f(x_0). \tag{1.1}$$

Such a crude approximation is likely to be inaccurate away from $x = x_0$, as illustrated in Figure 1.1(b) for the function $f(x) = \cos x$ with $x_0 = \frac{\pi}{3}$.

If the *derivative* of f is also known at $x = x_0$, that is $f'(x_0)$ is available[1] (see Figure 1.2(a)), then $f'(x_0)$ may be used to approximate the slope of f at points

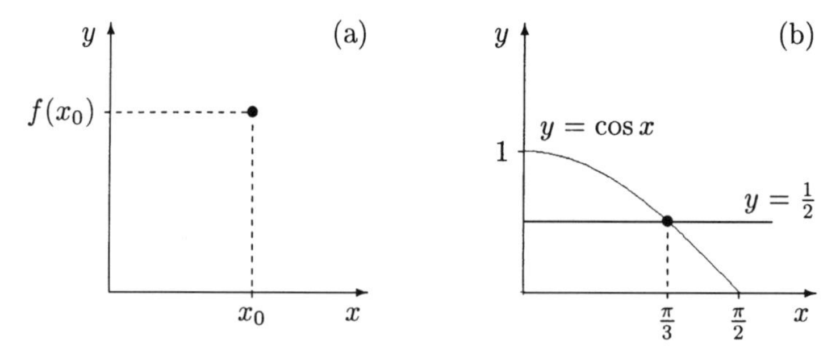

Figure 1.1 (a) A function defined by a single value $f(x)$ at $x = x_0$. (b) A constant approximation to $\cos x$.

[1] The notation $f^{(j)}$ denotes the jth-order derivative of f with respect to x. Commonly used alternative forms are $f' \equiv f^{(1)}$, $f'' \equiv f^{(2)}$ and $f''' \equiv f^{(3)}$.

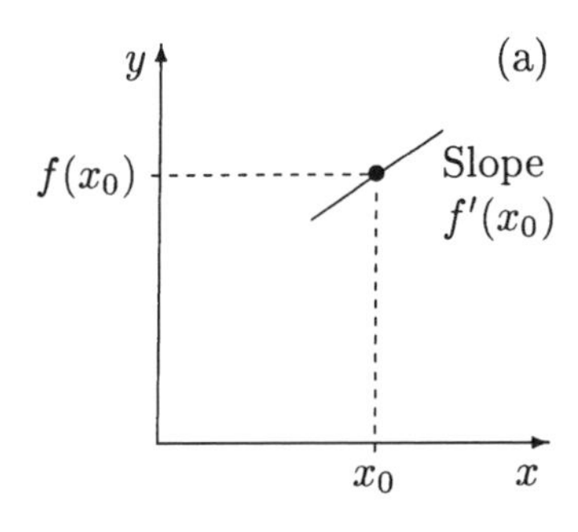

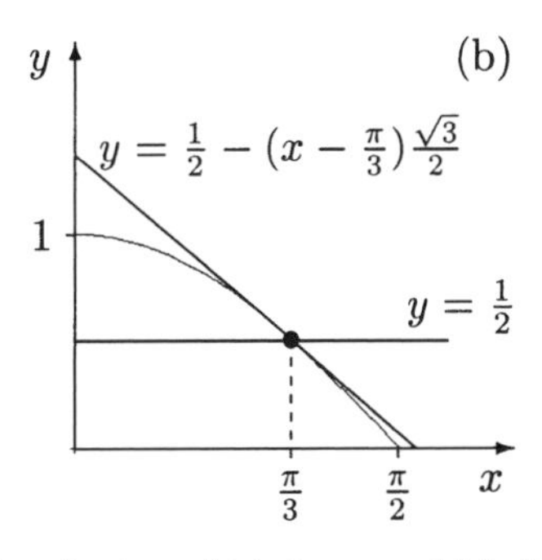

Figure 1.2 (a) A function f defined by a value $f(x)$ and a slope $f'(x)$ at $x = x_0$. (b) An illustration using $\cos x$.

close to $x = x_0$. In other words, in a *neighbourhood* of $x = x_0$, the graph of $y = f(x)$ may be approximated by a straight line passing through the point $(x_0, f(x_0))$ with slope $f'(x_0)$. The appropriate *linear* function is defined by

$$p_1(x) = f(x_0) + (x - x_0)f'(x_0). \tag{1.2}$$

This is easy to verify. Since $p_1'(x) = f'(x_0)$, then

$$\begin{aligned} p_1(x_0) &= f(x_0) + (x_0 - x_0)f'(x_0) = f(x_0), \\ p_1'(x_0) &= f'(x_0). \end{aligned}$$

Figure 1.2(b) illustrates eqn (1.2) using $f(x) = \cos x$ at $x_0 = \frac{\pi}{3}$. Here $f'(x_0) = -\sin(x_0) = -\sin\frac{\pi}{3} = -\frac{\sqrt{3}}{2}$. The range of values of x for which $p_1(x)$ is a satisfactory estimate to $f(x)$ is increased over that for the constant approximation (see Table 1.1).

If, finally, the second derivative of f is available at $x = x_0$, in other words $f''(x_0)$ is known (see Figure 1.3), then 'close to' $x = x_0$, f may be approximated by a *quadratic* function passing through the point $(x_0, f(x_0))$ with slope $f'(x_0)$ and second derivative $f''(x_0)$, of the form

$$p_2(x) = f(x_0) + (x - x_0)f'(x_0) + \frac{1}{2}(x - x_0)^2 f''(x_0). \tag{1.3}$$

Table 1.1 Range of values of x for which $p_0(x)$ and $p_1(x)$ predict the value of $\cos x$ to within 0.05.

	$p_0(x)$	$p_1(x)$
x_{min}	0.99	0.64
x_{max}	1.11	2.75

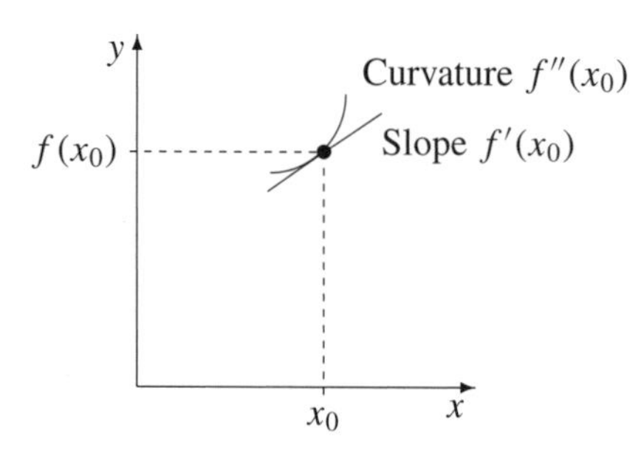

Figure 1.3 A function f defined by a value $f(x)$, a slope $f'(x)$ and a second derivative $f''(x)$ at $x = x_0$.

The properties of p_2 at $x = x_0$ are, again, not difficult to check. Since

$$p_2'(x) = f'(x_0) + (x - x_0)f''(x_0) \quad \text{and} \quad p_2''(x) = f''(x_0),$$

then

$$\begin{aligned} p_2(x_0) &= f(x_0) + (x_0 - x_0)f'(x_0) + \frac{1}{2}(x_0 - x_0)^2 f''(x_0) \\ &= f(x_0), \\ p_2'(x_0) &= f'(x_0) + (x_0 - x_0)f''(x_0) \\ &= f'(x_0), \\ p_2''(x_0) &= f''(x_0). \end{aligned}$$

In each of the cases (1.1), (1.2) and (1.3), f is approximated by a *polynomial*, that is $f(x) \simeq p_n(x)$ where p_n is a polynomial of *degree* at most n.

Example 1.1 Let the function f be defined by $f(x) = x^3$. Then $f'(x) = 3x^2$ and $f''(x) = 6x$, and with $x_0 = 1$, we have $f(1) = 1$, $f'(1) = 3$ and $f''(1) = 6$. The constant, linear and quadratic polynomial approximations to f, based upon (1.1), (1.2) and (1.3), are

$$\begin{aligned} p_0(x) &= f(1) \\ &= 1, \\ p_1(x) &= f(1) + (x - 1)f'(1) \\ &= p_0(x) + 3(x - 1) \\ &= -2 + 3x, \\ p_2(x) &= f(1) + (x - 1)f'(1) + \frac{1}{2}(x - 1)^2 f''(1) \\ &= p_1(x) + \frac{1}{2}(x - 1)^2 f''(1) \\ &= 1 - 3x + 3x^2. \end{aligned}$$

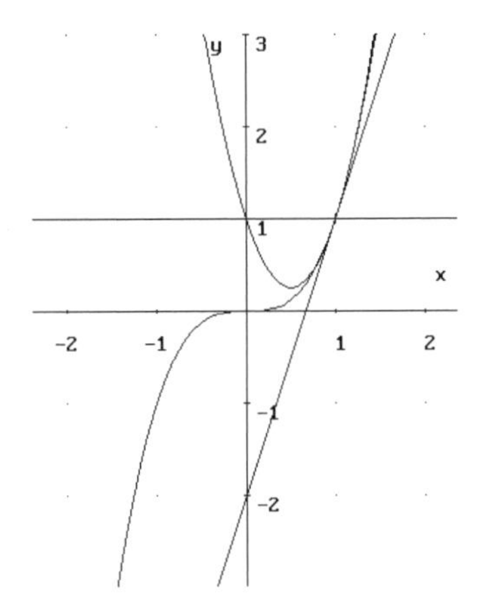

Figure 1.4 Local behaviour of higher-order polynomial approximations.

Graphs of $f(x)$, $p_0(x)$, $p_1(x)$ and $p_2(x)$ are shown in Figure 1.4. It appears that the higher-degree polynomials provide 'improved' approximations in the neighbourhood of $x = x_0$.

Continuing, it can be shown that if $f'''(x_0)$ is also available, then f may be approximated by the *cubic* polynomial

$$p_3(x) = f(x_0) + (x - x_0)f'(x_0) + \frac{1}{2}(x - x_0)^2 f''(x_0) + \frac{1}{6}(x - x_0)^3 f'''(x_0). \quad \mathbf{(1.4)}$$

It is left as an exercise (see Exercise 1.2) for the reader to show that $p_3^{(j)}(x_0) = f^{(j)}(x_0)$, $j = 0, 1, 2, 3$.

Example 1.2 Extending Example 1.1, determine a cubic polynomial approximation to f based on eqn (1.4). $f'''(x) = 6$ is a constant and hence $f'''(1) = 6$. Formula (1.4) gives

$$\begin{aligned} p_3(x) &= f(1) + (x - 1)f'(1) + \frac{(x - 1)^2}{2} f''(1) + \frac{(x - 1)^3}{6} f'''(1) \\ &= p_2(x) + \frac{(x - 1)^3}{6} f'''(1) \\ &= 1 - 3x + 3x^2 + \frac{1}{6} \times (x^3 - 3x^2 + 3x - 1) \times 6 \\ &= x^3. \end{aligned}$$

The original function, which itself was a polynomial, is recovered (see Exercise 1.3).

Example 1.3 Repeat Examples 1.1 and 1.2 with $x_0 = 2$.

Since $f(x) = x^3$, $f'(x) = 3x^2$, $f''(x) = 6x$, $f'''(x) = 6$ and $x_0 = 2$, then $f(2) = 8$, $f'(2) = 12$, $f''(2) = 12$ and $f'''(2) = 6$. The polynomial approximations to f based upon eqns (1.1) to (1.4) are

$$\begin{aligned}
p_0(x) &= f(2) \\
&= 8, \\
p_1(x) &= f(2) + (x-2)f'(2) \\
&= p_0(x) + 12(x-2) \\
&= -16 + 12x, \\
p_2(x) &= f(2) + (x-2)f'(2) + \frac{1}{2}(x-2)^2 f''(2) \\
&= p_1(x) + \frac{1}{2}(x-2)^2 f''(2) \\
&= 8 - 12x + 6x^2 \\
p_3(x) &= f(2) + (x-2)f'(2) + \frac{1}{2}(x-2)^2 f''(2) + \frac{1}{6}(x-2)^3 f'''(2) \\
&= p_2(x) + \frac{1}{6}(x-2)^3 f'''(2) \\
&= x^3.
\end{aligned}$$

Different 'construction' points x_0 give different polynomials unless f is a polynomial and the degree of the approximating polynomial matches that of f, in which case $p_n \equiv f$ for all values of x_0.

1.2 Taylor polynomials

Here we bring some formality to the rather 'hand-waving' approach adopted in Section 1.1 concerning the approximation of a function f by a polynomial. Specifically, for a polynomial of degree n defined by

$$\begin{aligned}
p_n(x) &= b_0 + b_1(x-x_0) + b_2(x-x_0)^2 + \cdots \\
&\quad + b_k(x-x_0)^k + \cdots + b_n(x-x_0)^n \qquad \textbf{(1.5)} \\
&= \sum_{k=0}^{n} b_k(x-x_0)^k,
\end{aligned}$$

what values of the constant coefficients $b_0, \ldots, b_n$ ensure that p_n and its first n derivatives coincide with f and its first n derivatives at the point $x = x_0$?

One response is to compute the first n derivatives of p_n from eqn (1.5) and equate each expression in turn with the appropriate derivative of f at $x = x_0$. From eqn (1.5), $p_n(x_0) = b_0$. Since $p_n(x)$ must equal $f(x)$ at $x = x_0$, so $b_0 = f(x_0)$. Differentiating eqn (1.5) gives

$$p_n'(x) = b_1 + 2b_2(x - x_0) + \cdots + kb_k(x - x_0)^{k-1} + \cdots + nb_n(x - x_0)^{n-1}.$$

Consequently, $p_n'(x_0) = b_1$ and since $p_n'(x)$ must equal $f'(x)$ at $x = x_0$ then $b_1 = f'(x_0)$.

To assist with the general case, three useful properties of the term $b_k(x - x_0)^k$ appearing in eqn (1.5) are

I the jth derivative is $k(k-1)\dots(k-j+1)b_k(x-x_0)^{k-j}$ for all $j < k$,

II the kth derivative is a constant with the value $k!b_k$, and

III derivatives of order $j > k$ are identically zero.

In the jth derivative of eqn (1.5), terms $k < j$ are zero (Property III), the jth term has the value $j!b_j$ (Property II), and terms $k > j$ are zero since (Property I) they contain a factor $(x - x_0)^{k-j}$ which equals zero at $x = x_0$.

In other words, the jth derivative of p_n at $x = x_0$ is equal to $j!b_j$ and must equal the jth derivative of f at $x = x_0$. Hence, for each $k = j$, $j = 0, \dots, n$,

$$k!b_k = f^{(k)}(x_0) \;\Rightarrow\; b_k = \frac{f^{(k)}(x_0)}{k!}$$

and the polynomial (1.5) assumes the form

$$\begin{aligned} p_n(x) &= f(x_0) + (x - x_0)f'(x_0) + \frac{(x-x_0)^2}{2!}f''(x_0) + \cdots \\ &\quad + \frac{(x-x_0)^k}{k!}f^{(k)}(x_0) + \cdots + \frac{(x-x_0)^n}{n!}f^{(n)}(x_0) \\ &= \sum_{k=0}^{n} \frac{(x-x_0)^k}{k!} f^{(k)}(x_0). \end{aligned} \qquad \textbf{(1.6)}$$

Formula (1.6) is the *Taylor polynomial* for f, of degree n, constructed at $x = x_0$. A useful alternative form is obtained via the transformation $x - x_0 = h$, that is $x = x_0 + h$, to give

$$\begin{aligned} p_n(x_0 + h) &= f(x_0) + hf'(x_0) + \frac{h^2}{2!}f''(x_0) + \cdots + \frac{h^n}{n!}f^{(n)}(x_0) \\ &= \sum_{k=0}^{n} \frac{h^k}{k!} f^{(k)}(x_0). \end{aligned} \qquad \textbf{(1.7)}$$

h is a small (incremental) step size – the distance between the 'evaluation' point x and the 'construction' point x_0. Typically, eqn (1.6) is used when the equation of a curve, in terms of x, is required. Equation (1.7) is used when the value of $p_n(x)$ is required for a specified value of x (and hence h).

Example 1.4 Construct the Taylor polynomial p_n for the function f defined by $f(x) = \ln(x-1)$, $2 < x < 3$, at $x_0 = 2$. From differential calculus

$$\begin{aligned} f(x) &= \ln(x-1), & f(2) &= 0, \\ f'(x) &= 1/(x-1), & f'(2) &= 1, \\ f''(x) &= -1/(x-1)^2, & f''(2) &= -1, \\ f'''(x) &= 2/(x-1)^3, & f'''(2) &= 2, \\ f''''(x) &= -6/(x-1)^4, & f''''(2) &= -6, \end{aligned}$$

and in general

$$f^{(k)}(x) = (-1)^{k-1}\frac{(k-1)!}{(x-1)^k}, \quad f^{(k)}(2) = (-1)^{k-1}(k-1)!$$

With $x_0 = 2$, and substituting for $f(2)$, $f'(2)$ etc. in eqn (1.6),

$$p_n(x) = (x-2) - \frac{(x-2)^2}{2} + \frac{(x-2)^3}{3} + \cdots + (-1)^{n-1}\frac{(x-2)^n}{n}.$$

DERIVE can be useful for 'side-stepping' all the calculus associated with the construction of Taylor polynomials.

DERIVE Experiment 1.1

The *DERIVE* function `taylor` provides expansions of the form (1.6). Its specification is

`taylor(f,x,x0,n)`

`f`	object function $f(x)$
`x`	independent variable x
`x0`	'construction' point x_0
`n`	degree of Taylor polynomial p_n

To reproduce the result of Example 1.4, `Author` the *DERIVE* expression `taylor(ln(x-1),x,2,3)` and `Simplify` to obtain

$$\frac{2x^3 - 15x^2 + 42x - 40}{6}.$$

It is not difficult to check that this expression is identical to the first three terms of Example 1.4, namely[2]

$$(x-2) - \frac{(x-2)^2}{2} + \frac{(x-2)^3}{3}.$$

[2] To use *DERIVE* simply `Author (x-2)-(x-2)^2/2+(x-2)^3/3` and `Simplify`.

The linear, quadratic and cubic Taylor polynomials may be plotted to illustrate the improvement gained by increasing the degree of p_n. Author `ln(x-1)` Plot Beside Plot and return to Algebra. Author the expression

```
vector(taylor(ln(x − 1), x, 2, n), n, 1, 3)
```

The approximation improves with the degree of the polynomial but, sufficiently far from the 'construction' point $x_0 = 2$, all the approximating curves are poor imitations of the graph of $\ln(x - 1)$.

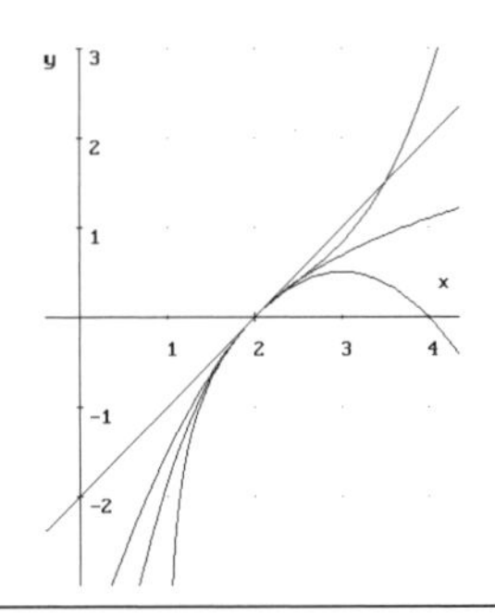

You can use the function `taylor` and the plotting facilities of *DERIVE* to experiment with alternative functions and investigate the effect of changing x_0 and/or n (see Exercises 1.7, 1.8 and 1.9).

1.3 Taylor's theorem

In Section 1.2 a polynomial p_n, of degree n, was constructed that matched a function f and its first n derivatives at $x = x_0$. For values of $x \neq x_0$ an error was observed (e.g. Figure 1.4), that is $f(x) - p_n(x) \neq 0$ for $x \neq x_0$. Here this error is quantified.

Taylor's theorem[3]

Let f be a function of one variable x for which $f(x)$, $f^{(1)}(x)$, ..., $f^{(n)}(x)$ exist and are continuous on the *closed interval* $[a, b]$, and where $f^{(n+1)}(x)$ exists for $a < x < b$. If $p_n(x)$ is the Taylor polynomial (1.6) for some $x_0 \in [a, b]$, then given an $x \in [a, b]$ there exists a ξ (dependent upon x) lying between x_0 and x such that

$$R_n(x) = f(x) - p_n(x) = \frac{(x - x_0)^{n+1}}{(n+1)!} f^{(n+1)}(\xi). \tag{1.8}$$

[3] For a proof of Taylor's theorem, see [18], pp. 41–3.

$R_n(x)$ is the *remainder term* and quantifies the error in using $p_n(x)$ to approximate $f(x)$ at $x \neq x_0$.

Observation *As x approaches x_0 so the error approaches zero. This is consistent with the construction of the Taylor polynomial for which the condition $p_n(x_0) = f(x_0)$ is enforced.*

The difficulty in using *Taylor's theorem* is identifying a suitable value for ξ. Its dependence upon the 'evaluation' point x makes this task virtually impossible. In practice eqn (1.8) is used to obtain an *upper bound* for the absolute value[4] of the error $R_n(x)$,

$$\begin{aligned} |R_n(x)| &= |f(x) - p_n(x)| \\ &\leq \max_{x_0<\xi<x} \left| \frac{(x-x_0)^{n+1}}{(n+1)!} f^{(n+1)}(\xi) \right|. \end{aligned} \tag{1.9}$$

An important point to note about eqn (1.9) is that for well-behaved functions (sufficient derivatives at $x = x_0$) $R_n(x)$ can be shown to approach zero as n increases[5]. In this case equality between $p_n(x)$ and $f(x)$ is achieved, formally termed *convergence*; that is

$$f(x) = \lim_{n\to\infty} \sum_{k=0}^{n} \frac{(x-x_0)^k}{k!} f^{(k)}(x_0). \tag{1.10}$$

Equation (1.10) is the *Taylor series* for f about $x = x_0$ and, in principle, implies that any function having sufficient continuity can be represented to an arbitrary precision by a polynomial of a suitably high degree. If f is a polynomial of a certain degree then the series will be finite (see Exercise 1.3).

Example 1.5 For Example 1.4, determine an error bound for $p_1(x)$ to $p_5(x)$ at $x = 2.4$ (given $x_0 = 2$). We have

$$\begin{aligned} R_n(x) &= \frac{(x-x_0)^{n+1}}{(n+1)!} f^{(n+1)}(\xi) \\ &= \frac{(x-2)^{n+1}}{(n+1)!} \frac{(-1)^n n!}{(\xi-1)^{n+1}} = \frac{(-1)^n}{n+1} \left(\frac{x-2}{\xi-1}\right)^{n+1}. \end{aligned}$$

Further

$$|R_n(2.4)| \leq \max_{2<\xi<2.4} \frac{1}{n+1} \left| \frac{0.4}{\xi - 1} \right|^{n+1}.$$

---

[4] Equation (1.9) assumes $x > x_0$. If $x < x_0$, the term $\max\limits_{x_0<\xi<x}$ is replaced by $\max\limits_{x<\xi<x_0}$.

[5] Details are given in [18], pp. 43–6.

The largest value of $|R_n(2.4)|$ occurs at $\xi = 2$ (by inspection) and equals $(0.4)^{n+1}/(n+1)$. At $x = 2.4$, $\ln(x - 1)$ is equal to $0.33647\ldots$.

Table 1.2 Error bounds for Taylor polynomial approximations to $\ln(x - 1)$.

	Approximation	**Observed error**	**Error bound**
n	$p_n(2.4)$	$\lvert f(x) - p_n(x)\rvert$ $= \lvert\ln(1.4) - p_n(2.4)\rvert$	$\max\lvert R_n(x)\rvert$ $= (0.4)^{n+1}/(n+1)$
1	0.4	0.06352	0.08
2	0.32	0.01647	0.02133
3	0.3413	0.00483	0.0064
4	0.33493	0.00154	0.00205
5	0.33698	0.00051	0.00068

In this case (see Table 1.2), $\max|R_n(x)|$ overestimates the magnitude of the observed error by approximately 30%. This is typical. An error bound computed from eqn (1.9) will usually be pessimistic (a considerable overestimate of the actual error). However, in the absence of $f(x)$, $x \neq x_0$, it is the best that can be achieved!

Observations *Example 1.5 provides experimental evidence for the 'convergence' statement expressed by eqn (1.10) – the error reduces as n increases. Further, this error is* controllable *– the addition of more terms in the Taylor polynomial reduces the error. Such an error is termed a* truncation error *(see Chapter 2), due here to a truncated series.*

1.4 Maclaurin series

A commonly used special case of Taylor series is that for which $x_0 = 0$. Here a *power series* for the function f is built up of the form $a_0 + \cdots + a_n x^n + \cdots$. Replacing x_0 by zero in the Taylor series (1.10) yields the *Maclaurin series*

$$f(x) = f(0) + xf'(0) + \frac{x^2}{2!}f''(0) + \cdots + \frac{x^n}{n!}f^{(n)}(0) + \cdots. \tag{1.11}$$

COLIN MACLAURIN
(1698–1746)
Scottish mathematician

LIFE

Maclaurin's father died when the boy was six months old, and his mother when he was nine years old. He was raised by an uncle who was a minister. Maclaurin entered Glasgow University as a divinity student, transferred to mathematics after one year and received his Master's degree at the age of 17. He was professor of mathematics at Marischal College (Aberdeen) from 1717 to 1725 and then at the University of Edinburgh until 1745. Maclaurin was elected a Fellow of the Royal Society in 1719 and was awarded prizes by the Académie des Sciences for his work on the impact of bodies (1724) and for the study of tides (1740). The latter prize was shared with Euler and Daniel Bernoulli. Maclaurin played an active role in the defence of Edinburgh during the Jacobite rebellion of 1745, and when the city fell he fled to York.

WORK

Maclaurin's first major work was *Geometrica Organica* (1720). In 1719 he met Newton in London and became his disciple. Some of Newton's analytic methods were bitterly attacked by major mathematicians and much of Maclaurin's work arose from his efforts to defend Newton's ideas geometrically. A *Treatise of Fluxions* (1742) was the first systematic formulation of Newton's methods, written as a reply to Berkeley's attack on the 'calculus' for its lack of rigorous foundations. Here Maclaurin used the special case of Taylor series now known as *Maclaurin series*. The treatise was a standard of mathematical rigour in calculus until the work of Cauchy in 1821. However, it encouraged British mathematicians to use geometry rather than the new analysis being developed in Europe.

Maclaurin was an outstanding experimentalist. He devised many ingenious mechanical devices, made important astronomical observations, performed actuarial computations for insurance societies, and helped to improve maps of the islands around Scotland. He did notable work in geometry (higher plane curves), wrote an important memoir on the theory of tides and gave the 'integral test' for the convergence of an infinite series. Maclaurin's *Treatise on Algebra* (1748) was published two years after his death.

Example 1.6 For the function f defined by $f(x) = \mathrm{e}^x$, $f^{(k)}(x) = \mathrm{e}^x$ for all $k > 0$. Hence $f^{(k)}(0) = 1$ and the Maclaurin series for e^x is

$$\mathrm{e}^x = 1 + x + \frac{1}{2!}x^2 + \frac{1}{3!}x^3 + \cdots = \sum_{k=0}^{\infty} \frac{x^k}{k!}. \qquad \textbf{(1.12)}$$

DERIVE Experiment 1.2

Author expx Plot Beside Plot defines and plots the exponential e^x. Now, return to the Algebra window.

To visualize the convergence of the Maclaurin series (1.12), Author the expression `m(n):=taylor(expx,x,0,n)`, and then (for values of $n = 0, 1, 2, 3$, etc.) Author Simplify and Plot Plot the expressions `m(0)`, `m(1)`, `m(2)`, `m(3)`, etc.

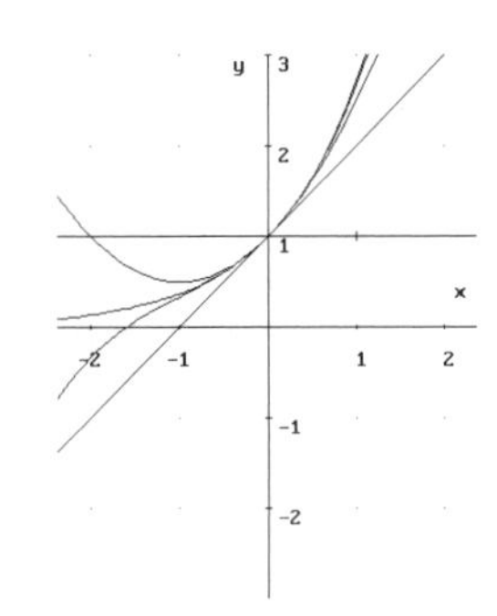

Notice how higher-order polynomials approximate e^x satisfactorily on an increasing interval about $x = 0$.

1.5 Rational (Padé) approximation

The Taylor series model for function approximation is based on polynomial forms. Here the use of *ratios* of low-order polynomials – so-called *rational approximants* – is introduced as a method of function representation. In place of the polynomial

$$p_n(x) = a_0 + a_1x + a_2x^2 + a_3x^3 + \cdots + a_nx^n$$

a polynomial ratio

$$\frac{p_m(x)}{q_r(x)} = \frac{a_0 + a_1x + \cdots + a_mx^m}{b_0 + b_1x + \cdots + b_rx^r} \tag{1.13}$$

is sought. Typically the values of m and r are significantly less than n; that is, a ratio of substantially lower-order polynomials is used. It is assumed that $b_0 \neq 0$, so eqn (1.13) can be normalised with respect to b_0 to give

$$R_{m,r}(x) = \frac{p_m(x)}{q_r(x)} = \frac{\alpha_0 + \alpha_1x + \cdots + \alpha_mx^m}{1 + \beta_1x + \cdots + \beta_rx^r}. \tag{1.14}$$

HENRI EUGÈNE PADÉ
(1863–1953)
French mathematician

LIFE

Padé was educated at the Lycée St Louis and École Normale Supérieure in Paris and then taught in secondary schools. He studied at Leipzig (1889) and Göttingen (1890) under Klein and Schwarz. He returned to France and obtained a doctorate under Hermite's supervision. He then taught at Lille, Poitiers and Bordeaux. In 1906 he received the Grand Prize of the French Academy and two years later became Rector of Besançon. In 1917 he became Rector of the Academy of Dijon and from 1923 until he retired in 1934 he was Rector at Aix-Marseille.

WORK

Padé wrote 41 papers, 29 of which were on continued fractions and Padé approximants. He wrote an elementary algebra book and translated Klein's Erlangen programme into French. His thesis and related papers on Padé approximants became well known after Borel included them in his 1901 book on divergent series.

The coefficients $\alpha_0, \ldots, \alpha_m, \beta_1, \ldots, \beta_r$ are chosen so that $R_{m,r}$ matches f and a specified number of its derivatives at the fixed point $x = 0$. In this way the process is similar to the construction of a Maclaurin series

$$f(x) = \gamma_0 + \gamma_1 x + \gamma_2 x^2 + \cdots + \gamma_n x^n + \cdots \tag{1.15}$$

in which $\gamma_i = f^{(i)}(0)/i!$ and $\gamma_i = 0$ if $i < 0$.

There are $m + 1 + r$ unknown coefficients in eqn (1.14) and matching the first $m+1+r$ derivatives of $R_{m,r}$ with respect to x, at $x = 0$, to the $m+1+r$ coefficients $\gamma_0, \ldots, \gamma_{m+r}$, the following equations are obtained[6]:

$$\begin{bmatrix} \gamma_m & \cdots & \gamma_{m-r+1} \\ \vdots & & \vdots \\ \gamma_{m+r-1} & \cdots & \gamma_m \end{bmatrix} \begin{bmatrix} \beta_1 \\ \vdots \\ \beta_r \end{bmatrix} = \begin{bmatrix} -\gamma_{m+1} \\ \vdots \\ -\gamma_{m+r} \end{bmatrix}, \tag{1.16}$$

$$\alpha_j = \sum_{k=0}^{j} \gamma_{j-k}\beta_k, \quad j = 0, \ldots, m, \tag{1.17}$$

where $\beta_0 = 1$. The system (1.16) can be solved for $\beta_1, \ldots, \beta_r$ and then eqn (1.17) can be used to evaluate $\alpha_0, \ldots, \alpha_m$. $R_{m,r}$ is the (m, r) *Padé approximation*[7] to f. If $r = 0$, $R_{m,0}$ is the Maclaurin expansion for f.

Note $\alpha_j = 0$ for $j > m$ and $\beta_j = 0$ for $j > r$.

Example 1.7 Construct the Maclaurin expansion of order 4 for the function f defined by $f(x) = \ln(1 + x)$. Use the coefficients to construct the Padé approximants $R_{3,1}$ and $R_{2,2}$.

The required derivatives of f are

$$f^{(j)}(x) = (-1)^{j-1}\frac{(j-1)!}{(1+x)^j}, \quad j = 1, 2, 3, 4.$$

Evaluation at $x = 0$ gives $f(0) = 0$, $f'(0) = 1$, $f''(0) = -1$, $f'''(0) = 2$ and $f''''(0) = -6$, and the required Maclaurin expansion is

$$x - x^2 + 2x^3 - 6x^4.$$

---

[6] For details, see [18], Section 6.5.

[7] A comprehensive treatment of Padé approximants is given in [2].

Here $\gamma_0 = 0$, $\gamma_1 = 1$, $\gamma_2 = -1$, $\gamma_3 = 2$ and $\gamma_4 = -6$.

$\boxed{R_{3,1}}$ Here $m = 3$ and $r = 1$, and eqn (1.16) takes the form

$$\gamma_3\beta_1 = -\gamma_4 \Rightarrow 2\beta_1 = 6 \Rightarrow \beta_1 = 3.$$

Using eqn (1.17)

$$\begin{aligned}
\alpha_0 &= \gamma_0\beta_0 = 0 \\
\alpha_1 &= \gamma_1\beta_0 + \gamma_0\beta_1 = 1 \\
\alpha_2 &= \gamma_2\beta_0 + \gamma_1\beta_1 = 2 \\
\alpha_3 &= \gamma_3\beta_0 + \gamma_2\beta_1 = -1,
\end{aligned}$$

and

$$R_{3,1} = \frac{x + 2x^2 - x^3}{1 + 3x}.$$

$\boxed{R_{2,2}}$ Here $m = 2$ and $r = 2$, and eqn (1.16) takes the form

$$\begin{bmatrix} \gamma_2 & \gamma_1 \\ \gamma_3 & \gamma_2 \end{bmatrix} \begin{bmatrix} \beta_1 \\ \beta_2 \end{bmatrix} = \begin{bmatrix} -\gamma_3 \\ -\gamma_4 \end{bmatrix} \Rightarrow \begin{bmatrix} -1 & 1 \\ 2 & -1 \end{bmatrix} \begin{bmatrix} \beta_1 \\ \beta_2 \end{bmatrix} = \begin{bmatrix} -2 \\ 6 \end{bmatrix}$$

$$\Rightarrow \beta_1 = 4, \quad \beta_2 = 2.$$

Using eqn (1.17)

$$\begin{aligned}
\alpha_0 &= \gamma_0\beta_0 = 0 \\
\alpha_1 &= \gamma_1\beta_0 + \gamma_0\beta_1 = 1 \\
\alpha_2 &= \gamma_2\beta_0 + \gamma_1\beta_1 + \gamma_0\beta_2 = 3,
\end{aligned}$$

and

$$R_{2,2} = \frac{x + 3x^2}{1 + 4x + 2x^2}.$$

The improvement shown by $R_{3,1}$ and $R_{2,2}$ over the Maclaurin polynomial is evident (see Figure 1.5). Of note is the fact that each Padé approximant contains a singularity.

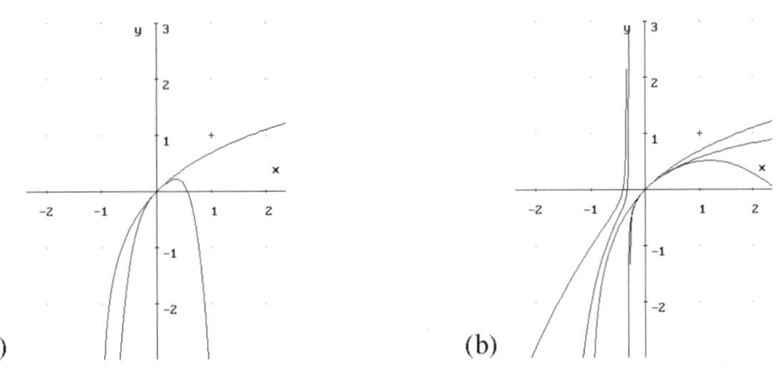

Figure 1.5 (a) Order 4 Maclaurin expansion for $\ln(1 + x)$. (b) $R_{3,1}$ and $R_{2,2}$ Padé approximants to $\ln(1 + x)$.

Summary

In this chapter we have seen how to approximate a function 'f of x' with a polynomial p_n based on information given at a single point $x = x_0$ (Section 1.1). The process was formalised in Section 1.2 to develop the *Taylor polynomial* (1.6). In Section 1.3 questions of accuracy were discussed and, in particular, an upper bound for the error $|R_n(x)| = |f(x) - p_n(x)|$ was developed via *Taylor's theorem* (1.9). Section 1.4 dealt with the special case in which $x_0 = 0$, the *Maclaurin series* (1.11), and in Section 1.5 the approximation was extended to rational functions (ratios of polynomials) leading to the *Padé approximant*.

Exercises

1.1 Use eqns (1.1)–(1.3) to obtain constant, linear and quadratic approximations to e^x constructed at the point $x_0 = 1$.

1.2 Show that the cubic polynomial p_3 of eqn (1.4) matches f and its first three derivatives at the point $x = x_0$.

1.3 If f is a polynomial of degree n, show that f is its own Taylor series.

1.4 Determine the first three *non-zero* terms of the Maclaurin series for $\sin x$ and for $\cos x$.

Determine the Taylor series for $\cos(a + h)$ about a and show that the series can be written in the form

$$\cos a\left[1 - \frac{h^2}{2!} + \frac{h^4}{4!} - \cdots\right] - \sin a\left[h - \frac{h^3}{3!} + \frac{h^5}{5!} - \cdots\right].$$

To which trigonometric identity does this correspond?

1.5 Show that if $\tan x = \frac{2}{5}$ and $\tan y = \frac{3}{7}$ then $\tan(x + y) = 1$. Deduce that $\tan^{-1}(2/5) + \tan^{-1}(3/7) = \pi/4$. Use $\mathcal{DERIVE}$ to obtain the Maclaurin series for $\tan^{-1} x$, and apply the series to evaluate $\tan^{-1}(2/5)$ and $\tan^{-1}(3/7)$ to 6 decimal places. Hence, evaluate π to 5 decimal places.

1.6 Obtain the Maclaurin series for e^x and e^{-x} and hence verify that the Maclaurin series for $\cosh x$ is

$$1 + \frac{x^2}{2!} + \frac{x^4}{4!} + \frac{x^6}{6!} + \cdots.$$

Find the smallest value of n for which

$$\max_{-1 \le x \le 1} |\cosh x - p_n(x)| \le 10^{-4}.$$

Deduce that if $x \geq 0$ then

$$\int_0^x \cosh\sqrt{t}\,\mathrm{d}t = x + \frac{x^2}{2\cdot 2!} + \frac{x^3}{3\cdot 4!} + \frac{x^4}{4\cdot 6!} + \frac{x^5}{5\cdot 8!} + \cdots,$$

and evaluate $\int_0^{\frac{1}{2}} \cosh\sqrt{t}\,\mathrm{d}t$ to 3 decimal places.

1.7 Use *DERIVE* to generate the first few terms of the Maclaurin series for $\tan x$. Hence determine the value of the expression

$$\lim_{x\to 0} \frac{\tan x}{x}.$$

Obtain the result directly by using *DERIVE* – `Author` and `Simplify` the expression `lim(tanx/x,x,0)`. You might plot $\tan x/x$ to confirm the analysis – `Author` `tanx/x` and `Plot`.

1.8 Use *DERIVE* to explore the Taylor series for $\sin x$ (a) about $x_0 = 0$ and (b) about $x_0 = \frac{\pi}{2}$. For each value of x_0, plot the Taylor polynomials up to degree $n = 9$ and comment upon the 'goodness of fit' on the interval $0 \leq x \leq \pi$. The *DERIVE* expression

```
vector(taylor(f,x,x0,n),n,a,b,c)
```

will generate the Taylor polynomials p_n for values of n from a to b in steps of c. The result is returned as a vector which can be plotted using `Plot`. The expression

```
vector(taylor(f,x,x0,n)-f,n,a,b,c)
```

returns the vector of errors, $f(x) - p_n(x)$.

Differentiate the polynomials obtained for $x_0 = 0$ (the *DERIVE* expression `dif(f,x)` returns the derivative $\mathrm{d}f/\mathrm{d}x$). What have you obtained? Confirm your answer.

1.9 Refer to the observation on p. 20, viz eqn (1.8) approaching zero as $x \to x_0$.

Obtain the Taylor series for $\sqrt{x}$, $x \geq 0$, about the point $x_0 = 1$. For $n = 3$, tabulate (a) $p_n(x)$, (b) $|\sqrt{x} - p_n(x)|$ and (c) $|R_n(x)|$ for values of $x = 0, 0.5, 0.75, 0.875$ and 0.9375. Comment on the behaviour of the values arising from (b) and (c).

1.10 Using the information provided in Example 1.7, construct the $R_{4,0}$, $R_{1,3}$ and $R_{0,4}$ Padé approximants to $\ln(1 + x)$.

1.11 Obtain the $R_{1,1}$ and $R_{2,2}$ Padé approximants to e^x and compare them (graph using *DERIVE*, for example) with the underlying Maclaurin expansions of order 2 and 4.

2 Numbers, errors and arithmetic

By the very nature of applied numerical analysis, many of the ideas discussed in this text must be implemented on a calculating machine, often a personal computer. Although computers are excellent at rapidly performing huge quantities of arithmetic, an inherent weakness is their tendency to introduce and propagate numerical errors.

In this chapter several issues relating to the computer representation and arithmetic of numbers will be discussed together with a simple assessment of the cause, measurement and control of arithmetic error.

2.1 Representation of numbers

We shall be concerned with two types of number, *integers* and *reals*.

Definition An integer N consists of an optional sign, $+$ or $-$, followed by a 'string' of *decimal* digits,

$$N = \pm d_1 d_2 d_3 \dots d_m \tag{2.1}$$

m is the *word length* of N and equals the number of digits in the string.

By convention the first digit d_1 is not zero[1]. For example, we do not write 034 to mean thirty-four, we write 34. Thus

$$d_1 \in \{1, 2, 3, 4, 5, 6, 7, 8, 9\}.$$

[1] An exception to the 'leading zeros' rule is the number zero itself; 0 *is* a valid integer.

The remaining digits $d_2, \ldots, d_m$ may take the value zero, that is

$$d_i \in \{0, 1, 2, 3, 4, 5, 6, 7, 8, 9\}, \quad i = 2, \ldots, m.$$

Example 2.1 1205, -100 and $+34$ are valid integers whereas 0093 (leading zeros) and -12.1 (decimal point) are invalid integers.

At this point it is useful to ask

what is the significance of the digits d_1 to d_m?

Example 2.2 34 is a concise representation of the quantity '3 times 10 plus 4', that is $3 \times 10 + 4$. This can be further expanded to $3 \times 10^1 + 4 \times 10^0$. In other words, 34 represents a specific combination of descending powers of 10.

The integer N, defined in eqn (2.1), can be expanded as

$$N = \pm d_1 \times 10^{m-1} + d_2 \times 10^{m-2} + \cdots + d_{m-1} \times 10^1 + d_m \times 10^0. \qquad \textbf{(2.2)}$$

The integer 10 is called the *base* and gives rise to the familiar *decimal number system*. The term involving 10^0 is called the *units*, the term with 10^1 the *tens*, the term with 10^2 the *hundreds*, etc.

Example 2.3 A convenient way of representing an integer in the form (2.2) is to use two rows of boxes. The upper row contains descending powers of the base and the lower row contains the multiple (coefficient) of each power appearing in the number N. Thus,

$$113 = 1 \times 10^2 + 1 \times 10^1 + 3 \times 10^0$$

can be represented as

Base 10

10^2	10^1	10^0
1	1	3

$N =$ 100 +10 +3 = 113

Reading from left to right, we have the familiar hundreds, tens and units columns.

The *base* used by a computer is usually 2 (the *binary number system*) and

$$N = \pm b_1 b_2 b_3 \ldots b_m \tag{2.3}$$

$$= \pm b_1 \times 2^{m-1} + b_2 \times 2^{m-2} + \cdots + b_{m-1} \times 2^1 + b_m \times 2^0. \tag{2.4}$$

The b_i are *binary digits* satisfying[2] $b_1 = 1$ and $b_i \in \{0, 1\}, i = 2, \ldots, m$.

Example 2.4 Repeating Example 2.3 with the binary number system,

$$113 = (1110001)_2$$

$$= 1 \times 2^6 + 1 \times 2^5 + 1 \times 2^4 + 0 \times 2^3 + 0 \times 2^2 + 0 \times 2^1 + 1 \times 2^0,$$

that is

Base 2

2^6	2^5	2^4	2^3	2^2	2^1	2^0
1	1	1	0	0	0	1

$N =$ 64 +32 +16 +0 +0 +0 +1 = 113

Definition A *real* number x has two standard forms. The everyday form consists of an optional sign, + or −, followed by two unsigned integers separated by a decimal point,

$$x = \pm d_1 d_2 d_3 \ldots d_j . d_{j+1} d_{j+2} \ldots d_m \tag{2.5}$$

This *fixed-point* form is so-named since it is the position of the *decimal point* that fixes the magnitude of the number, be it tens or tenths, etc.

$d_1 \ldots d_j$ is the *integer part* and $d_{j+1} \ldots d_m$ is the *fractional part*, where $d_i \in \{0, 1, 2, 3, 4, 5, 6, 7, 8, 9\}$, $i = 1, \ldots, m$, subject to $d_1 \neq 0$.

Example 2.5 Examples of valid fixed-point real numbers are

Number	Magnitude	Number	Magnitude
31.02	Tens	.93	Tenths
−2.6147	Units	429.	Hundreds
0.93	Tenths	0.003	Thousandths

[2] If $m = 1$ then $b_1 = 0$ *is* valid, i.e. 0 is a valid binary number.

Examples of invalid fixed-point real numbers are

Number	Reason	Valid form
031.02	Leading zero in integer part	31.02
429	No decimal point	429.

Observations *Either (but not both) the integer part or fractional part can be omitted (as in .93 and 429.). However, 429. is not necessarily the same as 429.0 (this is elaborated in Section 2.3.3) and the fractional part may have leading zeros (as in 0.003).*

Definition A *floating-point* real number x has an optional sign, + or −, a decimal point and an unsigned integer multiplied by a power of the base,

$$x = \pm .d_1 d_2 d_3 \ldots d_m \times 10^n. \tag{2.6}$$

The signed integer n is the *exponent* and $d_i \in \{0, 1, 2, 3, 4, 5, 6, 7, 8, 9\}$, $i = 1, \ldots, m$, subject to $d_1 \neq 0$. It is now the value of the exponent that fixes the magnitude of the number, and this form (in binary) is used by computers to represent real numbers,

$$x = \pm .b_1 b_2 b_3 \ldots b_m \times 2^n. \tag{2.7}$$

Example 2.6 Examples of valid floating-point real numbers are

Number	Magnitude	Number	Magnitude
$.3102 \times 10^2$	Tens	$.93 \times 10^0$	Tenths
$-.26147 \times 10^1$	Units	$.429 \times 10^3$	Hundreds
$.93 \times 10^0$	Tenths	$.3 \times 10^{-2}$	Thousandths

Examples of invalid floating-point real numbers are

Number	Reason	Valid form
$.03102 \times 10^3$	Leading zero	$.3102 \times 10^2$
42.9×10^1	Decimal point not in first position	$.429 \times 10^3$

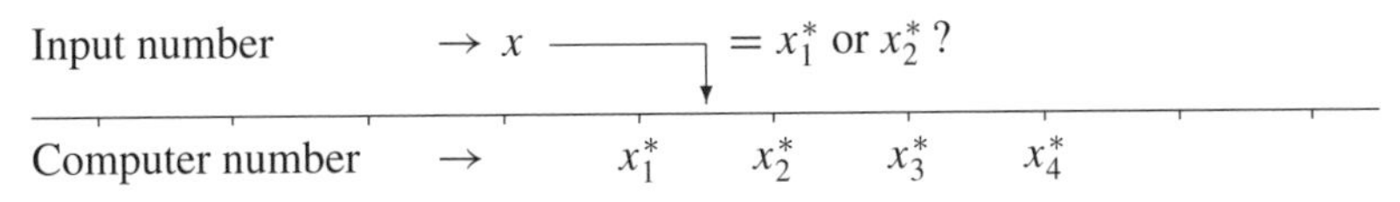

Figure 2.1 Typical number line for a finite-precision machine.

2.1.1 Storage errors

The action of entering a number (via the keyboard, for example) into a computer may introduce an error, even before any arithmetic has taken place! Here we illustrate this statement by looking at the method used by most modern computers to store numbers.

In decimal terms a stored computer number has the floating-point form[3]

$$x^* = \pm .d_1^* d_2^* d_3^* \ldots d_w^* \times 10^n, \qquad \textbf{(2.8)}$$

where $d_i^* \in \{0, 1, 2, 3, 4, 5, 6, 7, 8, 9\}$, $i = 1, \ldots, w$, and $d_1^* \neq 0$. w is the *computer word length*. Ignoring the exponent, eqn (2.8) provides for a total of $9 \times 10^{w-1}$ possible numbers (d_1^* can take one of nine values and the remaining $w - 1$ digits can each take one of ten values). It is known that there exists an infinity of real numbers. However, only $9 \times 10^{w-1}$ can be stored (recognised) exactly by the computer[4] and on entering an arbitrary real number x the chances are high that x will not be stored (recognised) exactly. Instead, x is stored as a 'nearby' number x^* which the computer does 'recognise' (see Figure 2.1).

Example 2.7

On a two-digit fixed-point machine, 100 numbers can be stored exactly, from .00 to .99 in steps of .01

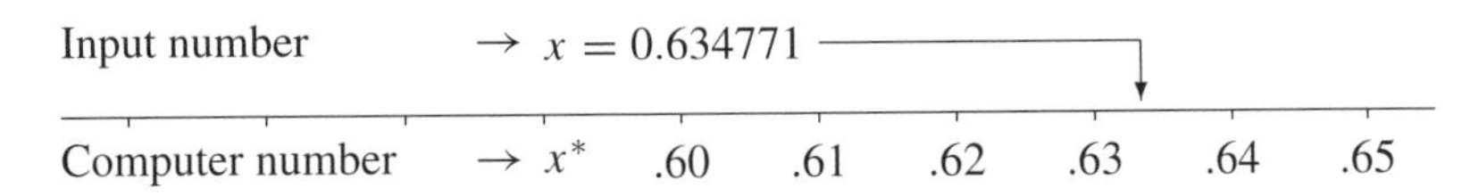

The question is what number x^* is stored by the computer given an input number such as $x = 0.634771$?

The mechanism used by a computer to get from x to x^* is discussed in Section 2.3, but first we need a means of quantifying the difference between two numbers, say x and x^*.

[3] The * denotes that the digits in the stored number x^* may not be the same as those in the entered number x.

[4] Such a computer is termed a *finite-precision machine*.

2.2 Error measurement

Numerical errors will almost certainly creep into computer arithmetic. Before an analysis of these errors can be made, a system of error measurement must be established.

2.2.1 Absolute error

Definition Let x^* be an estimate to the number x. The *absolute error* in x^* is defined as

$$\Delta = |x - x^*|. \tag{2.9}$$

Δ is the 'distance' between the numbers x and x^*. In general it is not important to know whether x^* is an overestimate or an underestimate, and by convention the *modulus* of $x - x^*$ is taken, so Δ is non-negative.

Example 2.8

Table 2.1 shows several values of an exact value x, an approximation x^* and the associated absolute error Δ.

Table 2.1 Examples of absolute error.

x	x^*	Δ
1	0.99	0.01
1	1.01	0.01
−1.5	−1.2	0.3
100	99.99	0.01
100	99	1

Observations *In Example 2.8 the estimates 0.99 and 1.01 as approximations to 1 both have the same absolute error, despite one being an underestimate and one an overestimate. Further, 99.99 as an approximation to 100 also has an absolute error of 0.01, yet intuitively we might consider this a better approximation than 0.99 is to 1. The absolute error does* not *distinguish between these two cases.*

An alternative interpretation of Δ is to define an *interval of certainty*. Given an estimate x^* to x with an associated absolute error Δ, then x lies in the interval $x^* - \Delta \leq x \leq x^* + \Delta$.

Example 2.9

If $x^* = 5.6372$ and $\Delta = 0.002$, $x \in [5.6352, 5.6392]$. Since the error is quantified to three decimal places it seems perverse to specify the value of x^* (and the interval of certainty for x) to four decimal places. Intuitively we might expect to be able to write $x^* = 5.64$ with some certainty; that is, the lower-order decimal digits (in the thousandths and ten-thousandths positions) are untrustworthy and hence should be ignored in 'some way' (discussed in Section 2.3).

2.2.2 Relative error

The weakness of Δ as a measure of numerical error is that it does *not* account for the *magnitude* of the numbers involved.

Definition The *relative error* is defined as

$$\delta = \frac{\Delta}{|x|} \simeq \frac{\Delta}{|x^*|} \tag{2.10}$$

and measures the error in x^* relative to the size of the number itself.

Example 2.10 Extending Table 2.1, Table 2.2 includes the computed relative error.

The relative error *does* distinguish the intuitive differences observed in the first and fourth approximations, that 99.99 is a better estimate to 100 than 0.99 is to 1. Further, 99 is as good an approximation to 100 as 0.99 is to 1.

Table 2.2 Examples of relative error.

x	x^*	Δ	δ
1	0.99	0.01	0.01
1	1.01	0.01	0.01
−1.5	−1.2	0.3	0.2
100	99.99	0.01	0.0001
100	99	1	0.01

Note As a rule of thumb, the relative error should be no greater than about 0.01, which implies a 1% error in x^*.

DERIVE Experiment 2.1

Activate approximate mode – `Options Precision Approximate` – and `Author a(x,y) :=abs(x-y)` to define the absolute error. `Author a(1,.99) Simplify` returns the value 0.01 and Author a(1,y) Plot Beside Plot Algebra plots the absolute error of approximating 1 by the number y – it is a function of y.

To 'see' the weakness of Δ, Author vector(a(x,y), x, .1, 1, .1) Plot Plot Algebra to define and plot a vector of values of the absolute error in approximating x by y. The slope is either 1 or -1, for *all* values of x. In other words, the variation (sensitivity) of Δ to changes in the approximation y is independent of the value x that is being approximated.

The relative error is defined by Author r(x,y):=a(x,y)/abs(x) and to 'see' the behaviour of δ for the data given above Author the expression vector(r(x,y), x, .1, 1, .1), and Simplify Plot Window Split Horizontal <Enter> Delete All Plot Algebra – notice how the slope of the relative error is greater for values of x closer to zero. In other words the variation (sensitivity) of δ to changes in y is increased for smaller values of x.

2.3 Errors in computation

In Section 2.1.1 it was stated that errors may occur during data input to a computer. Here these errors are quantified using the measures proposed in Section 2.2.

In order to form a basis for the analysis of computer arithmetic (see Section 2.4) we need to revisit the mechanism of storing a number and to quantify precisely the errors occurring in the process. In other words,

what are the values of Δ and δ when a number x is entered into a computer and stored as the number x^*?

In (decimal) floating-point form the real number

$$x = \pm .d_1 d_2 d_3 \ldots d_{m-1} d_m \times 10^n$$

is stored on a computer as the real number

$$x^* = \pm .d_1^* d_2^* d_3^* \ldots d_{w-1}^* d_w^* \times 10^n.$$

Figure 2.2 serves as a useful visual *aide-mémoire* and if, as is implied, the computer word length w is greater than or equal to the number word length m, then there is a 'box' in the computer to store each digit d_i of x. In other words, $d_1^* = d_1$,

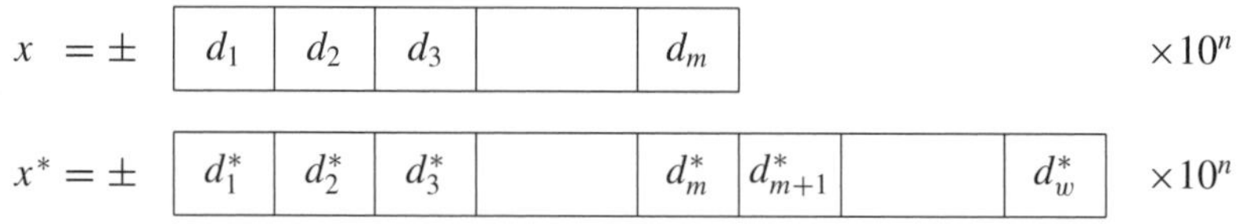

	d_1	d_2	d_3		d_m				
$x = \pm$	d_1	d_2	d_3		d_m				$\times 10^n$
$x^* = \pm$	d_1^*	d_2^*	d_3^*		d_m^*	d_{m+1}^*		d_w^*	$\times 10^n$

Figure 2.2 Storing a real number exactly.

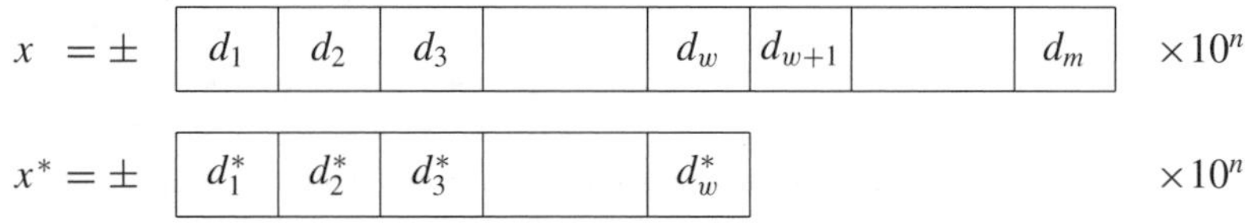

$x = \pm$	d_1	d_2	d_3		d_w	d_{w+1}		d_m	$\times 10^n$
$x^* = \pm$	d_1^*	d_2^*	d_3^*		d_w^*				$\times 10^n$

Figure 2.3 Storage error.

$d_2^* = d_2, \ldots, d_m^* = d_m$. The remaining computer digits are set to zero, that is $d_i^* = 0$, $i = m+1, \ldots, w$. In this case there is *no* error – the stored number is precisely equal to the input number, $x^* = x$.

Example 2.11 $x = 26.493$ (fixed-point form) has the floating-point form $x = .26493 \times 10^2$, that is $m = 5$ and $n = 2$. If $w = 6$ (usual PC word length) then x will be stored as the number $x^* = .264930 \times 10^2$, that is 26.4930 (note the introduction of the zero in the unused ten-thousandths position). Clearly $x - x^*$ equals zero.

If, on the other hand, $w < m$ (see Figure 2.3), an error will occur on storing the number x since just w of the m digits in x can be included, in some form, in the number x^*. In other words, $d_i^* \neq d_i$ and $x^* \neq x$.

Computers use one of two algorithms for storing an input number x as the approximation x^*, called *chopping* and *rounding*.

2.3.1 Chopping

The simplest mechanism of storing x when $w < m$ is to ignore all digits in x after d_w – called *chopping*. In this case $d_i^* = d_i, i = 1, \ldots, w$, and

$$x = \pm .d_1 d_2 d_3 \ldots d_{w-1} d_w \ldots d_{m-1} d_m \times 10^n$$

$$x^* = \pm .d_1 d_2 d_3 \ldots d_{w-1} d_w \qquad \times 10^n$$

The number x is *chopped to w digits*. From Figure 2.4, if $x_i^* \leq x < x_{i+1}^*$ then $x^* = x_i^*$. In other words, x^* equals the computer number that is less than or equal to x.

Example 2.12 To chop $x = 31.2357$ to 4 digits, write x in floating-point form, $x = .312357 \times 10^2$, 'throw away' all digits after the fourth to give $x^* = .3123 \times 10^2$, and return to fixed-point form $x^* = 31.23$ (if desired).

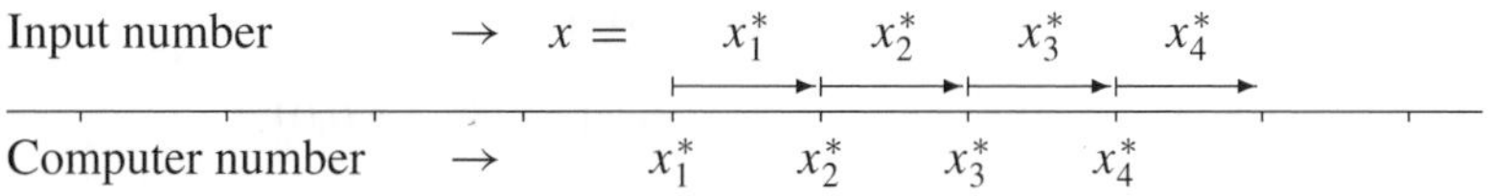

Input number $\rightarrow$ $x =$ x_1^* x_2^* x_3^* x_4^*

Computer number $\rightarrow$ x_1^* x_2^* x_3^* x_4^*

Figure 2.4 Chopping x on a finite-precision machine.

Computer number	$\rightarrow$	x_1^*	x_2^*	x_3^*	x_4^*
		Δ	Δ	Δ	

Figure 2.5 Maximum absolute chopping error equals the distance between adjacent computer numbers.

The *absolute chopping error* is

$$\Delta = |x - x^*| = .00\ldots00d_{w+1}\ldots d_m \times 10^n.$$

There are w zeros before the digit d_{w+1}, so

$$\Delta = .d_{w+1}\ldots d_m \times 10^{n-w}.$$

This provides a value for the absolute error in x^*. However, for every number x the computation needs to be repeated, since different numbers have different digits, and the solution to evaluating Δ is to compute the worst-case scenario – an *upper bound* in 'mathspeak'.

The largest value of Δ occurs when $d_{w+1} = \cdots = d_m = 9$, that is

$$\Delta \leq .9\ldots9 \times 10^{n-w}.$$

However, $.9\ldots9$ is (marginally) less than 1, so

$$\Delta < 1 \times 10^{n-w} = 10^{n-w}, \qquad \textbf{(2.11)}$$

shown in Figure 2.5. For example, the two-digit computer of Example 2.7 (p. 33) admits $w = 2$, and with $n = 0$ satisfies $\Delta < 10^{-2} = .01$ – precisely the distance between successive computer numbers.

The upper bound (2.11) holds for all numbers stored on a computer of word length w. All that is required is the exponent n, once x is written in floating-point form.

It follows that an upper bound for the *relative chopping error* δ is given by

$$\delta = \frac{\Delta}{|x|} < \frac{10^{n-w}}{.d_1d_2\ldots d_m \times 10^n}.$$

$.d_1d_2\ldots d_m$ is greater than or equal to .1 (remember that $d_1 \neq 0$), so

$$\delta < \frac{10^{n-w}}{.1 \times 10^n} = 10^{1-w}. \qquad \textbf{(2.12)}$$

Example 2.13 For Example 2.12, $w = 4$ and $n = 2$. Using eqn (2.11), the absolute chopping error is bounded by 10^{2-4}, that is 0.01. The *observed* error is

$$\Delta = |x - x^*| = |31.2357 - 31.23| = 0.0057$$

which *is* bounded by the theoretical maximum 0.01.

From eqn (2.12) the relative error is bounded by 10^{1-4}, that is 0.001. The observed relative error is

$$\delta = \frac{\Delta}{|x|} = \frac{0.0057}{31.2357} \simeq 0.00018$$

which *is* certainly bounded by 0.001.

The upper bounds (2.11) and (2.12) appear to hold in practice although it should be noted that they are very pessimistic!

2.3.2 Rounding

The usual mechanism for storing x when $w < m$ is to use the *closest* number x^* that can be stored exactly on the computer. Called *rounding*, this process entails adding 5 to the digit d_{w+1} and chopping the result to w digits.

Example 2.14 The number set of a certain computer consists of the integers 0, ±1, ±2, ±3, to ±9.

Given a number x, what is its closest neighbour in this set?

Here the digit d_w is the units column, and 5 is added to the tenths column, that is .5 is added to x.

If $x = 4.6$ then $x + .5 = 5.1$ which chops to 5 (4.6 is closer to 5 than 4). Hence, 4.6 is stored as 5 – to integer accuracy. The chopping algorithm stores 4.6 as 4, so the storage mechanism can affect the arithmetic.

On the other hand, if $x = 4.2$ then $x + .5 = 4.7$ which chops to 4 (chopping also stores 4.2 as 4).

So,

what are the error bounds for the rounding algorithm?

Adding 5 to the digit d_{w+1} gives an intermediate number x^+,

$$\begin{array}{rlcr} x & = \pm .d_1 d_2 d_3 & \ldots\ d_{w-1} d_w d_{w+1} d_{w+2}\ \ldots\ d_{m-1} d_m & \times 10^n \\ & \pm . & 5 & \times 10^n \\ \hline x^+ & = \pm .d_1^+ d_2^+ d_3^+ & \ldots\ d_{w-1}^+ d_w^+ d_{w+1}^+ d_{w+2}\ \ldots\ d_{m-1} d_m & \times 10^n \end{array}$$

From right to left, digits d_m to d_{w+2} are unchanged. Digit d_{w+1} is certainly altered (5 has been added) and digits d_w back to d_1 are possibly altered due to the 'carrying' process in addition. The final rounded number is

$$x^* = \pm .d_1^* d_2^* d_3^* \ldots d_{w-1}^* d_w^* \times 10^n,$$

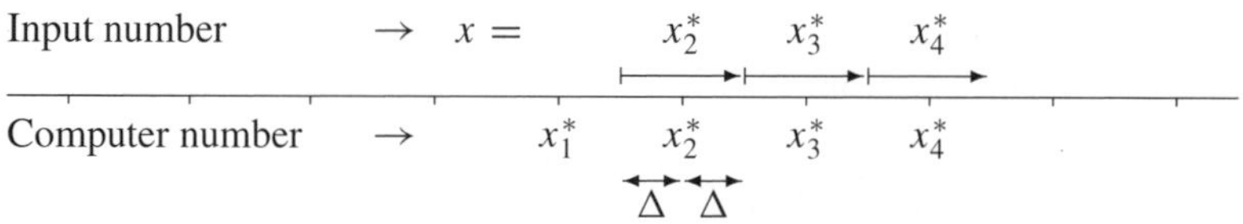

Figure 2.6 Rounding x on a finite-precision machine. The maximum absolute rounding error equals half the distance between adjacent computer numbers.

where $d_i^* = d_i^+$, $i = 1, \ldots, w$. To develop a bound for the error $|x - x^*|$ it is convenient to consider separately the cases $d_{w+1} \leq 4$ and $d_{w+1} \geq 5$.

If $d_{w+1} \in \{0, 1, 2, 3, 4\}$, adding 5 to d_{w+1} results in a digit no greater than 9, that is $d_{w+1}^+ \in \{5, 6, 7, 8, 9\}$ and no carrying is required to determine the intermediate number x^+. In other words, $d_i^+ = d_i$, $i = 1, \ldots, w$, and chopping x^+ to w digits results in the rounded number

$$x^* = \pm .d_1 d_2 d_3 \ldots d_{w-1} d_w \times 10^n.$$

The *absolute rounding error* is

$$\begin{aligned} \Delta = |x - x^*| &= .00 \ldots 00 d_{w+1} \ldots d_m \times 10^n \\ &= .d_{w+1} \ldots d_m \times 10^{n-w}. \end{aligned}$$

The largest value of Δ occurs when $d_{w+1} = 4$ and the remaining digits equal 9, that is

$$\Delta \leq .49 \ldots 9 \times 10^{n-w}.$$

Since $.49 \ldots 9$ is less than .5, then (see Figure 2.6)

$$\Delta < .5 \times 10^{n-w}. \tag{2.13}$$

This value is precisely half the value for chopping, eqn (2.11).

The *relative rounding error* δ is, by definition,

$$\delta = \frac{\Delta}{|x|} < \frac{.5 \times 10^{n-w}}{.d_1 d_2 d_3 \ldots d_m \times 10^n}.$$

$.d_1 d_2 d_3 \ldots d_m$ is greater than or equal to .1 ($d_1 \neq 0$) and

$$\delta < \frac{1}{2} \times 10^{1-w}. \tag{2.14}$$

The error bounds (2.13) and (2.14) also hold for the case where $d_{w+1} \in \{5, 6, 7, 8, 9\}$.

Example 2.15 To round $x = 31.2357$ to 4 digits, write down the floating-point form $x = .312357 \times 10^2$, add 5 to the fifth digit to give the intermediate value $x^+ = .312407 \times 10^2$, 'throw away' all digits after the fourth to give $x^* = .3124 \times 10^2$ and return to fixed-point form, $x^* = 31.24$ (if required). This is different from the chopped value of Example 2.12.

The *observed* absolute and relative errors are $\Delta = 0.0043$ and $\delta \simeq 0.00014$. Using eqns (2.13) and (2.14) with $w = 4$ and $n = 2$ the error bounds are $.5 \times 10^{2-4}$ (0.005) and $.5 \times 10^{1-4}$ (0.0005). Thus, the bounds (2.13) and (2.14) appear to hold in practice.

Observations *Although the error bounds for rounding are half those for chopping, the* observed *rounding error is not necessarily half the observed chopping error.*

DERIVE Experiment 2.2

To get *DERIVE* to chop and round 'to order' we need to 'manufacture' a few functions. `Author n(x):=floor(1.5+log(x,10))` defines a function n that returns the exponent of x when written in floating-point form. For example `Author n(23.478) Simplify` returns the value 2 and `Author n(100)` gives the value 3.

To chop a number x to w digits `Author`

```
chop(x,w):=floor(10^(w-n(x))x)/10^(w-n(x))
```

Thus, `Author chop(23.478,4) Simplify` gives 23.47. To round a number x to w digits `Author`

```
round(x,w):=chop(x+.5*10^(n(x)-w),w)
```

Thus, `Author round(23.478,4) Simplify` gives 23.48.

Experiment with these two functions. When do `chop` and `round` give the same result? What happens if w is greater than or equal to m (the number word length)? Make sure that the 'digits' value under `Options Precision` is set to a large value (say 20).

2.3.3 Significant digits

The phrase '... is rounded to ... significant digits' is often encountered in numerical analysis texts. Here the term *significant digit* is explained.

Definition *Significant digits* in a decimal number are defined as

(1) all digits from 1 to 9,

(2) zeros *not* fixing the position of the decimal point, and

(3) zeros *not* replacing decimal places of lower order.

Property 3 describes zeros appearing in a number to the right of the last non-zero digit that are not simply 'filling in' for unspecified digits.

Example 2.16 The following decimal numbers all have four significant digits:

Number	Explanation
5813.	All digits are non-zero (Property 1)
58.13	All digits are non-zero
58130.	The zero *is* fixing the decimal point (Property 2) – the magnitude *is* 'ten thousands'. Without the zero, 5813 has magnitude 'thousands'.
.0005813	The three zeros *are* fixing the decimal point (Property 2).

Leading zeros are not significant. However, trailing zeros (Property 3) are more difficult to classify and an example may be illuminating.

Example 2.17 In the number 2.30 the zero may mean that the multiplier for the 'hundredths' term is zero (the zero *is* significant). It may also mean that no digit is defined for the 'hundredths' position.

If, for example, 2.3 is entered on a computer with word length 3 then in the stored number 2.30, which is exact, the zero is meaningless – there is no digit in the original number in the 'hundredths' position and 2.30 only has two significant digits.

On the other hand, if 2.301 is entered on the same computer (using the rounding algorithm), the stored number will be 2.30 where the zero really does mean that there are no hundredths. In this case 2.30 has three significant digits.

If a (decimal) floating-point number is rounded to w digits then the following statements are equivalent:

(1) The number has w significant digits.

(2) $\Delta < .5 \times 10^{n-w}$.

(3) $\delta < .5 \times 10^{1-w}$.

(4) The number has $w - n$ decimal places of accuracy.

2.4 Computer arithmetic

Thus far, the error analysis has concentrated upon *storage* errors. Once a number x has been entered, it is likely to be used in an arithmetic process, such as computing the value of the expression $x^2 - 3x + 1$. In practice the stored number x^* is used and the computed value is $(x^*)^2 - 3x^* + 1$.

It is the aim of this section to provide some elementary ideas concerning the *propagation of errors* through an arithmetic process on using a stored approximation x^* in place of the entered data x.

Let x^* and y^* estimate the numbers x and y, with known upper bounds for the absolute (Δ_x and Δ_y) and relative (δ_x and δ_y) errors, that is $x = x^* \pm \Delta_x$ and $y = y^* \pm \Delta_y$. Arithmetic operations on x^* and y^* can be classified as combinations of addition, subtraction, multiplication and division.

2.4.1 Addition

In practice the number $x^* + y^*$ is computed with absolute error

$$\begin{aligned} \Delta_{x+y} &= |(x+y) - (x^* + y^*)| \\ &= |x - x^* + y - y^*| \\ &\le |x - x^*| + |y - y^*| = \Delta_x + \Delta_y, \end{aligned} \tag{2.15}$$

where use has been made of the *triangle inequality*. In other words, the absolute error in the sum of two numbers is bounded above by the sum of the individual absolute errors. The relative error bound is[5]

$$\delta_{x+y} = \frac{\Delta_{x+y}}{|x+y|} \le \frac{\Delta_x + \Delta_y}{|x+y|} \simeq \frac{\Delta_x + \Delta_y}{|x^* + y^*|}. \tag{2.16}$$

2.4.2 Subtraction

$x - y$ is approximated by $x^* - y^*$ with absolute error

$$\begin{aligned} \Delta_{x-y} &= |x - y - (x^* - y^*)| \\ &= |x - x^* - (y - y^*)| \\ &\le |x - x^*| + |y - y^*| = \Delta_x + \Delta_y, \end{aligned} \tag{2.17}$$

which is the same as for addition. This is hardly surprising since subtraction is the addition of a negative number. The relative error is

$$\delta_{x-y} = \frac{\Delta_{x-y}}{|x-y|} \le \frac{\Delta_x + \Delta_y}{|x-y|} \simeq \frac{\Delta_x + \Delta_y}{|x^* - y^*|}. \tag{2.18}$$

This last expression highlights a 'hidden' danger when performing arithmetic operations. If x and y are 'close' together then the difference $x - y$ will be small and, consequently, the relative error may be large. This implies that subtracting two numbers that are close together may result in a loss of significant digits in the result.

---

[5] In practice, one often only has access to the approximation x^* and associated maximum absolute error. In this case the relative error bound can only be estimated by $\Delta/|x^*|$.

Example 2.18 The numbers 0.1234 and 0.1236 both have 4 significant digits. If they represent rounded values of input data, then the maximum absolute error in each case is $\frac{1}{2} \times 10^{-4}$. The corresponding relative error bounds are 0.00041 and 0.00040, respectively.

Their difference equals 0.0002 which has just one sigificant digit – the three zeros simply fix the position of the decimal point.

Using eqns (2.17) and (2.18),

$$\Delta_{x-y} \leq \Delta_x + \Delta_y = \frac{1}{2} \times 10^{-4} + \frac{1}{2} \times 10^{-4} = 10^{-4},$$

and

$$\delta_{x-y} \leq \frac{\Delta_{x-y}}{|x^* - y^*|} = \frac{10^{-4}}{2 \times 10^{-4}} = \frac{1}{2}.$$

The magnitude of the absolute error is largely unaffected, whereas the relative error has increased by three orders of magnitude.

2.4.3 Multiplication

The propagation of errors in multiplication is less straightforward. In place of xy the product x^*y^* is computed with absolute error

$$\begin{aligned}\Delta_{xy} &= |xy - x^*y^*| \\ &= |(x^* + \Delta_x)(y^* + \Delta_y) - x^*y^*| \\ &= |x^*\Delta_y + y^*\Delta_x + \Delta_x\Delta_y| \\ &\leq |x^*\Delta_y| + |y^*\Delta_x| + |\Delta_x\Delta_y| \\ &\simeq |x^*|\Delta_y + |y^*|\Delta_x.\end{aligned} \tag{2.19}$$

To arrive at eqn (2.19) it is assumed that the magnitude of Δ is much smaller than that of x and y, and so the product $\Delta_x\Delta_y$ is much less than $|x^*|\Delta_y$ and $|y^*|\Delta_x$, and may be neglected. The relative error is

$$\delta_{xy} = \frac{\Delta_{xy}}{|xy|} \simeq \frac{\Delta_{xy}}{|x^*y^*|} = \delta_x + \delta_y. \tag{2.20}$$

A further 'hidden trap' is implied by eqn (2.19).

Example 2.19 The numbers 123456 and 21.84 have associated (maximum) abolute errors of .5 and .005, respectively. The corresponding relative errors are $.4 \times 10^{-5}$ and $.2 \times 10^{-3}$, respectively.

The product 2696279.04 has error bounds

$$\Delta_{xy} \leq 123456 \times .005 + 21.84 \times .5 = 628.2 \quad \text{and} \quad \delta_{xy} \leq .204 \times 10^{-3}.$$

The relative error of the product is no worse than the worst relative error of the individual numbers. However, the absolute error has soared. The 'cautionary note' is 'don't multiply by large numbers' – they also multiply the absolute errors!

2.4.4 Division

By means of an analysis similar to that of Section 2.4.3, the following formulae can be obtained for the propagation errors associated with division:

$$\Delta_{x/y} \leq \frac{|x^*|\Delta_y + |y^*|\Delta_x}{(y^*)^2}, \tag{2.21}$$

$$\delta_{x/y} \leq \delta_x + \delta_y. \tag{2.22}$$

Example 2.20 This example reinforces the caveats identified in Examples 2.18 and 2.19, and provides a summary of the pitfalls when implementing arithmetic on a finite-precision machine.

The following table lists several exact numbers and their corresponding stored (rounded) approximations, each having five significant digits.

	Exact	Rounded	Δ	δ
w	10/13	$.76923 \times 10^0$	$.8 \times 10^{-6}$	$.1 \times 10^{-5}$
x	.769201	$.76920 \times 10^0$	$.1 \times 10^{-5}$	$.13 \times 10^{-5}$
y	98765.9	$.98766 \times 10^5$	$.1 \times 10^0$	$.1 \times 10^{-5}$
z	.0000111111	$.11111 \times 10^{-4}$	$.1 \times 10^{-9}$	$.9 \times 10^{-5}$

The next table highlights several arithmetic sequences applied to the numbers above. The exact values are displayed to 5 or 8 digits.

Operation	Exact	Rounded	Δ	δ	Figures
$w-x$	$.29769 \times 10^{-4}$	$.30000 \times 10^{-4}$	$.23 \times 10^{-6}$	$.77 \times 10^{-2}$	2
$(w-x)/z$	$.26792 \times 10^1$	$.27000 \times 10^1$	$.21 \times 10^{-1}$	$.78 \times 10^{-2}$	2
$(w-x)y$	$.29402 \times 10^1$	$.29630 \times 10^1$	$.23 \times 10^{-1}$	$.78 \times 10^{-2}$	2
$x+y$	$.98766669 \times 10^5$	$.98767 \times 10^5$	.331	$.36 \times 10^{-5}$	5

Observations *Several points can be inferred from the above table.*

(1) $w - x$*:* large δ*, don't subtract numbers that are close togther.*

(2) $(w - x)/z$*:* large δ *and* Δ*, don't divide by small numbers.*

(3) $(w - x)y$*:* large δ *and* Δ*, don't multiply by large numbers.*

(4) $x + y$*:* large Δ*, don't add numbers of widely differing magnitudes.*

These points are quite general, although different numerical values may be obtained on machines having different word lengths.

A consequence of 'undesirable' arithmetic operations is *loss of significant digits*, which can often be avoided by reordering the calculations.

Example 2.21 If $x \gg 1$ the expression $f(x) = \sqrt{x+1} - \sqrt{x}$ represents the difference of two numbers that are close to each other, and loss of significance would be expected. However

$$f(x) = \frac{\left(\sqrt{x+1} - \sqrt{x}\right)\left(\sqrt{x+1} + \sqrt{x}\right)}{\sqrt{x+1} + \sqrt{x}} = \frac{1}{\sqrt{x+1} + \sqrt{x}}.$$

The final expression has a large denominator which is quite acceptable.

For example, if $x = 100000$ and a computer has a word length of $w = 5$, the exact value of $f(100000)$ is $.15811 \times 10^{-2}$ (to 5 digits) whereas the original formula $f(100000) = 0$ using 5-digit floating-point arithmetic. The modified formula gives $f(100000) = .15811 \times 10^{-2}$ using 5-digit floating-point arithmetic[6].

Such a result is difficult to 'arrange' with DERIVE since the package tends to internally reorganise calculations.

2.5 Truncation errors

A classical result given in Chapter 1 is the *Maclaurin series* for e^x,

$$e^x = \sum_{k=0}^{\infty} \frac{x^k}{k!} = 1 + x + \frac{1}{2}x^2 + \frac{1}{6}x^3 + \frac{1}{24}x^4 + \cdots \tag{2.23}$$

For $x = 1$ a series expansion for the number e is obtained, that is

$$\begin{aligned} 2.71828\ldots &= e \\ &= 1 + 1 + \frac{1}{2} + \frac{1}{6} + \frac{1}{24} + \cdots \end{aligned} \tag{2.24}$$

The series on the right-hand side of eqn (2.24) is infinite, which reduces its usefulness. In practice the first $K + 1$ terms are used. Table 2.3 shows the partial sums s_K of the first $K + 1$ terms of eqn (2.24), and the resulting absolute error when using s_K as an approximation to e. As the number of terms in the series (2.24) is increased so the absolute error in the partial sum s_K is reduced. The arithmetic involved in evaluating each s_K is without error, that is each s_K is an exact number in its own right. Yet s_K clearly only approximates e. Such an error is called a *truncation error*. In this case an infinite series is, literally, truncated after a finite number of terms.

A positive aspect of truncation error is that it is controllable. To improve the estimate s_K (reduce the error), more terms are included in the series. In Table 2.3, the partial

[6] In *w-digit floating-point arithmetic*, the result of every intermediate calculation is rounded to w significant digits *before* use in subsequent calculations.

Table 2.3 Truncation error, $\Delta = |e - s_K|$, and its controllability.

K	s_K	Δ
0	1	1.7182
1	2	0.7182
2	2.5	0.2182
3	2.6667	0.0516
4	2.7083	0.0099
5	2.71677	0.0016

sums s_K always underestimate the value of e, and the series converges *monotonically* from below. This is not always the case.

DERIVE Experiment 2.3

Use *DERIVE* to evaluate the first 11 partial Maclaurin sums for e^{-x} at $x = 1$. `Author m(x,n):=taylor(exp-x,x,0,n)` then `Author` and `Simplify` the expression `vector(m(x,n),n,0,10)`. Now `Manage` and `Substitute` the value 1 for x and `approX` the resulting expression. You should see values similar to those shown in column s_K of Table 2.4.

Table 2.4 Maclaurin approximations to e^{-1}.

K	s_K	$e^{-1} - s_K$
0	1	−0.63
1	0	0.37
2	0.5	−0.13
3	0.333333	0.35×10^{-1}
4	0.375	-0.71×10^{-2}
5	0.366667	0.12×10^{-2}
6	0.368056	-0.18×10^{-3}
7	0.367857	0.22×10^{-4}
8	0.367882	-0.25×10^{-5}
9	0.36787918	0.25×10^{-6}
10	0.36787946	-0.23×10^{-7}

The exact value of e^{-1} is 0.36787944 (to 8 decimal places) and successive values of s_K are alternately over- and underestimates. Nonetheless, the sequence $\{s_K\}$

still converges. The error column, $e^{-1} - s_K$, shows this clearly. To use *DERIVE* `Author` and `Simplify` the expression `vector(exp-x-m(x, n), n, 0, 10)`, `Manage Substitute` the value 1 for x and `approX` the resulting expression.

Summary Truncation errors *arise from what is excluded (terms excluded from a series) and* rounding errors *arise from what is included (evaluation of the terms included in a series). In other words, truncation errors are approximation errors, and arise in the analysis of a problem (the replacement of one problem by another approximating problem, such as a finite series), whereas rounding errors are arithmetic errors, and arise during the numerical computation of a problem on a computer. Thus, in the expressions*

$$\begin{aligned} s &= a_0 + a_1 + \cdots + a_n + a_{n+1} + a_{n+2} + \cdots \\ &\simeq a_0^* + a_1^* + \cdots + a_n^* \end{aligned}$$

the truncation error arises from ignoring *the terms a_{n+1} etc., and the rounding error arises from using approximations (due to finite machine precision) to the terms remaining in the series.*

Example 2.22 To illustrate these comments consider the problem of estimating the derivative of a function 'f of x' at $x = x_0$ by the slope of a chord joining the points $(x_0, f(x_0))$ and $(x_0 + h, f(x_0 + h))$, shown in Figure 2.7(a).

The exact derivative value is $f'(x_0)$ and the slope of the chord is

$$D_1 = \frac{f(x_0 + h) - f(x_0)}{h},$$

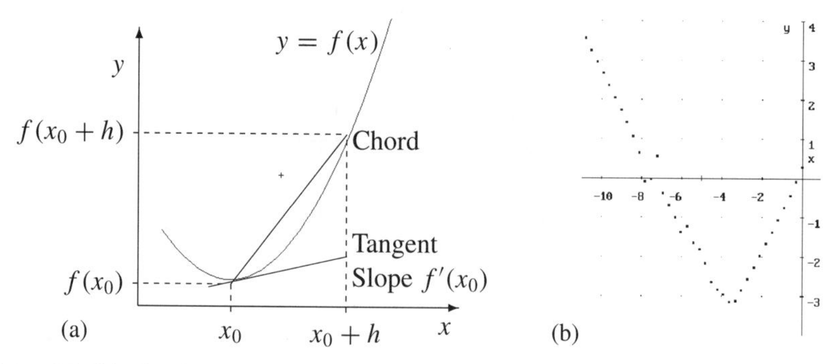

Figure 2.7 (a) Chordal approximation to a first derivative. (b) Applied to e^x at $x_0 = 1$ showing $\log_{10}|\Delta|$ vs. $\log_{10} h$.

with error $\Delta = f'(x_0) - D_1$. If $f(x) = \mathrm{e}^x$ and $x_0 = 1$ then $f'(x_0) = \mathrm{e}$ and

$$D_1 = \frac{\mathrm{e}^{1+h} - \mathrm{e}}{h}.$$

Figure 2.7(b) illustrates the effect of reducing the step size h. The graph shows $\log_{10}|\Delta|$ vs. $\log_{10} h$. As h is reduced (reading right to left) from 1 the error initially reduces, as we might expect. However, there clearly exists an 'optimum' step size, say h_{opt}, below which the error begins to increase (the analytical determination of h_{opt} is addressed in Section 7.6).

Observations *Towards the right of Figure 2.7(b) the truncation error is dominant. In other words the 'difference' between the exact model $f'(x)$ and the numerical model $(f(x+h) - f(x))/h$ is large, i.e. what has been left out of the formula for D_1 is significant. This error occurs even when $f'(x)$ and D_1 are evaluated exactly.*

Towards the left of the graph, rounding error is dominant. This occurs because as $h \to 0$ so $f(x+h) \to f(x)$ (implying subtraction of two numbers close together) and $1/h$ implies division by a small number. Both scenarios were highlighted as sources of error in Example 2.20, items 1 and 2, and result from the terms in the formula for D_1.

2.6 Error propagation

The notion of a computer-induced error has been developed. Simply stated, because the word length of a computer is finite, there exists an infinity of numbers 'out there' that cannot be stored exactly. Instead of storing the input number x, the machine stores a number x^* which is the closest representable number on the machine, with absolute error $\Delta = |x - x^*|$.

Here the potential effect of using x^* in place of x in a sequence of calculations is illustrated by means of an example.

Let I_n be a function of n defined by

$$I_n = \int_0^1 \mathrm{e}^{t-1} t^n \,\mathrm{d}t\,, \quad n = 0, 1, 2, \ldots \tag{2.25}$$

For small n the integral can be evaluated using repeated integration by parts. For n much larger than, say, 3 this becomes an enormous task! An alternative approach is to develop a *recurrence relation* which relates two successive terms in the sequence $\{I_n\}$.

Using integration by parts,

$$\begin{aligned} I_n = \left[t^n \mathrm{e}^{t-1}\right]_0^1 - \int_0^1 \mathrm{e}^{t-1} n t^{n-1} \,\mathrm{d}t &= 1 - n \int_0^1 \mathrm{e}^{t-1} t^{n-1} \,\mathrm{d}t \\ &= 1 - n I_{n-1}, \end{aligned} \tag{2.26}$$

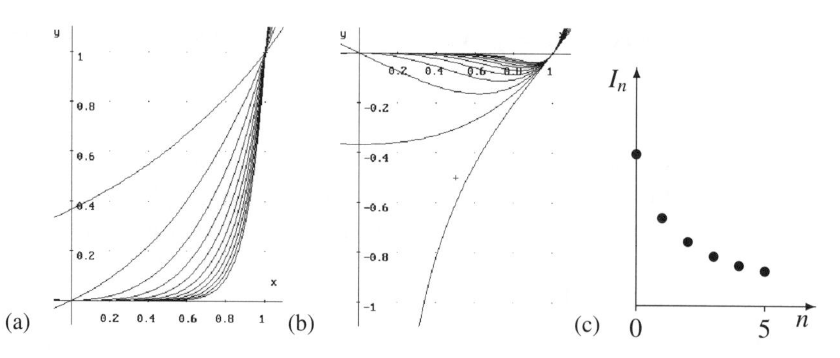

Figure 2.8 (a) Behaviour of the integrand in eqn (2.25), (b) behaviour of the integrand in eqn (2.27), (c) the sequence $\{I_n\}$.

and given the first term

$$I_0 = \int_0^1 e^{t-1}\, dt = [e^{t-1}]_0^1 = 1 - \frac{1}{e},$$

subsequent terms in the sequence $\{I_n\}$ can be generated from eqn (2.26), $I_n = 1 - nI_{n-1}$, $n = 1, 2, \ldots$

Before implementing this relation, two features of the expected behaviour of $\{I_n\}$ are identified.

I Since $e^{t-1}t^n$ is non-negative on the interval $0 \le t \le 1$ then $I_n > 0$, illustrated in Figure 2.8(a).

II $$I_n - I_{n-1} = \int_0^1 e^{t-1}(t^n - t^{n-1})\, dt = \int_0^1 e^{t-1}t^{n-1}(t-1)\, dt. \qquad \textbf{(2.27)}$$

On [0, 1], e^{t-1} is non-negative, $t^{n-1}(t-1)$ is non-positive and so the integrand is non-positive. That is, $I_n - I_{n-1} < 0$ (see Figure 2.8(b)), or $0 < I_n < I_{n-1}$. Consequently, $\{I_n\}$ is a monotonically decreasing positive sequence (see Figure 2.8(c)).

And so to the recurrence relation (2.26). e is *irrational* and the first value $I_0 = 1 - 1/e$ cannot be stored exactly. The value I_0^* is stored with absolute error $\Delta_0 = |I_0 - I_0^*|$ and, consequently, a sequence $\{I_n^*\}$ is generated from the relation $I_n^* = 1 - nI_{n-1}^*$.

The output shown in Table 2.5 is generated[7] from a rounded value of I_0 stored to 6 digits. This is the only error introduced – all subsequent arithmetic in the column headed I_n^* is exact. The output is displayed to 6 decimal places. The column headed I_n is exact (to 6 decimal places).

What has gone wrong?

---

[7] `Author iterates([1-(n+1)i,n+1],[i,n],[0.632121,0],12)` followed by `Simplify approX` will get $\mathcal{DERIVE}$ to reproduce the column I_n^*.

Table 2.5 An unstable arithmetic process.

n	I_n	I_n^*
0	0.632121	0.632121
1	0.367879	0.367879
2	0.264241	0.264242
3	0.207277	0.207274
4	0.170893	0.170904
5	0.145532	0.14548
6	0.126802	0.12712
7	0.112383	0.11016
8	0.100932	0.11872
9	0.091612	−0.06848
10	0.083877	1.6848
11	0.077352	−17.5328
12	0.071773	211.394

A simple analysis can be based upon the absolute error. The initial error is $\Delta_0 = |I_0 - I_0^*|$. After one iteration

$$\Delta_1 = |I_1 - I_1^*| = |(1 - I_0) - (1 - I_0^*)| = |I_0 - I_0^*| = \Delta_0.$$

After two iterations

$$\Delta_2 = |I_2 - I_2^*| = |(1 - 2I_1) - (1 - 2I_1^*)| = 2|I_1 - I_1^*| = 2\Delta_0.$$

After three iterations

$$\Delta_3 = |I_3 - I_3^*| = |(1 - 3I_2) - (1 - 3I_2^*)| = 3|I_2 - I_2^*| = 6\Delta_0.$$

It appears as though the error is following the pattern $\Delta_n = n!\Delta_0$. Assuming that this is the case,

$$\begin{aligned}\Delta_{n+1} = |I_{n+1} - I_{n+1}^*| &= |[1 - (n+1)I_n] - [1 - (n+1)I_n^*]| \\ &= (n+1)|I_n - I_n^*| \\ &= (n+1)n!\Delta_0 \\ &= (n+1)!\Delta_0.\end{aligned}$$

So, $\Delta_n = n!\Delta_0$ is a general result (proof by induction). If $\Delta_0 = \frac{1}{2} \times 10^{-6}$ (6 digits of accuracy in I_0^*) and $n = 12$ ($12! = 479001600$) then

$$\Delta_{12} = 479001600 \times \frac{1}{2} \times 10^{-6} = 239.5.$$

We can be a little more precise. The observed initial error Δ_0 is closer to 0.44×10^{-6}, which results in the value $\Delta_{12} = 210.8$ – not bad! The analysis shows that at each step the current error is multiplied by n, hence the catastrophic result for I_{12}^*.

Does this mean that computing I_n is 'out of bounds'? Certainly not! The trick is to 'think laterally'. As n approaches ∞ so I_n approaches 0. Starting with the approximation $I_{15}^* = 0$, the terms in the sequence can be determined in *reverse* order from the recurrence relation $I_{n-1}^* = (1 - I_n^*)/n$. In this case the error at each step is divided by n. The computed value of I_0^* is accurate to 6 decimal places!! (see Exercise 2.9)

Summary

This chapter addressed the storage and arithmetic of numbers on a computer. The representation of numbers such as $x = \pm.d_1 \ldots d_m \times 10^n$ (eqn (2.6)) was discussed in Section 2.1 together with the fact that the computer representation $x^* = \pm.d_1^* \ldots d_w^* \times 10^n$ (eqn (2.8)) may introduce errors if $w < m$. The absolute (2.9) and relative (2.10) errors were proposed as measures of the difference $x - x^*$ (Section 2.2). They were applied to the storage mechanisms of chopping and rounding (Section 2.3) to estimate the maximum error introduced on storing a number. The propagation of these errors via the basic arithmetic processes of addition, subtraction, multiplication and division was studied in Section 2.4. Section 2.5 dealt with truncation errors and, finally, in Section 2.6 we saw the potentially disastrous effect of the error build-up through repeated arithmetic operations, together with a possible 'remedy'.

Exercises

2.1 Convert $(1010)_2$, $(100101)_2$ and $(.1100011)_2$ to decimal.

2.2 Find the first five binary digits of $(.1)_{10}$. Obtain values for the absolute and relative errors in your result.

2.3 Write each of the following numbers in (decimal) floating-point form, stating the word length m and the exponent n.

$14.5, \quad -14.636, \quad .0093, \quad 2/125.$

2.4 Determine the absolute and relative errors when the value 0.4 is stored on a (binary) fixed-point machine having a word length of $w = 5$.

2.5 In each arithmetic expression assume that each number is a rounded estimate with associated 'worst-case' rounding error. If all subsequent arithmetic is exact (a) determine the value of each expression and (b) obtain bounds for the absolute and relative errors in your answers.

(i) 394.5×3.115
(ii) $465.4 + 0.03822$

(iii) $1.2450 \times 0.0036628 + 125.6$
(iv) $125.3/0.001252 - 12500$

For each answer (c) determine an interval that contains the true value of the expression and hence (d) write down the answer to the maximum number of *reliable* digits.

2.6 If $x = 1.29 \pm 0.005$ then determine the value and associated error bounds of the polynomial expression $x^2 - 3x + 1$. The coefficients in the polynomial are integers with *no* associated error.

2.7 In the following expressions, $x \simeq y$. Reorder each expression to obtain a form that is not sensitive to subtractive cancellation (x and y are positive integers in (i), (ii) and (iii)).

(i) $\sqrt{x} - \sqrt{y}$
(ii) $\sin x - \sin y$
(iii) $\ln x - \ln y$
(iv) $\sin(x + \alpha) - \sin \alpha$, where $x \approx 0$
(v) $x - \sqrt{x^2 - \alpha}$, where $x \gg \alpha$

2.8 Find the roots of the quadratic equation

$$x^2 + .4002 \times 10^0 x + .8 \times 10^{-4} = 0 \qquad \textbf{(2.28)}$$

by using the formula $x = (-b \pm \sqrt{b^2 - 4ac})/2a$. Work in 4-digit floating-point arithmetic (round *intermediate* results to 4 digits). The exact values of the roots are $-.0002$ and $-.4$, so what went wrong?

Use the substitution $x = 1/y$ to obtain a quadratic equation involving the variable y. Solve the equation to find the roots, again using 4-digit floating-point arithmetic, and hence determine an accurate estimate of the smaller of the two roots (in magnitude) of eqn (2.28).

2.9 Use $\mathcal{DERIVE}$ to reproduce the values of I_n^* in Table 2.5 (see the footnote on p. 50). Reorder the calculations (see Section 2.6) from I_{15}^* to obtain the value of I_0^* to 5 decimal places.

2.10 A sequence $\{x_n\}$ is generated by the recurrence relation $x_{n+1} = 3x_n + b$, $n = 0, 1, \ldots$ where $x_0 = 0$. If $\Delta_b = 0.05$, find (a) a recurrence relation and (b) an expression in terms of n for the error $\Delta_n = |x_n - x_n^*|$. If $\Delta_0 = \frac{1}{2} \times 10^{-2}$, use part (b) to determine the value of n for which x_n^* has just one decimal place of accuracy. [*Hint*: $\mathcal{DERIVE}$ might help!]

2.11 Use $\mathcal{DERIVE}$ to obtain the Maclaurin expansion of $\ln(x + 1)$ for $n = 0, \ldots, 10$. Evaluate the error of each expansion at $x = -0.5$, 0, 0.5, 1 and 1.5, and comment on the results.

2.12 The e^x key is broken on your calculator. With *DERIVE* obtain the Maclaurin polynomials of degree 0 to 20 for e^{-x} and use them to estimate e^{-5}. What is happening? Reorder the calculations to obtain an accurate estimate of e^{-5} (0.00673794 to 6 sig. figs.).

2.13 **'Linear or exponential' error growth?** The sequence $\{x_n\}$ is approximated by $\{x_n^*\}$ with error $e_n = |x_n - x_n^*|$. If the error growth is *linear*, $e_n \simeq kn\varepsilon$ and the algorithm generating the sequence $\{x_n^*\}$ is *stable*. If $e_n \simeq k^n\varepsilon$, the error growth is *exponential* and the generating algorithm is *unstable*.

The sequence $\frac{2}{3}, \frac{1}{3}, \frac{1}{6}, \frac{1}{12}, \ldots$ can be obtained by setting $x_0 = \frac{2}{3}$, $x_1 = \frac{1}{3}$ and computing subsequent terms from the relation $x_{n+1} = \frac{5}{2}x_n - x_{n-1}$, $n = 1, 2, \ldots$ Using 4-digit floating-point arithmetic, compute the terms x_2^* to x_{10}^* from the recurrence relation and compare the results with the direct evaluation of each term, $x_n = 2^{1-n}/3$.

Repeat the above procedures as applied to the alternative recurrence formula $x_{n+1} = x_n/2$, $n = 0, 1, \ldots$, where $x_0 = \frac{2}{3}$. For each case determine whether the error growth is linear or exponential.

3 Polynomial interpolation

The Taylor polynomial (Chapter 1) represents a function f in the 'vicinity' of a *point*, $x = x_0$, given values of f and its derivatives at $x = x_0$. Here the approximation is expanded to an *interval* $[a, b]$ on which a set of function values is known, akin to the puzzle-book problem of 'joining the dots'.

Example 3.1 Table 3.1 shows values of a material property, specific heat c, for a set of specified temperatures[1] U. A scientist may be interested in values of c at non-tabulated

Table 3.1 Measured values of the specific heat of a medium density polymer at specified temperatures.

Temperature U (°C)	Specific heat c (kJ/kg K)
20	1.750
40	2.200
60	2.500
80	2.750
100	3.500
120	6.000
126	11.200
140	2.600
179	2.700
200	2.725
220	2.750

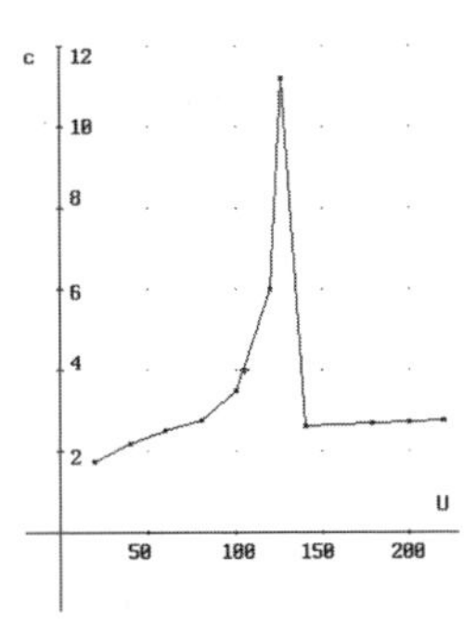

[1] Taken from A.S. Wood (1993). The butt fusion welding of polymers, *Chem. Engng. Sci.*, **48**, 3071–3.

temperatures. *Interpolation* is a mathematical technique which can furnish estimates of these unknown values.

The process of joining a set of data points by lines is called *interpolation*. In this chapter we aim to approximate the function f underlying the data by a *polynomial* p_n on an interval $[a, b]$.

Example 3.2 Figure 3.1(a) shows a quadratic approximation to $\sin x$, based upon three function values at $x = \frac{\pi}{2}$, $x = \pi$ and $x = 2\pi$. For comparison the quadratic Taylor polynomial constructed at $x_0 = \frac{\pi}{2}$ is also shown. The error curves (see Figure 3.1(b)) show that the Taylor polynomial is more accurate in the vicinity of $x = \frac{\pi}{2}$ whereas the interpolating polynomial is better over the interval $[\frac{\pi}{2}, 2\pi]$.

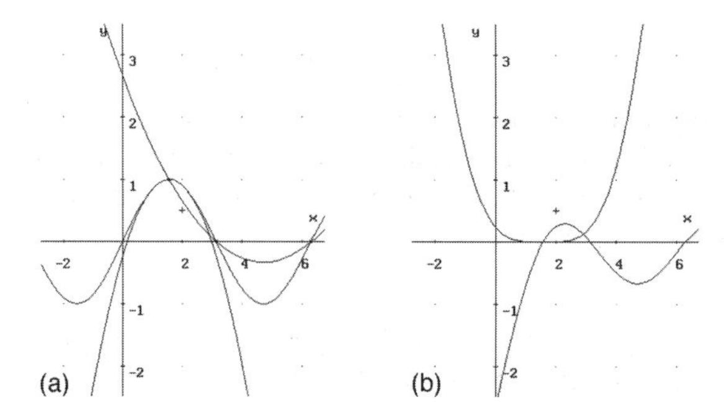

Figure 3.1 (a) Interpolation vs. Taylor approximation, (b) absolute errors.

The data required for this chapter consist of a set of nodes $x_0, \ldots, x_n$, and a set of function values $f_0, \ldots, f_n$, where $f_i = f(x_i)$.

3.1 Polynomial forms

To discuss interpolation it is helpful to introduce several polynomial forms. The usual description of a polynomial of degree n is a linear combination of the *monomials* $1, x, x^2, \ldots, x^n$

$$
\begin{aligned}
p_n(x) &= a_0 + a_1 x + a_2 x^2 + \cdots + a_n x^n \\
&= \sum_{k=0}^{n} a_k x^k, \qquad \textbf{(3.1)}
\end{aligned}
$$

called the *power form*. The a_k are constant coefficients. Algorithm 3.1 can be used to evaluate $p_n(x)$.

Algorithm 3.1
Function : Power form polynomial

Line	Description	
1	**Problem**	input $n, a_0, \ldots, a_n, x$
2	Initialize	$p_n := a_0$
3		for $k = 1$ to n
4	Compute	$t := a_k$
5	term k	for $j = 1$ to k
6		$t := t * x$
7		end
8	Add term k	$p_n := p_n + t$
9		end
10	**Solution**	output $p_n(x)$

Line 2 initializes the polynomial value to a_0 and (lines 3–9) for values of k from 1 to n updates the value (adds term k to $p_n(x)$). The form is convenient for analysis but is susceptible to loss of significance.

Example 3.3 The polynomial $p_1(x) = \frac{18001}{3} - x$ takes the value $\frac{1}{3}$ at $x = 6000$ and $-\frac{2}{3}$ at $x = 6001$. On a finite-precision machine (with a word length of 5 decimal digits) the coefficients are stored as $a_0^* = 6000.3$ and $a_1^* = -1$ (exact), and hence

$$\begin{aligned} p_1(6000) &= 6000.3 - 6000 = 0.3, \\ p_1(6001) &= 6000.3 - 6001 = -0.7. \end{aligned}$$

Only one digit of the exact value is recovered, yet the coefficients are accurate to 5 digits! 4 significant digits have been lost (in this case due to the subtraction of two near-neighbour large numbers).

The drawback seen in Example 3.3 can be alleviated by changing the origin of x to a non-zero value c and writing the polynomial (3.1) as

$$\begin{aligned} p_n(x) &= b_0 + b_1(x - c) + b_2(x - c)^2 + \cdots + b_n(x - c)^n \\ &= \sum_{k=0}^{n} b_k(x - c)^k, \end{aligned} \qquad \textbf{(3.2)}$$

called the *shifted power form*. c is a *centre* and the b_k are constant coefficients. Expressions (3.1) and (3.2) represent the *same* polynomial[2].

[2] In Algorithm 3.1, line 6 is replaced by $t := t * (x - c)$.

Example 3.4 Using Example 3.3 with centre $c = 6000$, then $b_0 = \frac{1}{3}$, $b_1 = -1$ and

$$p_1(x) = \frac{1}{3} - (x - 6000) = \frac{1}{3} - x + 6000 = \frac{18001}{3} - x,$$

which is the polynomial seen previously. In 5-digit fixed-point form, using the shifted power form, $p_1(x) = 0.33333 - (x - 6000)$ and

$$\begin{aligned} p_1(6000) &= 0.33333 - (6000 - 6000) = 0.33333, \\ p_1(6001) &= 0.33333 - (6000 - 6001) = -0.66667. \end{aligned}$$

These values are accurate to 5 digits – there is no loss of significance.

Can you spot the special significance of the coefficients b_k in eqn (3.2)? At $x = c$ it is not difficult to show that

$$b_k = \frac{p_n^{(k)}(c)}{k!}$$

and hence eqn (3.2) can be written as

$$p_n(x) = p_n(c) + (x - c)p_n'(c) + \frac{(x - c)^2}{2!}p_n''(c) + \cdots + \frac{(x - c)^n}{n!}p_n^{(n)}(c), \tag{3.3}$$

where $p^{(k)} \equiv \mathrm{d}^k p/\mathrm{d}x^k$. Equation (3.3) is the *Taylor polynomial* of degree n for the polynomial p_n (which is, of course, the polynomial itself as asserted in Example 1.2).

Equation (3.2) can be generalized by choosing n centres $c_1, c_2, \ldots, c_n$,

$$\begin{aligned} p_n(x) &= d_0 \\ &\quad + d_1(x - c_1) + d_2(x - c_1)(x - c_2) + \cdots \\ &\quad + d_n(x - c_1)\ldots(x - c_n) \\ &= d_0 + \sum_{k=1}^{n} d_k \prod_{j=1}^{k}(x - c_j), \end{aligned} \tag{3.4}$$

called the *Newton form*[3], which is particularly useful for *polynomial interpolation* (see Section 3.3). $c_1 = \cdots = c_n = c$ recovers the shifted power form and $c_1 = \cdots = c_n = 0$ recovers the power form.

[3] In Algorithm 3.1, line 6 is replaced by $t := t * (x - c_j)$.

3.2 Practical evaluation of polynomials

The kth term of the power form (3.1) is $a_k x^k$ and requires k multiplications. The number of flops[4] (cost) to evaluate all $n+1$ terms in eqn (3.1) is

$$\sum_{k=0}^{n} k = \sum_{k=1}^{n} k = \frac{n(n+1)}{2}.$$

n additions are required to sum the $n+1$ terms, giving a total cost of

$$\frac{n(n+3)}{2}. \tag{3.5}$$

The cost can be reduced by reordering the polynomial using factorization. The *factor* x appears in terms a_1 to a_n, and eqn (3.1) can be written as

$$p_n(x) = a_0 + x(a_1 + a_2 x + \cdots + a_{n-1}x^{n-2} + a_n x^{n-1}).$$

Within the parentheses, terms a_2 to a_n include a factor x and the polynomial can be further reorganized to

$$p_n(x) = a_0 + x(a_1 + x(a_2 + \cdots + a_{n-1}x^{n-3} + a_n x^{n-2})).$$

Factorizing a further $n-3$ steps gives the *nested power form*

$$p_n(x) = a_0 + x(a_1 + x(a_2 + \cdots + x(a_{n-2} + x(a_{n-1} + a_n x))\ldots)). \tag{3.6}$$

Equation (3.6) requires just $2n$ flops for evaluation. The polynomial is evaluated 'inside out'.

Algorithm 3.2

Function : Nested power form polynomial

Line	Description	
1	**Problem**	`input` $n, a_0, \ldots, a_n, x$
2	Initialize	$p_n := a_n$
3		`for` $k = n-1$ `to` 0 `step` -1
4		$p_n := a_k + p_n * x$
5		`end`
6	**Solution**	`output` $p_n(x)$

[4] A *flop* is a floating-point operation (addition, subtraction, multiplication, division, exponentiation).

Line 2 initializes $p_n(x)$ to a_n. Lines 3–5 loop back through the n remaining coefficients (line 4), adding a_k to x times the 'current' value of $p_n(x)$ using 2 flops (1 multiplication, 1 addition), giving a total cost of $2n$ flops.

Example 3.5 For the cubic polynomial

$$p_3(x) = a_0 + a_1x + a_2x^2 + a_3x^3, \tag{3.7}$$

the nested algorithm proceeds as follows:

Step k	p_n at step k
$k = n = 3$	$p_3 = a_3$
$k = 2$	$p_3 = a_2 + p_3x = a_2 + a_3x$
$k = 1$	$p_3 = a_1 + p_3x = a_1 + (a_2 + a_3x)x$ $= a_1 + a_2x + a_3x^2$
$k = 0$	$p_3 = a_0 + p_3x = a_0 + (a_1 + a_2x + a_3x^2)x$ $= a_0 + a_1x + a_2x^2 + a_3x^3$

The final form of $p_3(x)$ is identical to eqn (3.7).

The Newton form (3.4) can also be written in a nested form,

$$p_n(x) = d_0 + (x - c_1)(d_1 + (x - c_2)(d_2 + \cdots + (x - c_{n-2})(d_{n-2} + (x - c_{n-1})(d_{n-1} + d_n(x - c_n)))\ldots)). \tag{3.8}$$

In place of $2n$ flops, the subtraction of the centre c_k at each pass requires an additional flop, making $3n$ flops in all[5].

Example 3.6 Nested evaluation of the polynomial

$$\begin{aligned} p_3(x) &= 1 + 2(x-1) + 3(x-1)(x-2) + 4(x-1)(x-2)(x-3) \\ &= 1 + (x-1)(2 + (x-2)(3 + 4(x-3))) \end{aligned}$$

at $x = 4$ gives

$$\begin{aligned} p_3(4) &= 1 + 3(2 + 2(3 + 4 \times 1)) \\ &= 1 + 3(2 + 2 \times 7) \\ &= 1 + 3 \times 16 \\ &= 49. \end{aligned}$$

[5] In Algorithm 3.2, line 2 becomes $p_n := d_n$ and line 4 is replaced by $p_n := d_k + p_n * (x - c_{k+1})$.

Table 3.2 Arithmetic costs of power form and nested power form.

n	1	10	100	1000
Power	2	65	5150	501500
Nested power	2	20	200	2000
Ratio	1	3.25	25.75	250.75

Table 3.2 compares the costs of the 'basic' and 'nested' polynomial forms.

3.3 Interpolation

3.3.1 Linear interpolation

The simplest interpolation problem is that of joining two points with a line. As shown in Figure 3.2 there are many lines that pass through two specified points. To develop a process for constructing the line requires an idea of *uniqueness* – what line passing through two distinct points is unique? Thus, when the 'process' finds this line, *the* line is found (as opposed to *a* line).

The answer is a straight line. To see this, place a transparent ruler on Figure 3.3, which shows two *distinct* points (x_0, f_0) and (x_1, f_1). Move the edge around until it passes through the two points. In how many ways[6] can this be done?

The general form of a straight line is

$$y = a + bx, \tag{3.9}$$

where a and b are constants to be determined. Interpolation requires that the line must coincide with the function values f_0 and f_1 at the two given points. Substituting the coordinates into eqn (3.9) gives

$$\begin{matrix} f_0 = a + bx_0 & & a + x_0 b = f_0 \\ & \Rightarrow & \\ f_1 = a + bx_1 & & a + x_1 b = f_1 \end{matrix} \tag{3.10}$$

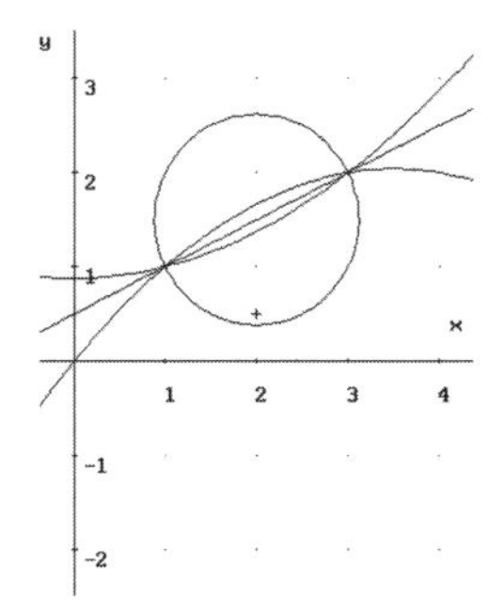

Figure 3.2 Joining the points (1, 1) and (3, 2): $y = (x + 1)/2$, $(x - 2)^2 + (y - 1.5)^2 = 1.25$, $y = (x^2 + 7)/8$, $y = (7x - x^2)/6$.

[6] One. You are not allowed to turn the line over, and the line has no direction.

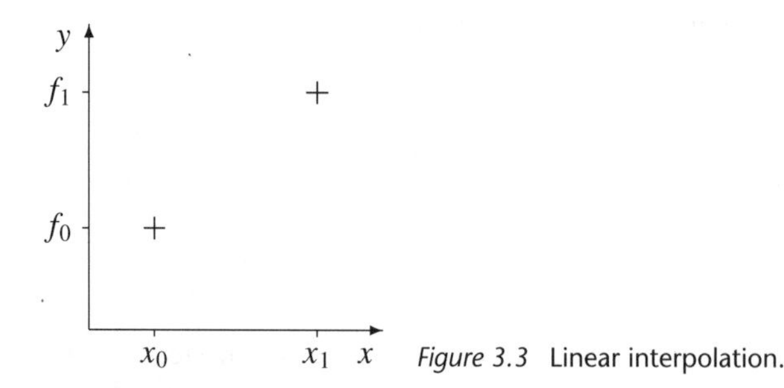

Figure 3.3 Linear interpolation.

a pair of linear equations with solution (x_0, x_1, f_0 and f_1 are known)

$$b = \frac{f_1 - f_0}{x_1 - x_0}, \qquad a = f_0 - x_0 b. \tag{3.11}$$

The equation of the straight line passing through the two points is then

$$p_1(x) = a + bx = f_0 + b(x - x_0) = f_0 + \left(\frac{f_1 - f_0}{x_1 - x_0}\right)(x - x_0). \tag{3.12}$$

p_1 is the *linear interpolating polynomial.* On substituting the values $x = x_0$ and $x = x_1$ it is easy to show that $p_1(x_0) = f_0$ and $p_1(x_1) = f_1$ – the line *does* pass through the two given points. $p_n(x)$ can be used to estimate the value of $f(x)$ at an intermediate point $x_0 < x < x_1$.

Example 3.7 If $\log_{10} 3 = 0.4771$ and $\log_{10} 4 = 0.6021$, use linear interpolation to estimate $\log_{10} 3.4$. Here $x_0 = 3$, $x_1 = 4$, $f_0 = 0.4771$ and $f_1 = 0.6021$. Hence

$$\begin{aligned} \ln x \simeq p_1(x) &= 0.4771 + \left(\frac{0.6021 - 0.4771}{4 - 3}\right)(x - 3) \\ &= 0.4771 + 0.125(x - 3) \end{aligned}$$

At $x = 3.4$

$$\log_{10} 3.4 \approx p_1(3.4) = 0.4771 + 0.125(4 - 3) = 0.5271.$$

The exact value is 0.5315 (to 4 decimal places).

The need for *distinct* points ($x_0 \neq x_1$) is clear. Since the factor $x_1 - x_0$ appears in the denominator of b, then $x_0 = x_1 \Rightarrow x_1 - x_0 = 0$, and b and a are undefined. Further, the system (3.10) has a unique solution if the determinant of the coefficient matrix is non-zero, that is

$$\begin{vmatrix} 1 & x_0 \\ 1 & x_1 \end{vmatrix} \neq 0 \quad \Rightarrow \quad x_1 - x_0 \neq 0. \tag{3.13}$$

a and b exist (and are unique) if condition (3.13) is satisfied. The distinct nature of the points therefore implies the uniqueness of the line p_1.

> *DERIVE Experiment 3.1*
> `p1(x0,f0,x1,f1):=f0+(f1-f0)(x-x0)/(x1-x0)` defines a straight line p_1 interpolating the data (x_0, f_0) and (x_1, f_1). The data (1, 1), (3, 2) is plotted with `Author [[1,1],[3,2]] Plot Beside Plot Algebra` and the straight line $(x+1)/2$ with `Author p1(1,1,3,2) Simplify Plot Plot Algebra`.
> Use the function p_1 to experiment with different data. In particular, what happens if $x_0 = x_1$, or $f_0 = f_1$?

3.3.1.1 Error of linear interpolation

In Example 3.7 there is an error in using the value $p_1(x)$ to replace $f(x)$, except at $x = x_0$ and $x = x_1$. An estimate of $f(x) - p_1(x)$ is required in order to attach some worth to the value $p_1(x)$. The approach uses Taylor's theorem (see Section 1.3).

Let the error at x be

$$\begin{aligned} e_1(x) &= f(x) - p_1(x) \\ &= f(x) - f(x_0) - \left(\frac{x - x_0}{x_1 - x_0}\right)(f(x_1) - f(x_0)). \end{aligned} \tag{3.14}$$

To enumerate $e_1(x)$, terms on the right-hand side are evaluated at x. From Taylor's theorem

$$f(x_0) = f(x) + (x_0 - x)f'(x) + \frac{(x_0 - x)^2}{2} f''(x) + \cdots$$

$$f(x_1) = f(x) + (x_1 - x)f'(x) + \frac{(x_1 - x)^2}{2} f''(x) + \cdots$$

and eqn (3.14) can be written as

$$\begin{aligned} e_1(x) &= -(x_0 - x)f'(x) - \frac{(x_0 - x)^2}{2} f''(x) - \cdots \\ &\quad -(x - x_0)\left[f'(x) + \frac{(x_1 + x_0 - 2x)}{2} f''(x) + \cdots\right] \\ &= \frac{(x - x_0)}{2}(x - x_1)f''(\xi), \quad x_0 < \xi < x_1. \end{aligned} \tag{3.15}$$

The value of ξ is unknown and a worst-case scenario is adopted in which the value $M_2 = \max |f''(x)|$, $x_0 < x < x_1$, is used to give the bound

$$|e_1(x)| \leq \frac{1}{2}|(x - x_0)(x - x_1)|M_2. \tag{3.16}$$

Example 3.8 For Example 3.7, $3.1 \leq x \leq 3.2$, $f(x) = \ln x$, $f''(x) = -1/x^2$ and $M_2 = \max|-1/x^2| = 1/(3.1)^2$. From eqn (3.16)

$$|e_1(3.16)| \leq \frac{1}{2}\frac{|(3.16-3.1)(3.16-3.2)|}{(3.1)^2} \approx 0.000125.$$

This implies that about 3 decimal places of accuracy are guaranteed (the observed error is 0.0001). Of course, this calculation assumes that the data f_0 and f_1 are sufficiently accurate!

3.3.2 Quadratic interpolation

The next interpolation problem is that of finding a *unique* polynomial passing through *three* distinct points, (x_0, f_0), (x_1, f_1) and (x_2, f_2), shown in Figure 3.4. Applying the interpolation conditions at $x = x_0, x_1$ and x_2, to the quadratic form $y = a + bx + cx^2$, three equations are obtained:

x	y	$f(x)$
x_0	$a + bx_0 + cx_0^2 =$	f_0
x_1	$a + bx_1 + cx_1^2 =$	f_1
x_2	$a + bx_2 + cx_2^2 =$	f_2

The system has a *unique* solution vector $[a, b, c]^{\mathrm{T}}$ if the determinant of the coefficient matrix is non-zero, that is

$$\begin{vmatrix} 1 & x_0 & x_0^2 \\ 1 & x_1 & x_1^2 \\ 1 & x_2 & x_2^2 \end{vmatrix} \neq 0 \Rightarrow (x_2 - x_1)(x_2 - x_0)(x_1 - x_0) \neq 0. \tag{3.17}$$

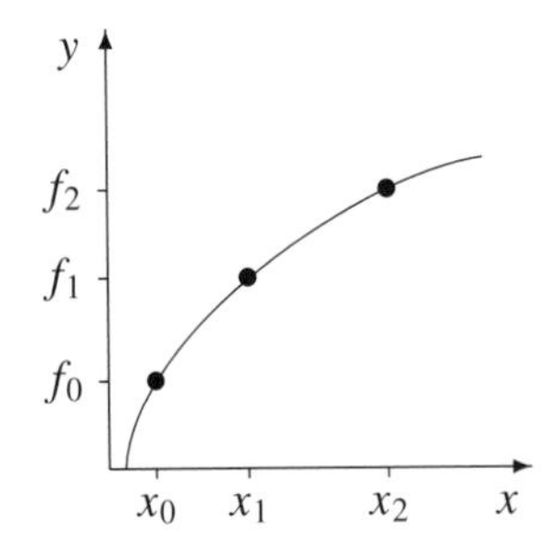

Figure 3.4 Quadratic interpolation.

Thus, no two nodes can be equal – *distinct* interpolation points – and

$$c = \frac{1}{x_2 - x_1}\left[\frac{f_2 - f_0}{x_2 - x_0} - \frac{f_1 - f_0}{x_1 - x_0}\right], \quad b = \frac{f_1 - f_0}{x_1 - x_0} - c(x_1 + x_0),$$

$$a = f_0 - bx_0 - cx_0^2.$$

DERIVE Experiment 3.2

DERIVE has its own interpolation functions. However, the quadratic case can be simulated with user-defined functions as follows: `Author`

```
c(x0,f0,x1,f1,x2,f2):=
   ((f2-f0)/(x2-x0)-(f1-f0)/(x1-x0))/(x2-x1)
b(x0,f0,x1,f1,x2,f2):=
   (f1-f0)/(x1-x0)-c(x0,f0,x1,f1,x2,f2)(x1+x0)
a(x0,f0,x1,f1,x2,f2):=
   f0-b(x0,f0,x1,f1,x2,f2)x0-c(x0,f0,x1,f1,x2,f2)(x0)^2
```

defines the constants c, b and a, and `Author`

```
p2(x0,f0,x1,f1,x2,f2):=a(x0,f0,x1,f1,x2,f2)+
   b(x0,f0,x1,f1,x2,f2)x+c(x0,f0,x1,f1,x2,f2)x^2
```

gives the polynomial interpolating (x_0, f_0), (x_1, f_1) and (x_2, f_2).

Plot the data $(1, 1)$, $(2, 4)$, $(3, 2)$ with `Author [[1,1],[2,4],[3,2]] Plot Beside Plot Algebra` and the polynomial expression $p_2(x)$ with `Author p2(1,1,2,4,3,2) Simplify Plot Plot Algebra`.

Use the function p_2 to experiment with different data. In particular, what happens if any pair of x values are equal, or any pair of f values?

3.3.3 Lagrange interpolation

The algebraic process of constructing an interpolating polynomial is becoming complicated. Nonetheless, higher-order polynomials generally lead to enhanced approximations, and here a systematic approach is introduced.

JOSEPH-LOUIS LAGRANGE (1736–1813)
French mathematician

LIFE

Lagrange's interest in mathematics began when he read a text by Halley. Lagrange was professor of geometry at the Royal Artillery School in Turin (1755–66) and helped to found the Royal Academy of Science there in 1757. In 1764 the Paris Academy awarded him a prize for his essay on the libration of the Moon. In 1766 Lagrange succeeded Euler as director of mathematics at the Berlin Academy of Science. In 1787 he left Berlin to become a member of the Paris Academy of Science and teach at the École Polytechnique, which he helped to found. Lagrange survived the French Revolution while others did not, and said of the death of the chemist Lavoisier: *'It took only a moment to cause this head to fall and a hundred years will not suffice to produce its like'.* Napoleon named him to the Legion of Honour and Count of the Empire in 1808.

WORK

During the 1790s Lagrange worked on the metric system and advocated a decimal base. He excelled in all fields of analysis and number theory and analytical and celestial mechanics. He published *Mécanique Analytique* (1786) which summarized all the work done in the field of mechanics since Newton and is notable for its use of the theory of differential equations. In it he transformed mechanics into a branch of mathematical analysis. His early work on the theory of equations led Galois to the idea of a group of permutations. In 1797 he published the first theory of functions of a real variable although he failed to give enough attention to matters of convergence.

Polynomial interpolation can be described as follows. Given $n+1$ *distinct* nodes $x_0, \ldots, x_n$ with associated function values $f_0, \ldots, f_n$, a polynomial p_n is sought such that

(1) $\text{degree}(p_n) \le n$ and

(2) $p_n(x)$ takes the function values $f_0, \ldots, f_n$ at the nodes $x_0, \ldots, x_n$.

Consider the linear polynomial (3.12) written in a *symmetric* form

$$\begin{aligned} p_1(x) &= f_0 + \left(\frac{f_1 - f_0}{x_1 - x_0}\right)(x - x_0) \\ &= \left(\frac{x - x_1}{x_0 - x_1}\right) f_0 + \left(\frac{x - x_0}{x_1 - x_0}\right) f_1 \\ &= \ell_0(x) f_0 + \ell_1(x) f_1. \end{aligned} \tag{3.18}$$

ℓ_0 and ℓ_1 are linear polynomials in x satisfying

$$\ell_k(x_i) = \begin{cases} 0, & i \ne k, \\ 1, & i = k. \end{cases} \tag{3.19}$$

Generalizing to $n+1$ points, the approach is to seek a polynomial

$$p_n(x) = \ell_0(x) f_0 + \ell_1(x) f_1 + \cdots + \ell_n(x) f_n. \tag{3.20}$$

Each function $\ell_0, \ldots, \ell_n$ is a polynomial of degree at most n and $p_n \in P_n$ (satisfying Condition 1). ℓ_0 and ℓ_1 are *not* the polynomials appearing in the linear form (3.18). To satisfy Condition 2, at $x = x_0$, $p_n(x_0) = f_0$ and so $\ell_0(x_0) = 1$. The remaining polynomials are zero, that is $\ell_k(x_0) = 0$, $k = 1, \ldots, n$. At the nodes $x_1, \ldots, x_n$ it is required that $\ell_0(x_i) = 0$ (since f_0 must not appear at these nodes). Thus, ℓ_0 is a polynomial of degree n with n zeros at $x = x_1, \ldots, x_n$,

$$\ell_0(x) = C_0(x - x_1)(x - x_2)\ldots(x - x_{n-1})(x - x_n),$$

where C_0 is a constant. The condition $\ell_0(x_0) = 1$ gives

$$C_0 = \frac{1}{(x_0 - x_1)(x_0 - x_2)\ldots(x_0 - x_n)}$$

and

$$\ell_0(x) = \frac{(x - x_1)(x - x_2)\ldots(x - x_n)}{(x_0 - x_1)(x_0 - x_2)\ldots(x_0 - x_n)} = \prod_{i=1}^{n} \frac{x - x_i}{x_0 - x_i}.$$

Similarly, the polynomials $\ell_1, \ldots, \ell_n$ are

$$\ell_k(x) = \frac{(x - x_0)\ldots(x - x_{k-1})(x - x_{k+1})\ldots(x - x_n)}{(x_k - x_0)\ldots(x_k - x_{k-1})(x_k - x_{k+1})\ldots(x_k - x_n)}$$

$$= \prod_{\substack{i=0 \\ i \neq k}}^{n} \frac{x - x_i}{x_k - x_i}, \qquad k = 1, \ldots, n. \tag{3.21}$$

The ℓ_k are the *Lagrange polynomials* and it is readily seen that they satisfy condition (3.19). Thus, the polynomial (3.20) satisfies Condition 2 and takes the form[7]

$$p_n(x) = \sum_{k=0}^{n} f_k \prod_{\substack{i=0 \\ i \neq k}}^{n} \frac{x - x_i}{x_k - x_i}. \tag{3.22}$$

This is the *Lagrange interpolating polynomial* of degree n. It is clear just why the nodes $x_0, \ldots, x_n$ must be distinct. If any pair of nodes, say x_i and x_k, were equal then n of the $n + 1$ Lagrange polynomials would have a zero factor in the denominator, rendering the polynomial undefined!

Example 3.9 For the data below, obtain the quadratic interpolating polynomial and use it to estimate $f(0.5)$.

x	1	−1	2
$f(x)$	0	−2	3

[7] Establishing the uniqueness of the interpolating polynomial p_n is best tackled by using 'proof by contradiction', as opposed to showing the the determinant of a certain coefficient matix is non-zero. Details can be found in [18], Section 4.2.

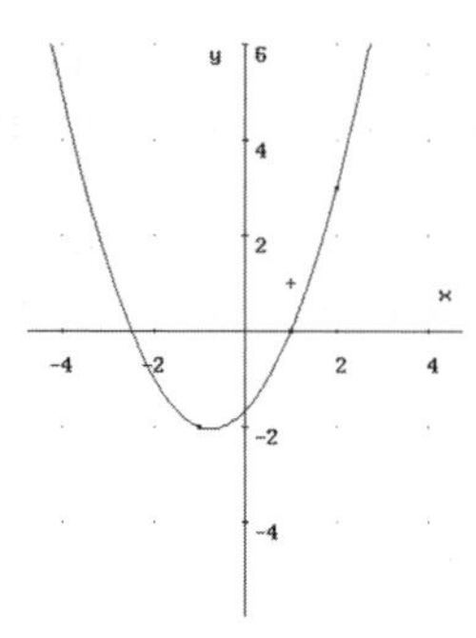

Figure 3.5 Quadratic interpolation.

The quadratic Lagrange polynomials are

$$\ell_0(x) = \frac{(x-x_1)(x-x_2)}{(x_0-x_1)(x_0-x_2)} = \frac{(x+1)(x-2)}{(2)(-1)} = \frac{-x^2+x+2}{2},$$

$$\ell_1(x) = \frac{(x-x_0)(x-x_2)}{(x_1-x_0)(x_1-x_2)} = \frac{(x-1)(x-2)}{(-2)(-3)} = \frac{x^2-3x+2}{6},$$

$$\ell_2(x) = \frac{(x-x_0)(x-x_1)}{(x_2-x_0)(x_2-x_1)} = \frac{(x-1)(x+1)}{(1)(3)} = \frac{x^2-1}{3}.$$

The quadratic interpolating polynomial is (see Figure 3.5)

$$p_2(x) = \sum_{k=0}^{2} \ell_k(x) f_k = 0\ell_0(x) - 2\ell_1(x) + 3\ell_2(x) = \frac{2x^2+3x-5}{3}.$$

$p_2(x)$ clearly passes through the three data points. At $x = 0.5$, $f(0.5) \approx p_2(0.5) = -1$.

Note The order of the nodes is unimportant. An ascending or descending ordering simply makes the set $\{x_0, \ldots, x_n\}$ easier to scan.

DERIVE Experiment 3.3

The function `fit` can interpolate the data $(x_0, y_0), \ldots, (x_n, y_n)$ with the polynomial $y(x) = a_n x^n + \cdots + a_1 x + a_0$. The specification is

```
fit(p,d)
```

p vector $[x, \underline{a}]$ where x is the independent variable and $\underline{a}$ is a polynomial in x (e.g. $ax^2 + bx + c$)

d $(n+1) \times 2$ matrix of data pairs (x, y)

For Example 3.9, Author `p:=[x,ax^2+bx+c]` (polynomial form) and `d:=[[1,0], [-1,-2],[2,3]]` (data set), followed by Author `fit(p,d)`

Simplify. The result (identical to Example 3.9) is

$$\frac{2x^2}{3} + x - \frac{5}{3}.$$

The data and the polynomial are easily graphed. Highlight the expression for d, then Plot Beside (make sure that Options State is set to Discrete) and Plot. Return to the Algebra window, highlight the quadratic polynomial expression, and Plot Plot (see Example 3.9).

'Extreme' data points can significantly affect the aesthetic nature of the interpolating polynomial, particularly when a moderate to large number of data points are used, which leads to a high-order 'oscillatory' polynomial.

Example 3.10 The data of Table 3.1 are fully interpolated poorly (see Figure 3.6(a)). The extreme value of $c = 11.2$ at $U = 126$ 'destroys' the polynomial. Ignoring this point gives a more acceptable polynomial (see Figure 3.6(b)). The graph is further enhanced by ignoring the data point (120, 6), shown in Figure 3.6(c).

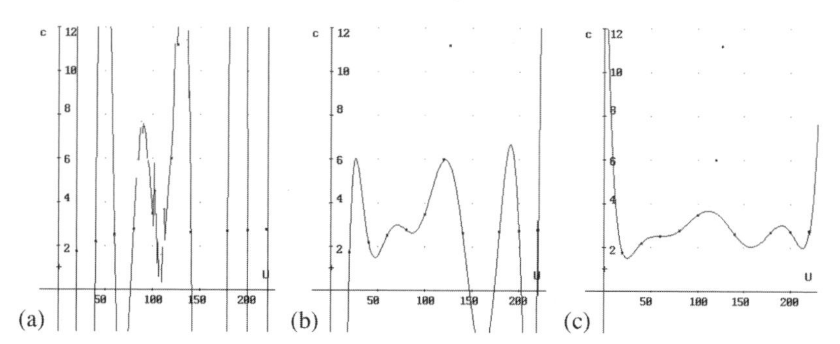

Figure 3.6 Interpolating the data of Table 3.1: (a) full interpolation, (b) ignoring the extreme point, (c) ignoring the next-most extreme point.

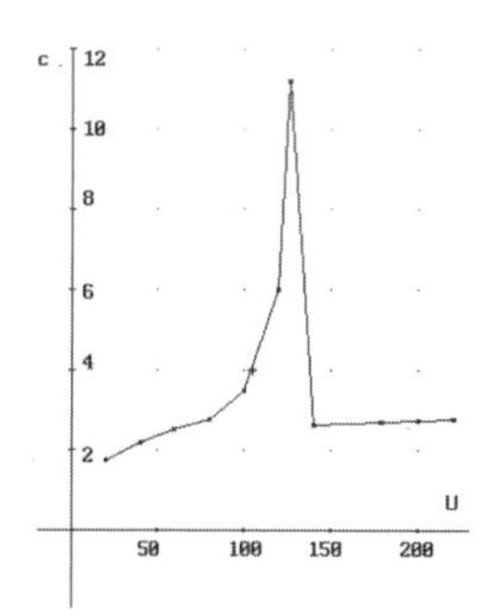

Figure 3.7 Piecewise linear interpolation.

Clearly, care must be taken with high-order interpolation and it may be appropriate to use *piecewise* linear interpolation between adjacent data points (see Figure 3.7).

The derivation of an error expression is beyond the scope of this text. For completeness an error bound is given by

$$|e_n(x)| = |f(x) - p_n(x)| \leq |(x - x_0)\dots(x - x_n)|\frac{M_{n+1}}{(n+1)!}, \tag{3.23}$$

where $M_{n+1} = \max|f^{((n+1)}(x)|$, $x_0 < x < x_n$ (for details see [18], Section 4.3).

3.4 Divided difference interpolation

The Lagrange interpolating polynomial is useful for analysis, but is not the ideal formula for evaluating the polynomial. Here the groundwork is laid for the development of efficient forms of the unique interpolating polynomial p_n (a) by simplifying the construction and (b) by reducing the computational effort required to evaluate the polynomial.

3.4.1 Divided differences

It is well known that the first derivative of a function f, with respect to x, is defined by the limiting process

$$\frac{\mathrm{d}f(x)}{\mathrm{d}x} = f'(x) = \lim_{h\to 0}\frac{f(x+h) - f(x)}{h}.$$

$[f(x+h) - f(x)]/h$ is the slope of the line joining $(x, f(x))$ and $(x+h, f(x+h))$ on the graph of $y = f(x)$, shown in Figure 3.8. As h is reduced to zero the line approaches the tangent to the curve $y = f(x)$ at x, that is the slope of the line tends to the slope of the curve, $f'(x)$. For small $h > 0$, $[f(x+h) - f(x)]/h$ is an approximation to the slope of f at x.

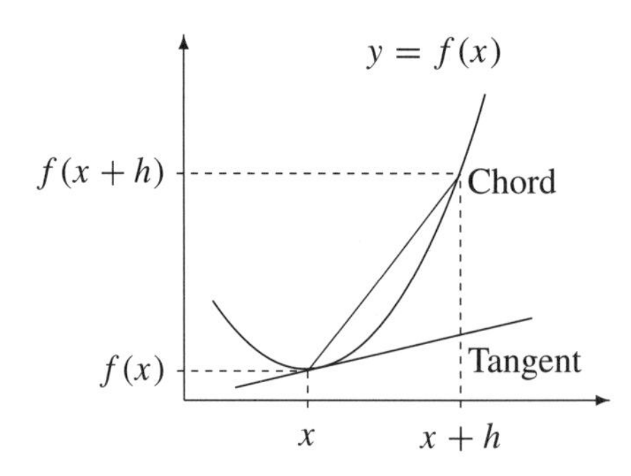

Figure 3.8 Approximating the first derivative by a chord.

Table 3.3 Function values and first-order divided differences.

x_i	f_i	$f[x_i, x_{i+1}]$
x_0	f_0	
		$f[x_0, x_1]$
x_1	f_1	
		$f[x_1, x_2]$
x_2	f_2	
$\vdots$		$\vdots$
x_{n-1}	f_{n-1}	
		$f[x_{n-1}, x_n]$
x_n	f_n	

Extending this to the data set $(x_0, f_0), \ldots, (x_n, f_n)$, where $f_i = f(x_i)$, $(f_{i+1} - f_i)/(x_{i+1} - x_i)$ is an estimate to the slope of f at $x = x_i$,

$$f'(x_i) \approx f[x_i, x_{i+1}] = \frac{f(x_{i+1}) - f(x_i)}{x_{i+1} - x_i}. \tag{3.24}$$

$f[x_i, x_{i+1}]$ is the *first-order divided difference* of f at $x = x_i$. The $n + 1$ function values define n first-order differences, $f[x_0, x_1], \ldots, f[x_{n-1}, x_n]$, that estimate $f'(x_0), \ldots, f'(x_{n-1})$, conveniently displayed as a *divided difference table* (see Table **??**). The $f[x_i, x_{i+1}]$ are obtained from the two values to the left, in columns x_i and f_i (for example, $f[x_0, x_1] = (f_1 - f_0)/(x_1 - x_0)$).

Example 3.11 The data below give the distance of a particle from a fixed point as a function of time[8]. Use the data to estimate the speed v of the particle at appropriate times where $v = \mathrm{d}s/\mathrm{d}t$.

Time t (s)	Distance s (m)
0	1
1	1
2	0.294
2.5	0.181
3	0.122

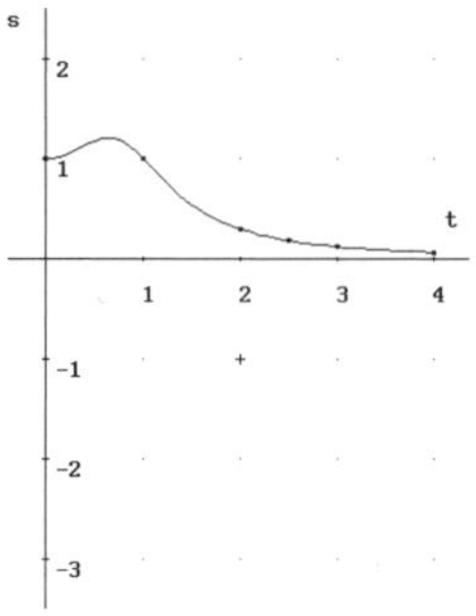

[8] In fact, $s(t) = (t^2 + 1)/(t^4 + 1)$.

At $t = 0$ an estimate of the speed is given by

$$v(0) = \left.\frac{\mathrm{d}s}{\mathrm{d}t}\right|_{t=0} \approx \frac{s_1 - s_0}{t_1 - t_0} = \frac{1-1}{1-0} = 0.$$

Completing the (first-order) divided difference table results in

Time t (s)	Distance s (m)	Speed v (m/s)
0	1	
		0
1	1	
		−0.706
2	0.294	
		−0.226
2.5	0.181	
		−0.118
3	0.122	

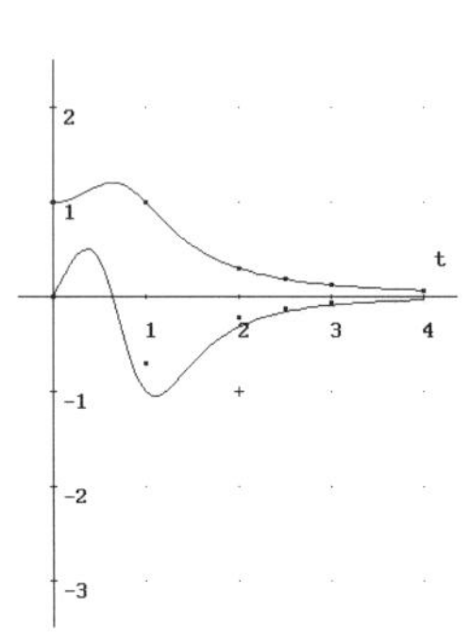

The additional column (which you should be able to reproduce) gives estimates to $v(0)$, $v(1)$, $v(2)$ and $v(2.5)$ which, in some cases, lie some distance from the exact 'time–speed curve'.

DERIVE Experiment 3.4

Author `dd(x0,f0,x1,f1):=(f1-f0)/(x1-x0)` simulates the divided difference $(f_1 - f_0)/(x_1 - x_0)$. Then Author `dd(2,.294,2.5,.181)` approX gives -0.226 (see Example 3.11).

To compute $f[x, x+h]$ for the function f defined by $f(x) = \sin x$, Author `ds(x,h):=dd(x,sinx,x+h,sin(x+h))`. Now investigate the behaviour of $f[x, x+h]$ at $x = 0$ as h approches zero (try $h = 1, 1/2, 1/4, 1/8$, etc., and approX each expression). What do you observe?

The *second-order divided difference* of f at $x = x_i$ is the difference of two first-order divided differences,

$$f[x_i, x_{i+1}, x_{i+2}] = \frac{f[x_{i+1}, x_{i+2}] - f[x_i, x_{i+1}]}{x_{i+2} - x_i}. \qquad \textbf{(3.25)}$$

$n + 1$ values $f_0, \ldots, f_n$ support $n - 1$ second-order divided differences, from $f[x_0, x_1, x_2]$ to $f[x_{n-2}, x_{n-1}, x_n]$, adding a column to Table 3.3 (see Table 3.4). For example, $f[x_0, x_1, x_2] = (f[x_1, x_2] - f[x_0, x_1])/(x_2 - x_0)$.

Table 3.4 Function values and two orders of divided difference.

x_i	f_i	$f[x_i, x_{i+1}]$	$f[x_i, x_{i+1}, x_{i+2}]$
x_0	f_0		
		$f[x_0, x_1]$	
x_1	f_1		$f[x_0, x_1, x_2]$
		$f[x_1, x_2]$	
x_2	f_2		$f[x_1, x_2, x_3]$
		$f[x_2, x_3]$	
x_3	f_3		
$\vdots$			$\vdots$
x_{n-1}	f_{n-1}		$f[x_{n-2}, x_{n-1}, x_n]$
		$f[x_{n-1}, x_n]$	
x_n	f_n		

Example 3.12 Appending a column of second-order divided differences to the table in Example 3.11 produces the table shown below:

Time t (s)	Distance s (m)	Speed v (m/s)	$f[x_i, x_{i+1}, x_{i+2}]$
0	1		
		0	
1	1		−0.353
		−0.706	
2	0.294		0.32
		−0.226	
2.5	0.181		0.108
		−0.118	
3	0.122		

Why does the final column (which you should be able to reproduce) *not* give estimates to the particle acceleration? After all, $a = \mathrm{d}v/\mathrm{d}t$.

Higher-order divided differences are given recursively by

$$f[x_0, x_1, \ldots, x_{k-1}, x_k] = \frac{f[x_1, \ldots, x_k] - f[x_0, \ldots, x_{k-1}]}{x_k - x_0}, \quad k = 1, \ldots, n. \tag{3.26}$$

For consistency, the function value $f(x_i) = f_i = f[x_i]$ is defined as the *divided difference of order zero* of f at $x = x_i$.

DERIVE Experiment 3.5
Author the expressions

```
xi(n):=vector(floor(100random(1))/100,i,0,n)
fi(f,x,xi):=vector(f,x,xi)
```

to define a vector of $n+1$ nodes $x_0, \ldots, x_n$ and $n+1$ function values $f(x_0), \ldots, f(x_n)$. The x_i are randomly distributed in [0, 1].

```
d(xi,i,j):=element(xi,i)-element(xi,j)
dd1(xi,fi):=vector(d(fi,i+1,i)/d(xi,i+1,i),i,1,
   dimension(xi)-1)
dd2(xi,dd1):=vector(d(dd1,i+1,i)/d(xi,i+2,i),i,1,
   dimension(dd1)-1)
dd3(xi,dd2):=vector(d(dd2,i+1,i)/d(xi,i+3,i),i,1,
   dimension(dd2)-1)
```

define the first-, second- and third-order divided differences, and

```
ddt(xi,fi):=[xi',fi',dd1(xi,fi)',dd2(xi,dd1(xi,fi))',
   dd3(xi,dd2(xi,dd1(xi,fi)))']
```

generates the associated divided difference table.

For random data sets taken from the interval $0 \leq x \leq 1$ for the polynomial expression x^2, generate the divided difference table to order 4. What do you notice? Test your 'hypothesis' for x and x^3 (you will need to define a function dd4 and modify the function ddt).

3.4.2 Divided difference polynomial

Divided differences are now utilized to write the Lagrange interpolating polynomial more efficiently (requiring less flops). It is stressed that an interpolating polynomial is unique and the task is one of reordering the formula (see Section 3.1).

The *constant* interpolating polynomial $p_0(x) = f_0$ can be written as

$$p_0(x) = f[x_0]. \tag{3.27}$$

The *linear* interpolating polynomial is

$$\begin{aligned} p_1(x) &= \left(\frac{x - x_1}{x_0 - x_1}\right) f_0 + \left(\frac{x - x_0}{x_1 - x_0}\right) f_1 \\ &= f_0 + \left(\frac{f_1 - f_0}{x_1 - x_0}\right)(x - x_0) \\ &= f[x_0] + f[x_0, x_1](x - x_0). \end{aligned} \tag{3.28}$$

With a little more work the *quadratic* interpolating polynomial is

$$p_2(x) = f[x_0] + f[x_0, x_1](x - x_0) + f[x_0, x_1, x_2](x - x_0)(x - x_1). \quad \textbf{(3.29)}$$

The first efficiency is already evident. From eqns (3.27) to (3.29)

$$\begin{aligned} p_1(x) &= p_0(x) + f[x_0, x_1](x - x_0), \\ p_2(x) &= p_1(x) + f[x_0, x_1, x_2](x - x_0)(x - x_1). \end{aligned}$$

That is, the interpolating polynomial of degree n makes *full* use of the polynomial of degree $n - 1$, simply adding one extra term to p_{n-1},

$$p_n(x) = p_{n-1}(x) + f[x_0, \ldots, x_n](x - x_0) \ldots (x - x_{n-1}). \quad \textbf{(3.30)}$$

A significant feature of this form is the ease of including an extra data point (x_{n+1}, f_{n+1}), where $x_{n+1} \notin \{x_0, \ldots, x_n\}$. The interpolating polynomial p_{n+1} is obtained from eqn (3.30) with n replaced by $n + 1$,

$$p_{n+1}(x) = p_n(x) + f[x_0, \ldots, x_n, x_{n+1}](x - x_0) \ldots (x - x_n).$$

Expanding eqn (3.30), the Lagrange interpolating polynomial can be written in the form

$$\begin{aligned} p_n(x) &= f[x_0] \\ &\quad + f[x_0, x_1](x - x_0) \\ &\quad + f[x_0, x_1, x_2](x - x_0)(x - x_1) \\ &\quad \vdots \\ &\quad + f[x_0, x_1, x_2, \ldots, x_n](x - x_0)(x - x_1) \ldots (x - x_{n-1}) \\ &= f[x_0] + \sum_{k=1}^{n} f[x_0, \ldots, x_k] \prod_{i=0}^{k-1} (x - x_i). \end{aligned} \quad \textbf{(3.31)}$$

This is the *Newton divided difference polynomial*. There appears to be little advantage over the Lagrange form, but there is!

Firstly, eqn (3.30) illustrates the ease of including an extra data point, at a cost of $5n + 6$ flops. For the Lagrange form, *all* $n + 2$ Lagrange polynomials $\ell_0, \ldots, \ell_{n+1}$ must be reconstructed at a cost of $4n^2 + 10n + 1$ flops!

Secondly, a *nested* form can be deduced from factorization (the factor $x - x_0$ is common to all terms but the first, the factor $x - x_1$ is common to all terms but the first two, etc.), namely

$$\begin{aligned} p_n(x) &= f[x_0] \\ &\quad + (x - x_0)\{f[x_0, x_1] \\ &\quad + (x - x_1)\{f[x_0, x_1, x_2] \\ &\quad + \cdots \\ &\quad + (x - x_{n-2})\{f[x_0, \ldots, x_{n-1}] \\ &\quad + (x - x_{n-1}) f[x_0, \ldots, x_n]\}\}\}. \end{aligned} \quad \textbf{(3.32)}$$

This form minimizes the operation count to evaluate $p_n(x)$ and further reduces the cost of adding an extra data point to $3n + 4$ flops.

Example 3.13 Repeating Example 3.9 using the polynomial form (3.32) requires a divided difference table.

x_i	$f[x_i]$	$f[x_i, x_{i+1}]$	$f[x_i, x_{i+1}, x_{i+2}]$
1	0		
		1	
−1	−2		$\frac{2}{3}$
		$\frac{5}{3}$	
2	3		

Make sure that you can reproduce the values appearing in columns $f[x_i, x_{i+1}]$ and $f[x_i, x_{i+1}, x_{i+2}]$. The quadratic interpolant is

$$\begin{aligned} p_2(x) &= f[x_0] + (x - x_0) f[x_0, x_1] + (x - x_0)(x - x_1) f[x_0, x_1, x_2] \\ &= 0 + (x - 1) + \frac{2}{3}(x - 1)(x + 1) \\ &= \frac{2x^2}{3} + x - \frac{5}{3}. \end{aligned}$$

To include the data point (0, 1), that is $x_3 = 0$ and $f_3 = 1$, the cubic Lagrange polynomals are

$$\begin{aligned} \ell_0(x) &= \frac{(x - x_1)(x - x_2)(x - x_3)}{(x_0 - x_1)(x_0 - x_2)(x_0 - x_3)} = \frac{-x^3 + x^2 + 2x}{2}, \\ \ell_1(x) &= \frac{(x - x_0)(x - x_2)(x - x_3)}{(x_1 - x_0)(x_1 - x_2)(x_1 - x_3)} = \frac{-x^3 + 3x^2 - 2x}{6}, \\ \ell_2(x) &= \frac{(x - x_0)(x - x_1)(x - x_3)}{(x_2 - x_0)(x_2 - x_1)(x_2 - x_3)} = \frac{x^3 - x}{6}, \\ \ell_3(x) &= \frac{(x - x_0)(x - x_1)(x - x_2)}{(x_3 - x_0)(x_3 - x_1)(x_3 - x_2)} = \frac{x^3 - 2x^2 - x + 2}{2}, \end{aligned}$$

and the cubic interpolating polynomial is

$$\begin{aligned} p_3(x) = \sum_{k=0}^{3} \ell_k(x) f_k &= 0\ell_0(x) - 2\ell_1(x) + 3\ell_2(x) + 1\ell_3(x) \\ &= \frac{4x^3 - 6x^2 - x + 3}{3}. \end{aligned}$$

$p_3(x)$ clearly passes through the four data points (note the change in the polynomial shape on adding an extra data point). The value of $f(0.5)$ is now estimated by $p_3(0.5)$, in other words by $\frac{1}{2}$. This is quite different from the 'quadratic' estimate of -1.

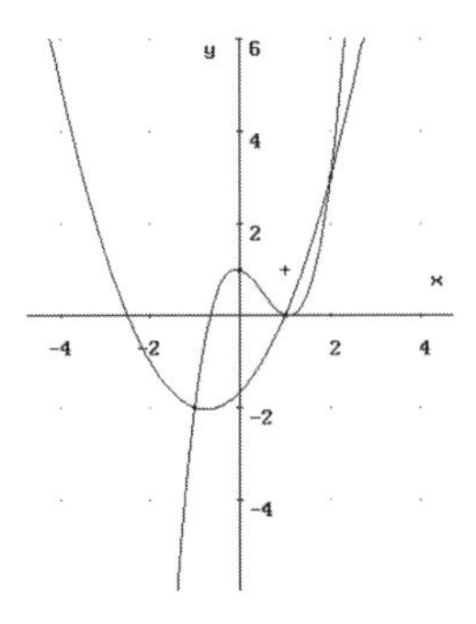

The divided difference approach requires an extra diagonal in the difference table

x_i	$f[x_i]$	$f[x_i, x_{i+1}]$	$f[x_i, x_{i+1}, x_{i+2}]$	$f[x_i, \ldots, x_{i+3}]$
1	0			
		1		
-1	-2		$\frac{2}{3}$	
		$\frac{5}{3}$		$\frac{4}{3}$
2	3		$-\frac{2}{3}$	
		1		
0	1			

and the cubic polynomial is

$$\begin{aligned} p_3(x) &= p_2(x) + (x - x_0)(x - x_1)(x - x_2) f[x_0, \ldots, x_3] \\ &= p_2(x) + \frac{4}{3}(x - 1)(x + 1)(x - 2) \\ &= \frac{4x^3 - 6x^2 - x + 3}{3}. \end{aligned}$$

3.5 Equispaced interpolation

Often the nodes $x_0, \ldots, x_n$ are spaced at a *fixed* interval h (for example, census data may be taken every 10 years, or the telemetry from a rocket may be transmitted every second). We ask:

What savings can be made when interpolating equispaced data?

Table 3.5 A table of forward differences.

x	$f(x)$	$\Delta f(x)$	$\Delta^2 f(x)$	$\Delta^3 f(x)$
x_0	f_0			
		Δf_0		
x_1	f_1		$\Delta^2 f_0$	
		Δf_1		$\Delta^3 f_0$
x_2	f_2		$\Delta^2 f_1$	
		Δf_2		
x_3	f_3			
$\vdots$	$\vdots$			

3.5.1 Difference operators and difference tables

Differences are similar to divided differences but work with equispaced data. The forward difference operator Δ is defined by[9]

$$\Delta^0 f(x) = f(x), \tag{3.33}$$

$$\Delta f(x) = \Delta^1 f(x) = f(x+h) - f(x), \tag{3.34}$$

$$\begin{aligned} \Delta^k f(x) &= \Delta(\Delta^{k-1} f(x)) \\ &= \Delta^{k-1}(\Delta f(x)) \\ &= \Delta^{k-1} f(x+h) - \Delta^{k-1} f(x), \quad k \geq 1. \end{aligned} \tag{3.35}$$

The Δ^k are conveniently displayed in a *difference table* (see Table 3.5).

Example 3.14 The polynomial $p_3(x) = x^3 - 6x^2 + 11x - 3$ gives rise to the following difference table at $x = 2, 4, 6, 8, 10$.

x	$p_3(x)$	$\Delta p_3(x)$	$\Delta^2 p_3(x)$	$\Delta^3 p_3(x)$
2	3			
		6		
4	9		48	
		54		48
6	63		96	
		150		48
8	213		144	
		294		
10	507			

Third-order differences are constant and higher-order differences are zero. This can

[9] Note that the interval size h does not appear in the definition.

be used to evaluate $p_3(x)$ at other equispaced points by 'working from right to left' to create additional diagonals in the difference table. To compute $p_3(12)$, follow the tables from A to D.

A				
2	3			
		6		
4	9		48	
		54		48
6	63		96	
		150		48
8	213		144	
		294		**48**
10	507			
12				

B				
2	3			
		6		
4	9		48	
		54		48
6	63		96	
		150		48
8	213		144	
		294		**48**
10	507		**192**	
12				

C				
2	3			
		6		
4	9		48	
		54		48
6	63		96	
		150		48
8	213		144	
		294		**48**
10	507		**192**	
		486		
12				

D				
2	3			
		6		
4	9		48	
		54		48
6	63		96	
		150		48
8	213		144	
		294		**48**
10	507		**192**	
		486		
12	**993**			

$p_3(12) = 993$.

The consequence of Example 3.14 is that for a set of function values on equispaced data points, a difference table displaying constant differences in 'column n' suggests that the function underlying the data is a polynomial of degree n (the nth derivative of a polynomial of degree n is constant). Interpolation provides the polynomial, and a technique for equispaced data is developed in this section.

For later use, two further difference operators are mentioned:

Backward difference operator: ∇

$$\nabla^0 f(x) = f(x), \tag{3.36}$$

$$\nabla f(x) = \nabla^1 f(x) = f(x) - f(x-h), \tag{3.37}$$

$$\nabla^k f(x) = \nabla^{k-1} f(x) - \nabla^{k-1} f(x-h), \quad k \geq 1. \tag{3.38}$$

Table 3.6 A table of backward differences.

x	$f(x)$	$\nabla f(x)$	$\nabla^2 f(x)$	$\nabla^3 f(x)$
x_0	f_0			
		∇f_1		
x_1	f_1		$\nabla^2 f_2$	
		∇f_2		$\nabla^3 f_3$
x_2	f_2		$\nabla^2 f_3$	
		∇f_3		
x_3	f_3			
$\vdots$	$\vdots$			

The ∇^k are conveniently displayed in a *difference table* (see Table 3.6).

Example 3.15 The second-order backward difference of f is

$$\begin{aligned}\nabla^2 f(x) &= \nabla f(x) - \nabla f(x-h)\\ &= [f(x) - f(x-h)] - [f(x-h) - f(x-2h)]\\ &= f(x) - 2f(x-h) + f(x-2h). \qquad (3.39)\end{aligned}$$

Shift operator: E

$$E^0 f(x) = f(x), \qquad (3.40)$$

$$Ef(x) = E^1 f(x) = f(x+h), \qquad (3.41)$$

$$E^{-1} f(x) = f(x-h), \qquad (3.42)$$

$$\begin{aligned}E^k f(x) &= f(x+kh)\\ &= E(E^{k-1} f(x)), \quad k = \pm 1, \pm 2, \ldots \qquad (3.43)\end{aligned}$$

E shifts the data point a number of intervals to the left or right.

There are many relationships between the three difference operators, of which two will be useful for the ensuing discussion:

$$\begin{aligned}&\Delta f(x) = f(x+h) - f(x) = Ef(x) - f(x) = (E-1)f(x)\\ &\quad \Rightarrow \quad \Delta \equiv E - 1, \quad E \equiv 1 + \Delta, \qquad (3.44)\end{aligned}$$

and

$$\begin{aligned}&\nabla f(x) = f(x) - f(x-h) = f(x) - E^{-1} f(x) = (1 - E^{-1}) f(x)\\ &\quad \Rightarrow \quad \nabla \equiv 1 - E^{-1}, \quad E \equiv (1 - \nabla)^{-1}. \qquad (3.45)\end{aligned}$$

DERIVE Experiment 3.6

Author `f(u,x,x0,h,n):=vector(lim(u,x,x0+ih),i,0,n)` defines f as a vector of $n+1$ values of $u(x)$ from $x = x_0$ to $x = x_0 + nh$. Then Author `v:=f(sinx,x,0,.1,5)` approX defines $\boldsymbol{v}$ to be a vector of 6 values of $\sin x$ from $x = 0$ to 0.5 in steps of 0.1

```
[0, 0.0998333, 0.198669, 0.295520, 0.389418, 0.479425]
```

The operators Δ, ∇, E and 1 may be defined by authoring

```
d(v,i):=element(v,i+1)-element(v,i)
n(v,i):=element(v,i)-element(v,i-1)
e(v,i):=element(v,i+1)
i(v,i):=element(v,i)
```

where $\boldsymbol{v}$ is a vector of function values. Use these definitions to verify the operator relations, as applied to the sample vector $\boldsymbol{v}$.

3.5.2 Forward difference polynomial

That a set of data is equispaced can be used to further 'streamline' the interpolating polynomial. It is assumed that the nodes $x_0, \ldots, x_n$ are in ascending order and may be described by an *index* i and an *interval* h,

$$x_i = x_0 + ih, \quad i = 0, \ldots, n. \tag{3.46}$$

i is the number of intervals between the data point x_i and the *origin* x_0. For a real number s,

$$x = x_0 + sh, \quad 0 \le s \le n. \tag{3.47}$$

If $s \in \{0, 1, 2, \ldots, n\}$, x corresponds to a data point. Otherwise x corresponds to a point lying between two adjacent data points.

There are two ways of incorporating the equispaced nature of the data. The first is based upon the operator relations of Section 3.5.1. Using the shift operator E and eqn (3.44), where $s = (x - x_0)/h$,

$$\begin{aligned} f(x) &= f(x_0 + sh) \\ &= E^s f(x_0) \\ &= (1+\Delta)^s f(x_0) \\ &= \left[1 + s\Delta + \frac{s(s-1)}{2!}\Delta^2 + \frac{s(s-1)(s-2)}{3!}\Delta^3 + \cdots\right] f(x_0). \end{aligned}$$

The binomial expansion of $(1+\Delta)^s$ is used and the coefficient of Δ^k is

$$\frac{s(s-1)\dots(s-(k-1))}{k!}\,\frac{(s-k)\dots 1}{(s-k)\dots 1}=\frac{s!}{k!(s-k)!}=\binom{s}{k},$$

the *binomial coefficient*. Thus

$$f(x_0+sh)=\sum_{k=0}^{\infty}\binom{s}{k}\Delta^k f_0.$$

This infinite series is exact but impractical. Truncating the series at $\Delta^n f_0$ gives a polynomial approximation p_n, of degree n, to the function f,

$$f(x_0+sh)\simeq p_n(x_0+sh)=\sum_{k=0}^{n}\binom{s}{k}\Delta^k f_0. \qquad \textbf{(3.48)}$$

This is the *Newton–Gregory forward difference polynomial.*

SIR ISAAC NEWTON
(1642–1727)
English mathematician/scientist

LIFE

Newton's life had three distinct periods: boyhood to graduation (1642–69), a highly productive period (1669–87) as Lucasian Professor at Cambridge, and finally as a well-paid government official in London with little further interest in mathematics.

Newton's family were farmers. His mother remarried and left him with his grandmother. Upon the death of his stepfather (1656), Newton's mother removed him from grammar school in Grantham, where he had shown little academic promise (school reports described him as 'idle' and 'inattentive'). An uncle decided that he should be prepared for university and he entered Trinity College, Cambridge, in June 1661.

Instruction at Cambridge was dominated by the philosophy of Aristotle (Newton also studied Descartes, Gassendi and Boyle). The new algebra and analytical geometry of Viète, Descartes and Wallis, and the mechanics of Galileo's Copernican astronomy, attracted him. His talent began to emerge on the arrival of Barrow to the Lucasian chair at Cambridge.

After suffering a nervous breakdown (1693) Newton left research for a position in London as Warden of the Royal Mint (1696) and Master (1699). In 1703 he was elected president of the Royal Society and was re-elected each year until his death. He was knighted in 1708 by Queen Anne, the first scientist to be so honoured.

WORK

Newton's scientific genius emerged suddenly when the plague closed the University in 1665. In less than two years he began major advances in mathematics, optics, physics and astronomy. He laid the foundation for differential and integral calculus several years before its independent discovery by Leibniz. The 'method of fluxions', as he termed it, was based on his crucial insight that the integration of a function is the inverse of differentiation. With differentiation as the basic operation, Newton produced simple analytical methods that unified many separate techniques previously developed to solve seemingly unrelated problems (areas, tangents, lengths of curves, the maxima and minima of functions). Newton's *De Methodis Serierum et Fluxionum* (1671) was only published in 1736 when John Colson produced an English translation.

Barrow resigned the Lucasian chair in 1669 recommending that Newton (still only 27) be appointed in his place. Newton's first work as Professor was on optics. He had reached the conclusion during the plague years that white light is not a simple entity. Scientists since Aristotle had believed that white light was a basic single entity, but the chromatic aberration in a telescope lens convinced Newton otherwise. When he passed a thin beam of sunlight through a glass prism Newton noted the spectrum of colours that was formed. He argued that white light is a mixture of many different types of rays which are refracted at slightly different angles, and that each type of ray produces a different spectral colour.

In 1672 Newton was elected a Fellow of the Royal Society, after donating a reflecting telescope, and published his first scientific paper on light and colour in the *Phil. Trans. Roy. Soc.* The paper was well received, but Hooke and Huygens objected to Newton's attempt to prove that light consists of the motion of small particles rather than waves. His corpuscular theory reigned until the wave theory was revived in the 19th century.

Newton's relations with Hooke deteriorated and he turned away from the Royal Society. He delayed the publication of a full account of his optical researches until after the death of Hooke in 1703. Newton's *Opticks* (1704) discussed the theory of light and colour. To explain some of his observations he had to use a wave theory of light in conjunction with his corpuscular theory.

Newton's greatest work was in physics and celestial mechanics, culminating in the theory of universal gravitation. By 1666 Newton had early versions of his laws of motion. He had also discovered the law giving the centrifugal force on a body moving uniformly in a circular path. However he did not have a correct understanding of the mechanics of circular motion. Newton's novel idea of 1666 was to imagine that the Earth's gravity influenced the Moon, counter-balancing its centrifugal force. From his law of centrifugal force and Kepler's third law of planetary motion, Newton deduced the inverse square law. In 1679 Newton applied his mathematical skill to proving a conjecture of Hooke's, that if a body obeys Kepler's second law then the body is being acted upon by a centripetal force.

In 1684 Halley, tired of Hooke's boasting, asked Newton whether he could prove Hooke's conjecture. At Halley's urging Newton reproduced earlier proofs and expanded them into a paper on the laws of motion and problems of orbital mechanics. Halley persuaded Newton to write a full treatment of his new physics and its application to astronomy. Over a year later (1687) Newton published the *Philosophiae Naturalis Principia Mathematica* or *Principia* as it is known.

The *Principia* is the greatest scientific book ever written. Newton analysed the motion of bodies in resisting and non-resisting media under the action of centripetal forces. The results were applied to orbiting bodies, projectiles, pendulums and free-fall near the Earth. He demonstrated that the planets were attracted toward the Sun by a force varying as the inverse square of the distance and generalized that all heavenly bodies mutually attract one another. Further generalization led Newton to the law of universal gravitation, 'all matter attracts all other matter with a force proportional to the product of their masses and inversely proportional to the square of the distance between them'.

Newton explained many previously unconnected phenomena: the eccentric orbits of comets, the tides and their variations, the precession of the Earth's axis and the motion of the Moon as perturbed by the gravity of the Sun. He discovered a method for finding roots of equations which is still in use (Newton–Raphson method). He worked on many curves (cartesian ovals, cissoid, conchoid, cycloid, epicycloid, epitrochoid, hypocycloid, hypotrochoid, kappa curve, serpentine) and classified cubic curves.

JAMES GREGORY
(1638–75)
Scottish mathematician

LIFE

Gregory was appointed professor at St Andrews in 1668 and became the first Professor of Mathematics at Edinburgh in 1674.

WORK

Gregory published *Optica Promota* (1663) which described the first practical reflecting telescope (the Gregorian telescope). A primary concave parabolic mirror converges the light to one focus of a concave ellipsoidal mirror. Reflected light rays from its surface converge to the ellipsoid's second focus (behind the main mirror). There is a central hole in the main mirror through which the light passes. The tube of the Gregorian telescope is thus shorter than the sum of the focal lengths of the two mirrors.

In 1665 Gregory went to the University of Padua and worked on using infinite convergent series to find the areas of the circle and hyperbola. He published *Geometricae Pars Universalis* (1668) which found the areas under curves and the volumes of their solids of revolution.

While at St Andrews he made important contributions in mathematics in addition to those described above. His mathematical work included infinite series expansions for $\tan^{-1} x$, $\tan x$ and $\sec^{-1} x$, and he was one of the first to distinguish convergent and divergent series. His series for $\tan^{-1} x$ (1671) yields, for $x = 1$,

$$\frac{\pi}{4} = 1 - \frac{1}{3} + \frac{1}{5} - \frac{1}{7} + \cdots$$

This expansion for $\pi/4$ was discovered independently by Leibniz in 1673.

That eqn (3.48) does interpolate f is seen by expanding the sum

$$p_n(x_0 + sh) = f_0 + s\Delta f_0 + \frac{s(s-1)}{2!}\Delta^2 f_0 + \cdots + \frac{s(s-1)\ldots(s-n+1)}{n!}\Delta^n f_0.$$

At $x = x_0$, $s = 0$. All terms from Δf_0 onwards contain s and so $p_n(x_0) = f_0$. At $x = x_1$, $s = 1$. All terms from $\Delta^2 f_0$ onwards contain $s - 1$, which is zero at $s = 1$, and $p_n(x_1) = f_0 + \Delta f_0 = f_1$, etc.

Example 3.16 Construct a difference table for the function f where $f(0.5) = 1$, $f(0.6) = 2$ and $f(0.7) = 5$, and use quadratic interpolation to estimate $f(0.63)$. The difference table is

x	$f(x)$	Δ	Δ^2				
0.5	1			x_0	f_0		
		1				Δf_0	
0.6	2		2	x_1	f_1		$\Delta^2 f_0$
		3				Δf_1	
0.7	5			x_2	f_2		

The quadratic polynomial $p_2(x_0+sh) = f_0+s\Delta f_0+\frac{1}{2}s(s-1)\Delta^2 f_0$ uses difference terms from the *upper* diagonal of the difference table to give $p_2(x_0 + sh) = 1 + s + s(s-1) = 1+s^2$. At $x = 0.63$, $s = 1.3$ and $f(0.63) \approx p_2(0.63) = 1+(1.3)^2 = 2.69$.

A more formal development of eqn (3.48) is to modify each term

$$f[x_0, \ldots, x_k](x - x_0)(x - x_1)\ldots(x - x_{k-1}) \tag{3.49}$$

in the Newton divided difference polynomial (3.31) to account for the equispaced data (see Exercise 3.16 for a guided analysis).

An alternative form of p_n uses the backward difference operator ∇,

$$p_n(x) = p_n(x_n + sh) = \sum_{k=0}^{n}(-1)^k \binom{-s}{k} \nabla^k f_n, \tag{3.50}$$

which is the *Newton–Gregory backward difference polynomial* (see Exercise 3.17 for a step-by-step analysis). Here the origin is moved to the 'last' point in the tabulated data, $x = x_n$.

Example 3.17 Repeat Example 3.16 using the backward difference formula (3.50). The difference table is identical to that of Example 3.16.

x	$f(x)$	∇	∇^2				
0.5	1			x_0	f_0		
		1				∇f_1	
0.6	2		**2**	x_1	f_1		$\nabla^2 f_2$
		3				∇f_2	
0.7	**5**			x_2	f_2		

The quadratic polynomial $p_2(x_2+sh) = f_2+s\nabla f_2+\frac{1}{2}s(s+1)\nabla^2 f_2$ uses difference terms from the *lower* diagonal of the difference table to give $p_2(x_2+sh) = 5+3s+s(s+1) = 5+4s+s^2$. At $x = 0.63$, $s = -0.7$ and $f(0.63) \simeq p_2(0.63) = 5-3\times 0.7+(0.7)^2 = 2.69$.

Equations (3.48) and (3.50) are different forms of the *same* polynomial, although this is not evident when written in terms of s.

Example 3.18 Examples 3.16 and 3.17 gave the quadratic polynomials $p_f(s) = 1+s^2$ and $p_b(s) = 5+4s+s^2$. The interpolated estimate at $x = 0.63$ was the same for both polynomials.

In p_f and p_b the meaning of s is different. To establish uniqueness both polynomials must be transformed to the x domain. In the first case $x = x_0 + sh$ from which $s = (x - x_0)/h$. With $x_0 = 0.5$ and $h = 0.1$, $s = 10x - 5$. Consequently, $p_b(x) = 1 + (10x - 5)^2 = 26 - 100x + 100x^2$. In the second case, $x = x_2 + sh$ and $s = (x - x_2)/h$. With $x_2 = 0.7$ and $h = 0.1$, $s = 10x - 7$. Hence $p_b(x) = 5 + 4(10x - 7) + (10x - 7)^2 = 26 - 100x + 100x^2$. Uniqueness!

The need for several interpolation formulae is due to the occurrence of rounding errors. On a finite-precision machine each formula may give a slightly different estimate to $f(x)$ due (a) to the different value of s and (b) to a slightly different calculation ordering.

DERIVE Experiment 3.7

This experiment lets you define a set of four equispaced nodes and associated function values and then determine the forward and backward difference polynomials. Uniqueness (to machine precision) should be your conclusion.

First set the working precision to `Options Precision Approximate` then use `Author` to define the data set

```
xi(x0,h):=[x0,x0+h,x0+2h,x0+3h]
fi(f,x,xi):=vector(f,x,xi)
```

To specify values of e^x on $x \in \{1, 1.1, 1.2, 1.3\}$ ($x_0 = 1$, $h = 0.1$) `Author` `xx:=xi(1,.1)` and `f:=fi(expx,x,xx)`. You can see the values by using `Simplify` on each of these expressions.

To define the first-, second- and third-order differences `Author`

```
d(xi,i,j):=element(xi,i)-element(xi,j)
d1(fi):=vector(d(fi,i+1,i),i,1,dimension(fi)-1)
d2(fi):=d1(d1(fi))
d3(fi):=d1(d2(fi))
```

A difference table to order Δ^3 is obtained with the function

```
dt(xi,fi):=[xi',fi',d1(fi)',d2(fi)',d3(fi)']
```

For the 'exponential' data set specified above, `Author` and `Simplify` the expression `dt(xx,f)` to generate the table:

x_i	f_i	Δ	Δ^2	Δ^3
1	2.71828			
		0.285883		
1.1	3.00416		0.0300672	
		0.315951		0.00316155
1.2	3.32011		0.0332287	
		0.349179		
1.3	3.66929			

The cubic forward and backward difference polynomials are

```
pf(d0,d1,d2,d3,s):=d0+s(d1+(s-1)(d2+(s-2)d3/3)/2)
pb(d0,d1,d2,d3,s):=d0+s(d1+(s+1)(d2+(s+2)d3/3)/2)
```

where $d_k = \Delta^k f_0$ in p_f and $d_k = \nabla^k f_3$ in p_b. Finally

```
s(x,xo,h):=(x-xo)/h
```

defines the transformation from s-space to x-space. $x_o = x_0$ in p_f and $x_o = x_3$ in p_b.

The forward difference polynomial for the data shown above is then given by `pf(2.71828,0.285883,0.0300672,0.00316155,z)`. To convert the polynomial to one in terms of x, first Author and Simplify `s(x,1,.1)` to obtain the transformation (from s) $10(x-1)$, and then Manage Substitute this expression for z in the expression for p_f. Now Simplify and Expand to obtain the result

$$0.526925x^3 - 0.235492x^2 + 1.60924x + 0.817605.$$

A plot of $p_f(x)$ and e^x shows remarkably good agreement in the interval $[1, 1.3]$ – Plot Beside Plot Algebra Author expx Plot Plot Algebra (see Figure 3.7). A similar construction for p_b, using $s(x, 1.3, .1)$, leads to the polynomial expression

$$0.526924x^3 - 0.235494x^2 + 1.60924x + 0.817598.$$

In this case there is a very small difference in the coefficients of x^3, x^2 and 1 (due to rounding) and it is quite plain that the two polynomials *are* the *same*.

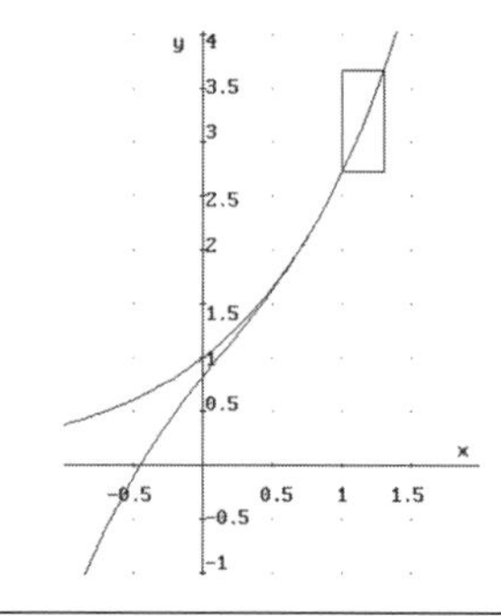

Figure 3.9 Forward and backward difference interpolation of e^x.

Finally, we note that there are many other interpolation techniques (one of which is introduced in Section 4.1). A common modification is to interpolate both function values and derivatives at a set of nodes. This leads to *Hermite interpolation.*

CHARLES HERMITE
(1822–1901)
French mathematician

LIFE

Hermite found formal examinations difficult and had to spend five years working for his BSc (1848). He held posts at the École Polytechnique, Collège de France, École Normale Supérieure and the Sorbonne. Poincaré is the best known of his students.

WORK

Hermite's work in the theory of functions includes the application of elliptic functions to the general equation of the fifth degree, the quintic equation. He published the first proof that e is a transcendental number (1873). Using methods similar to those of Hermite, Lindemann established (1882) that π was also transcendental.

Hermite is known for several mathematical entities bearing his name: Hermite polynomials, Hermite's differential equation, Hermite's interpolation formula and Hermitian matrices.

Summary

The material in this chapter dealt with the development of polynomial expressions whose graphs pass through a specified data set $\{(x_i, f_i)\}_{i=0}^{n}$. Several useful polynomial forms were introduced in Section 3.1 and efficient evaluation of them was discussed in Section 3.2. The construction, eqn (3.12), and uniqueness, eqn (3.13), of a straight line interpolating two points was discussed in Section 3.3.1, and a bound for the error of interpolation (3.16) was developed in Section 3.3.1.1. The ideas were generalised to three points (quadratic interpolation) and to $n+1$ points in Sections 3.3.2 and 3.3.3, leading to the Lagrange interpolating polynomial p_n of eqn (3.21). It was shown in Section 3.4 how divided differences could produce an efficient form of p_n, Newton's divided difference polynomial (3.31). The 'efficiency' was taken one step further in Section 3.5 in which the data was based upon equispaced nodes, leading to the Newton–Gregory forward difference polynomial (3.48).

Exercises

3.1 Write the polynomial $p(x) = 3 - (x-1)(4-(x+1)(5-x(6-(x+2))))$ in power form.

3.2 For each of the following power series exploit nested multiplication to find an efficient means for their valuation.

(a) $e^x = \sum_n \frac{x^n}{n!}$

(b) $\ln x = 2\sum_{n_{\text{odd}}} \frac{1}{n}\left(\frac{x-1}{x+1}\right)^n$

3.3 `Author` the *DERIVE* expression `p(x):=(x-99π)(x-100π)(x-101π)` and evaluate the expression at $x = 314.15$ using `Simplify approX` (set `Options Precision Exact` to 6 digits). Expand the expression for $p(x)$ to its power form and set this to equal the polynomial expression $q(x)$. Evaluate $q(314.15)$ and compare the answer with $p(314.15)$. What has happened?

Set `Digits` to 12 and re-evaluate $p(314.15)$ and $q(314.15)$. Comment on the answers.

3.4 Show that the first two polynomial forms in eqn (3.18) define the same polynomial.

3.5 Use eqn (3.12):

(a) to find the straight line passing through the data $f(x) = 2$ and 2.5 at $x = 1$ and 4, and
(b) to show that $p_1(x_0 + sh) = \ln[x_0(1 + \frac{h}{x_0})^s]$ interpolates $\ln x$ at $x = x_0$ and $x = x_0 + h$, where $x = x_0 + sh$.

3.6 Use linear interpolation to estimate the value of $\sin 40°$ given that $\sin 30° = \frac{1}{2}$ and $\sin 60° = \frac{\sqrt{3}}{2}$ (0.866 to three decimal places). Estimate the accuracy of your answer and compare this with the error bound obtained from eqn (3.16). [*Note*: Use radian measure.]

3.7 Use linear interpolation at $x = 1$ and $x = 2$ on the function defined by $f(x) = x^{1/3}$ to estimate the cube root of $\frac{3}{2}$.

3.8 The function f is tabulated at intervals of size h. If $|f''(x)| \leq M_2$ (a positive constant) throughout the table, show that the modulus of the error due to linear interpolation between any adjacent entries is bounded above by

$$\frac{M_2 h^2}{8}.$$

At what *practical* step size h ought $\cos x$ to be tabulated so that linear interpolation between adjacent entries produces an error no greater than $\frac{1}{2} \times 10^{-4}$?

3.9 Verify the 3×3 determinant result of eqn (3.17).

3.10 Use eqn (3.22) to find the interpolating polynomial that agrees with the data

x	-2	0	1
$f(x)$	5	1	2

3.11 Prove that the polynomial of degree at most n which interpolates f at $n + 1$ distinct points is f itself when $f \in P_n$. [*Hint*: Use the interpolation error (3.23).]

Find the interpolating polynomial for the function f defined by $f(x) = 1 - x^2$ at the 10 points $x \in \{0, 1, \ldots, 9\}$.

3.12 At what practical step size h ought $\cos x$ to be tabulated so that *quadratic* interpolation produces an error no greater than $\frac{1}{2} \times 10^{-4}$? [*Hint*: Use the error term (3.23) and refer to Exercise 3.8.]

3.13 Repeat Exercise 3.10 using the divided difference polynomial (3.31).

3.14 A certain function in x gives rise to the following data:

x	-2	0	1	2	5
$f(x)$	-15	1	-3	-7	41

Use divided differences to conjecture that the data satisfies a cubic relationship, and find this polynomial using formula (3.31).

3.15 Construct a difference table for the polynomial $p(x) = x^2 - x + 2$ at $x = -1$, 0 and 1. Evaluate $p(x)$ at $x = -2$ and 3 by extending the difference table 'from right to left'.

3.16 For equispaced data use eqns (3.46) and (3.47) to show that the k-term product of linear factors in eqn (3.49) can be written as

$$(x - x_0) \dots (x - x_{k-1}) = h^k \prod_{i=0}^{k-1} (s - i) = h^k k! \binom{s}{k}, \tag{3.51}$$

and use proof by induction to show that

$$f[x_0, \dots, x_k] = \frac{\Delta^k f_0}{k! h^k}. \tag{3.52}$$

Substitute eqns (3.51) and (3.52) into (3.49) and hence deduce the Newton–Gregory forward difference polynomial (3.48).

3.17 Use the E–∇ relation (3.45) to produce the Newton–Gregory backward difference polynomial (3.50).

3.18 For the data

x	0.0	0.2	0.4	0.6	0.8	1.0
$f(x)$	0.55	0.82	1.15	1.54	1.99	2.50

(a) construct a difference table,
(b) obtain the Newton–Gregory forward and backward difference interpolating polynomials in s and
(c) show that the two polynomials are identical when expressed in terms of the variable x.

3.19 In the table

n	1	2	3	4	5	6	7	8
$S(n)$	1	5	14	30	55	91	140	204

$S(n)$ is the sum of the squares of the first n positive integers. From a table of differences form a conjecture that $S(n)$ can be represented exactly by a polynomial of some degree in n. Use the Newton–Gregory forward difference formula to construct this polynomial.

3.20 Construct the polynomials p_1 and p_2 that interpolate f at 0 and h, and at 0, h and $2h$, respectively. Integrate these polynomials over the intervals $[0, h]$ and $[0, 2h]$, respectively, and comment on the results of your integration.

3.21 This experiment indicates the importance of numerical precision when computing polynomial interpolation coefficients. The material is based upon Exercise 3.19 and uses the *DERIVE* function `fit`. `Author`

```
s(n):=vector([i,sum(j^2,j,1,i)],i,1,n)
z:=[a,b,c,d,e,f,g,h]
q(n,x,y):=sum(element(y,i+1)x^i,i,0,n)
p(n,x,y):=fit([x,q(n,x,y)],s(n+1))
```

`s` defines the data of Exercise 3.19, `z` is a vector of 8 elements, `q` defines the polynomial $q(x) = z_0 + z_1 x + \cdots + z_n x^n$ and `p` computes the interpolating polynomial p_n passing through the data generated by $s(n+1)$ (using the function `fit`).

Determine the minimum number of digits – use `Options Precision` – required to obtain the cubic polynomial that interpolates the data for $n = 3, 4, 5, 6$ and 7. Comment on your results.

4 Other approximation methods

Chapter 3 concentrated on interpolation with a *single* polynomial satisfying the data $\{(x_i, f_i)\}_{i=0}^n$. For large data sets a single approximant will be a high-order polynomial, p_n. High-order polynomials naturally oscillate (a polynomial of degree n has $n - 1$ turning points) which may not represent physically acceptable behaviour. The evaluation of a high-order polynomial uses many floating-point operations. For example, the Lagrange polynomial (3.22) requires $3n(n + 1)$ flops.

Both drawbacks can be alleviated by using low-order polynomials either (a) in a piecewise fashion or (b) in a non-interpolatory manner.

4.1 Spline interpolation

In this section interpolation is applied in a *piecewise* manner where a different low-order polynomial is used to interpolate successive pairs of data.

The simplest approach uses linear interpolants (see Section 3.3.1). Given $n+1$ items of data in ascending order by x, the data (x_{i-1}, f_{i-1}) and (x_i, f_i) are interpolated by the straight line

$$p_{1,i}(x) = f_{i-1} + \left(\frac{f_i - f_{i-1}}{x_i - x_{i-1}}\right)(x - x_{i-1}), \quad i = 1, \ldots, n. \tag{4.1}$$

Successive approximants, $p_{1,i}$ and $p_{1,i+1}$, are *continuous* at the common point $x = x_i$, that is $p_{1,i}(x_i) = p_{1,i+1}(x_i)$. A piecewise linear interpolant is called a *linear spline* S_1, where $S_1 \in C^0[x_0, x_n]$.

This process works well for the data of Table 3.1 (see Figure 3.7). However, there are some 'sharp corners' which, physically, are unrealistic. The problem is the lack of 'smoothness' of the interpolants at the data points – continuity is assured, but there is a step change in the first derivative.

DERIVE Experiment 4.1

The linear spline (piecewise linear interpolation) is quite easy to set up in *DERIVE* – we have

$$p_{i,1} = f_{i-1} + a_i(x - x_{i-1}), \quad i = 1, \ldots, n,$$

where $a_i = \nabla f_i / h_i$ and $h_i = x_i - x_{i-1}$. A few auxiliary functions need to be defined first. `Author`

```
e1(v,i):=element(v,i)
e2(v,i,j):=element(v,i,j)
dim(v):=dimension(v)
xx(v):=vector(e1(v,i),i,1,dim(v)-1)
```

`e1` (`e2`) extracts element v_i ($v_{i,j}$) from the vector (matrix) v. `dim` returns the length of a vector, and `xx` chops the last element of a vector.

The data set is a $(n+1) \times 2$ matrix, say `y`. `Author`

```
x(y):=vector(e2(y,i,1),i,1,dim(y))
f(y):=vector(e2(y,i,2),i,1,dim(y))
h(x):=vector(e1(x,i)-e1(x,i-1),i,2,dim(x))
d(f):=vector(e1(f,i)-e1(f,i-1),i,2,dim(f))
```

`x` extracts the values $x_0, \ldots, x_n$ from `y`, `f` extracts the values $f_0, \ldots, f_n$ from `y`, `h` computes the intervals $h_1, \ldots, h_n$, and `d` computes the differences ∇f_i, $i = 1, \ldots, n$.

The coefficient vector $\boldsymbol{a}$ and coefficient set $[\boldsymbol{x}, \boldsymbol{f}, \boldsymbol{a}]$ are obtained from

```
a(d,h):=vector(e1(d,i)/e1(h,i),i,1,dim(d))
lco(y):=[xx(x(y)),xx (f(y)),a(d(f(y)),h(x(y)))]'
```

To construct the spline `Author`

```
p(z,z0,z1):=if(z<=z0,0,1)-if(z<=z1,0,1)
ls(c,z,x):=sum(p(z,e1(x,i),e1(x,i+1))
    (e2(c,i,2)+e2(c,i,3)(z-e2(c,i,1))),i,1,dim(c))
```

Once you have defined these functions[1], save them in the *DERIVE* file `lins.mth` for later use.

To compute the linear spline for the data of Table 3.1, `Author`

```
y:=[[20,1.75],[40,2.2],[60,2.5],[80,2.75],[100,3.5],
    [120,6],[126,11.2],[140,2.6],[179,2.7],[200,2.725],
    [220,2.75]]
u:=lco(y)
p1:=ls(u,z,x(y))
```

[1] The `if` statement has the form `if(test,then,else)`. If `test` is true, `then` is executed, otherwise `else` is executed.

and `Plot` – easy! After a little scaling, the picture should look like Figure 3.7. Use the function `ls` to experiment with other data sets.

A solution is to use splines[2] having greater 'smoothness' (measured in terms of the highest continuous derivative). In this section we discuss the most common spline which preserves first- and second-derivative continuity at the internal *knots* $x_1, \ldots, x_{n-1}$ – the *cubic spline*.

Definition For the data $\{(x_i, f_i)\}_{i=0}^{n}$, S_3 is a cubic spline in $[x_0, x_n]$ if

(1) S_3 restricted to $[x_{i-1}, x_i]$ is a polynomial of degree at most 3, and
(2) $S_3 \in C^2[x_0, x_n]$.

If $s_{3,i}$ and $s_{3,i+1}$ are cubic interpolants on adjacent sub-intervals then in addition to the condition $s_{3,i}(x_i) = s_{3,i+1}(x_i) = f_i$, $s'_{3,i}(x_i) = s'_{3,i+1}(x_i)$ and $s''_{3,i}(x_i) = s''_{3,i+1}(x_i)$ are also enforced.

The consequence of this definition is that individual interpolants can no longer be constructed in isolation. The piecewise interpolants $s_{3,1}, \ldots, s_{3,n}$ are interdependent through the derivative continuity conditions.

On $[x_{i-1}, x_i]$,

$$s_{3,i}(x) = f_{i-1} + a_i(x - x_{i-1}) + b_i(x - x_{i-1})^2 + c_i(x - x_{i-1})^3, \tag{4.2}$$

leaving $3n$ constants to be determined, namely $a_1, b_1, c_1, \ldots, a_n, b_n, c_n$. The form of $s_{3,i}$ ensures that the function is interpolated at $x = x_{i-1}$, that is $s_{3,i}(x_{i-1}) = f_{i-1}$.

Functional continuity is established by enforcing $s_{3,i}(x_i) = f_i$, $i = 1, \ldots, n$, in other words

$$f_{i-1} + a_i h_i + b_i h_i^2 + c_i h_i^3 = f_i, \quad i = 1, \ldots, n, \tag{4.3}$$

where $h_i = x_i - x_{i-1}$. This removes n constants and ensures that $s_{3,i}(x_i) = s_{3,i+1}(x_i) = f_i$. For first-derivative continuity,

$$\begin{aligned} s'_{3,i}(x) &= a_i + 2b_i(x - x_{i-1}) + 3c_i(x - x_{i-1})^2, \\ s'_{3,i+1}(x) &= a_{i+1} + 2b_{i+1}(x - x_i) + 3c_{i+1}(x - x_i)^2, \end{aligned}$$

and at $x = x_i$,

$$a_i + 2b_i h_i + 3c_i h_i^2 = a_{i+1}, \quad i = 1, \ldots, n-1, \tag{4.4}$$

removing a further $n - 1$ constants. For second-derivative continuity,

$$\begin{aligned} s''_{3,i}(x) &= 2b_i + 6c_i(x - x_{i-1}) \\ s''_{3,i+1}(x) &= 2b_{i+1} + 6c_{i+1}(x - x_i). \end{aligned}$$

[2] The term spline comes from the draughtsman's flexible implement which is used to draw 'smooth' curves through a series of points.

At $x = x_i$,

$$b_i + 3c_i h_i = b_{i+1}, \quad i = 1, \ldots, n-1, \tag{4.5}$$

removing a further $n - 1$ constants, leaving two unknowns to be determined. It is common practice to apply the conditions

$$s''_{3,1}(x_0) = s''_{3,n}(x_n) = 0,$$

in other words

$$b_1 = 0 \quad \text{and} \quad b_n + 3c_n h_n = 0, \tag{4.6}$$

leading to the *natural cubic spline*. Equations (4.3)–(4.6) represent a system of $3n$ linear equations in $3n$ unknowns. A little work can reduce the size of the system to n. From eqn (4.6), $b_1 = 0$, and the remaining equations (4.3)–(4.6) be be used to develop a further $n - 1$ linear equations for the unknowns $b_2, \ldots, b_n$.

From eqns (4.3), (4.5) and the second of (4.6)

$$a_i = \frac{f_i - f_{i-1}}{h_i} - b_i h_i - c_i h_i^2, \quad i = 1, \ldots, n, \tag{4.7}$$

$$c_i = \frac{b_{i+1} - b_i}{3h_i}, \quad i = 1, \ldots, n-1, \tag{4.8}$$

$$c_n = -\frac{b_n}{3h_n}. \tag{4.9}$$

Substituting these expressions into eqn (4.4) leads to the $n \times n$ system of linear equations

$$\boldsymbol{Hb} = \boldsymbol{g} \tag{4.10}$$

where

$$\boldsymbol{H} = \begin{bmatrix} 1 & & & & & \\ & 2(h_1 + h_2) & h_2 & & & \\ & h_2 & 2(h_2 + h_3) & h_3 & & \\ & & & \ddots & & \\ & & h_{n-2} & & 2(h_{n-2} + h_{n-1}) & h_{n-1} \\ & & & & h_{n-1} & 2(h_{n-1} + h_n) \end{bmatrix},$$

$g_1 = 0$, $g_i = 3(\nabla f_i / h_i - \nabla f_{i-1}/h_{i-1})$, $i = 2, \ldots, n$, and $\boldsymbol{b} = [b_1, \ldots, b_n]^{\mathrm{T}}$.

Example 4.1 The following data are to be interpolated using a natural cubic spline.

x	1	2	5	7	8
y	2	1	3	3	2

Here $n = 4$, $h_1 = 1$, $h_2 = 3$, $h_3 = 2$ and $h_4 = 1$. Further, $\nabla f_1 = -1$, $\nabla f_2 = 2$, $\nabla f_3 = 0$ and $\nabla f_4 = -1$, leading to $g_1 = 0$, $g_2 = 5$, $g_3 = -2$ and $g_4 = -3$. The system (4.10) becomes

$$\begin{bmatrix} 1 & 0 & 0 & 0 \\ 0 & 8 & 3 & 0 \\ 0 & 3 & 10 & 2 \\ 0 & 0 & 2 & 6 \end{bmatrix} \begin{bmatrix} b_1 \\ b_2 \\ b_3 \\ b_4 \end{bmatrix} = \begin{bmatrix} 0 \\ 5 \\ -2 \\ -3 \end{bmatrix}$$

with solution $b_1 = 0$, $b_2 = \frac{149}{197}$, $b_3 = -\frac{69}{197}$ and $b_4 = -\frac{151}{394}$. Equations (4.7)–(4.9) then give

$$c_1 = \frac{149}{591}, \quad c_2 = -\frac{218}{1773}, \quad c_3 = -\frac{13}{2364}, \quad c_4 = \frac{151}{1182},$$

and

$$a_1 = -\frac{740}{591}, \quad a_2 = -\frac{293}{591}, \quad a_3 = \frac{427}{591}, \quad a_4 = -\frac{440}{591}.$$

Consequently, $s_{3,i}(x)$, $i = 1, \ldots, 4$, can be constructed. Figure 4.1 compares the cubic spline with the quartic interpolating polynomial

$$p_4(x) = \frac{x^4}{120} - \frac{13x^3}{60} + \frac{211x^2}{120} - \frac{293x}{60} + \frac{16}{3}.$$

For this data there is little to be gained by using a spline. The graphs of $p_4(x)$ and $S_3(x)$ are very similar.

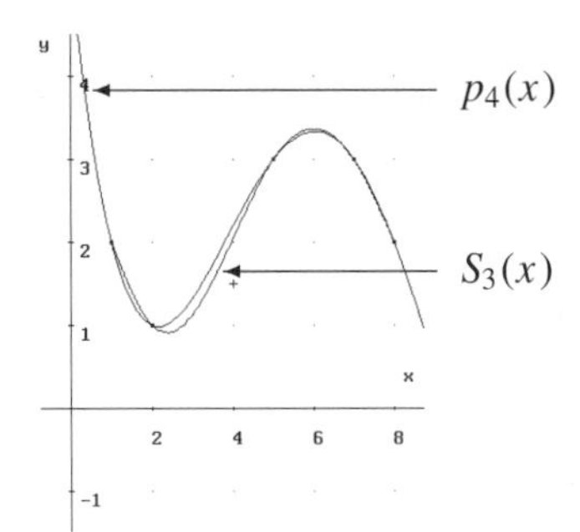

Figure 4.1 Cubic spline vs. interpolating polynomial.

Figure 4.2 shows the result of fitting a natural cubic spline to the data of Table 3.1. While the resulting approximant is significantly better than the polynomial interpolant in Figure 3.6(a), the extreme value at $U = 126$ still has the effect of introducing unsatisfactory oscillations into the spline interpolant. Within the scope of this text, piecewise linear interpolation appears to be the solution for this particular set of data.

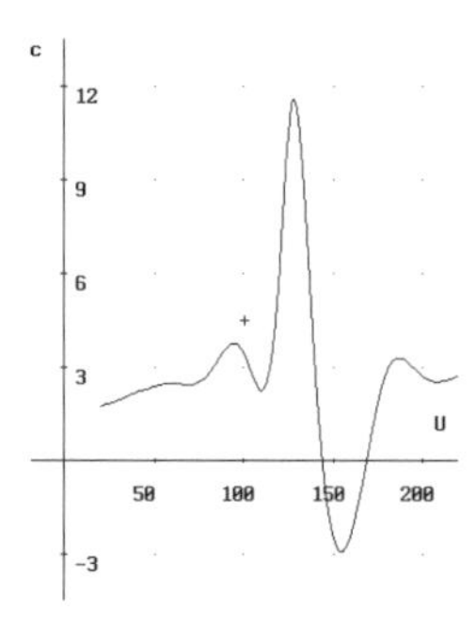

Figure 4.2 Natural cubic spline interpolation of data in Table 3.1.

DERIVE Experiment 4.2

The effort required to construct a natural spline for $\{(x_i, f_i)\}_{i=0}^{n}$ is substantial. Nonetheless, *DERIVE* can be made to do just that! There are quite a large number of pre-processing functions – be patient! They may be saved for later use.

First, some useful auxiliary functions (note that several functions used here are defined in Experiment 4.1) – Author

```
q(v,w,i):=e1(v,i)/e1(w,i)
s(v,i,m,j):=e1(v,i)+m*e1(v,j)
kd(i,j):=if(i=j,1,0)
```

q computes the quotient of two corresponding vector elements, v_i/w_i, s computes the sum ($m = +1$) or difference ($m = -1$) of two vector elements, and kd is the Kronecker quantity δ_{ij} (1 if $i = j$, otherwise 0).

The data set is an $(n + 1) \times 2$ matrix, say y, and functions x, f, h and d from Experiment 4.1 can be used to process this data.

Now set up the vector $\boldsymbol{g}$ and matrix $\boldsymbol{H}$ – Author

```
g(d,h):=append([0],vector(3(q(d,h,i)-q(d,h,i-1)),
   i,2,dim(h)))
n1(h):=vector(vector(kd(i,j)if(i>1,2s(h,j-1,1,j),1),
   i,1,dim(h)),j,1,dim(h))`
n2(h):=vector(vector(if(i>1,if(j>1,(kd(i+1,j)+kd(i,j+1))
   e1(h,min(i,j)),0),0),i,1,dim(h)),j,1,dim(h))`
hh(h):=n1(h)+n2(h)
```

then solve the system for the vectors $\boldsymbol{b}$, $\boldsymbol{c}$ and $\boldsymbol{a}$ – Author

```
b(y):=hh(h(x(y)))^(-1).g(d(f(y)),h(x(y)))
c(b,h):=append(vector(s(b,i+1,-1,i)/(3e1(h,i)),
   i,1,dim(b)-1),[-q(b,h,dim(b))/3])
a(d,h,b,c):=vector(q(d,h,i)-
   e1(h,i)(e1(b,i)+e1(c,i)e1(h,i)),i,1,dim(h))
```

Finally, set up an array of spline coefficients, `nco`, and construct the spline, `ns3` – `Author`

```
nco(b,y):=[xx(x(y)),xx(f(y)),
   a(d(f(y)),h(x(y)),b,c(b,h(x(y)))),b,c(b,h(x(y)))]`
ns3(c,z,x):=sum(p(z,e1(x,i),e1(x,i+1))
   (e2(c,i,2)+sum(e2(c,i,j)(z-e2(c,i,1))^(j-2),j,3,5)),
   i,1,dim(c))
```

Once you have defined these functions, save them in a *DERIVE* file, for example `nats3.mth`

Now define the data set (see Example 4.1) – `Author`

```
y:=[[1,2],[2,1],[5,3],[7,3],[8,2]]
```

compute the coefficient set – `Author`

```
u:=nco(b(y),y)
```

and construct the spline – `Author` and `Simplify`

```
s3:=ns3(u,z,x(y))
```

Once you have defined (and saved) the functions from q to `ns3`, you need *only* execute these last three steps to compute the natural spline! Highlight and `Plot Beside Plot` the data y and `Algebra` highlight `Simplify` and `Plot Plot` the spline `s3`.

You can check the calculations by using `Simplify` on the matrix `u` – you should recover the values computed by hand in Example 4.1.

To compute the natural spline for the data of Table 3.1, `Author`

```
y:=[[20,1.75],[40,2.2],[60,2.5],[80,2.75],[100,3.5],
   [120,6],[126,11.2],[140,2.6],[179,2.7],[200,2.725],
   [220,2.75]]
u:=nco(b(y),y)
s3:=ns3(u,z,x(y))
```

and `Plot` – easy!

Alternative conditions at $x = x_0$ and $x = x_n$ are

$$s'_{3,1}(x_0) = f'(x_0), \quad s'_{3,n}(x_n) = f'(x_n).$$

The slope of the curve is specified to give the equations

$$a_1 = f'(x_0) \quad \text{and} \quad a_n + 2b_n h_n + 3c_n h_n^2 = f'(x_n). \tag{4.11}$$

The resulting curve is a *clamped cubic spline*. The stretched vibrating surface of a drum is one example, whereby the skin is clamped around the perimeter of the kettle. Following an analysis similar to that for the natural cubic spline, a system of equations can be developed[3].

4.2 Least-squares approximation

Although not strictly interpolation, in that the data $\{(x_i, f_i)\}$ cannot be reconstructed from the approximating function, the method of *least squares* does seek a relationship between x and f. The 'trend' of the x–f relationship is assumed, for example linear, exponential or trigonometric, and the method identifies the 'best' coefficients to use in the relation.

Example 4.2 A classical physics experiment studies the connection between a mass suspended from the end of a spring to the extension of the spring (Hooke's law). For a set of masses m, measured values of extension x are recorded, such as those given in Table 4.1. In such an experiment we are looking for a relationship between the measured quantities, in this case mass and spring extension. To assist in this it is useful to plot the data, shown in Figure 4.3(a). It is known that the extension x of an elastic spring is proportional to the attached mass m, that is $x = km$ where k is the spring constant. To determine k for a particular spring, a straight line is fitted to the data (see Figure 4.3(b)), the slope of which gives an estimate to k.

Table 4.1 Data taken from a Hooke's law experiment.

Mass m	Extension x
0.1	9
0.05	5
0.15	16
0.2	21

[3] See [7], Section 3.6 for full details of the system, and a discussion of the uniqueness of cubic spline interpolants.

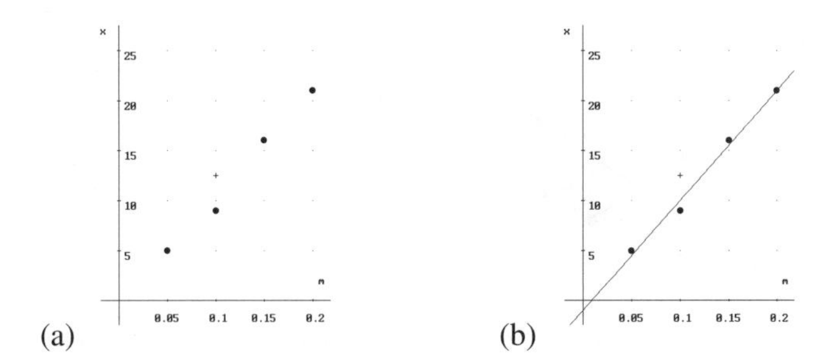

Figure 4.3 (a) Data and (b) straight-line fit to Hooke's law.

In this section the equation of the 'best' straight line is found (experimental error ensures that the measured data will not lie precisely on a straight line).

So, what is 'best'? Take a transparent ruler and try to visually improve on the straight-line fit shown in Figure 4.3(b) by moving the ruler around. Difficult, yes?

The classical approach to defining 'best' constructs a measure of error and then minimises this error. Let e_i be the *point error* in using the straight line $y = a + bx$ to estimate the observed data f_i at $x = x_i$, that is (see Figure 4.4)

$$e_i = f_i - (a + bx_i). \tag{4.12}$$

A first guess at 'best' might be to choose a and b such that the sum of the point errors

$$S = \sum_{i=0}^{n} e_i$$

is minimised. This is a poor model. Both straight lines shown in Figure 4.5 yield $S = 0$, yet clearly one line is better than the other. The difficulty in using S is that errors of opposite sign tend to cancel each other.

A simple remedy is to square the point errors. In this way the point error measure is non-negative and the modified problem is to find constants a and b that minimise the

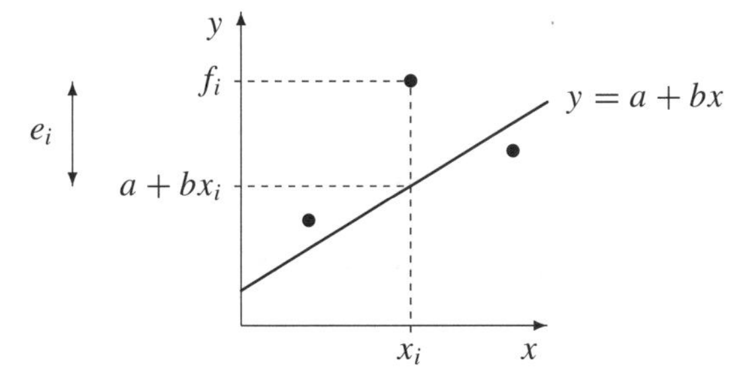

Figure 4.4 Point error, $e_i = f_i - (a + bx_i)$.

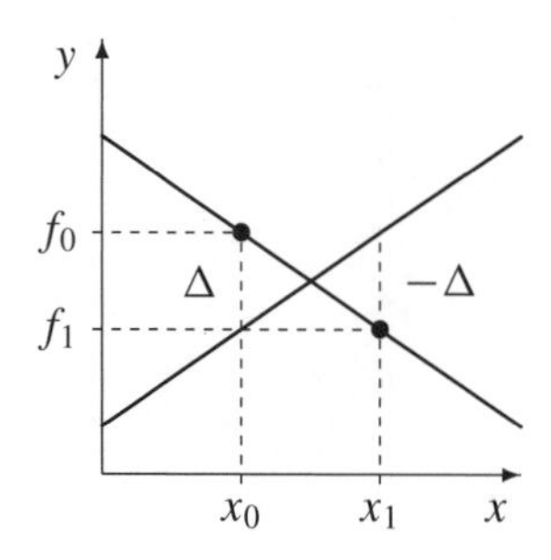

Figure 4.5 Sum of linear point errors.

sum of the squares of the point errors – *least-squares approximation*

$$\min_{a,b}\left[E(a,b)=\sum_{i=0}^{n}(f_i-a-bx_i)^2\right]. \tag{4.13}$$

E is a function of the two variables a and b (x_i and f_i are known) and, from differential calculus, has turning points satisfying the equations

$$\frac{\partial E}{\partial a}=\frac{\partial E}{\partial b}=0. \tag{4.14}$$

From eqn (4.13)

$$\frac{\partial E}{\partial a}=-2\sum_{i=0}^{n}(f_i-a-bx_i) \quad\text{and}\quad \frac{\partial E}{\partial b}=-2\sum_{i=0}^{n}x_i(f_i-a-bx_i),$$

and applying conditions (4.14) leads to

$$\begin{aligned}(n+1)a+\sum_{i=0}^{n}x_ib&=\sum_{i=0}^{n}f_i\\ \sum_{i=0}^{n}x_ia+\sum_{i=0}^{n}x_i^2b&=\sum_{i=0}^{n}x_if_i\end{aligned} \tag{4.15}$$

Equation (4.15) represents a pair of simultaneous equations – the *normal equations* – with solution

$$b=\frac{\overline{x}\,\overline{f}-\overline{xf}}{(\overline{x})^2-\overline{x^2}},\qquad a=\overline{f}-\overline{x}\,b \tag{4.16}$$

where $\overline{}$ signifies *arithmetic mean* (average).

Example 4.3

Find a least-squares straight line for the data

x	0.0	0.1	0.3	0.4	0.5
$f(x)$	0.5	1.4	2.0	2.5	3.1

For hand calculations the following table is useful in evaluating the coefficients required for the solution.

		x_i	f_i	x_i^2	$x_i f_i$
		0.0	0.5	0.0	0.0
		0.1	1.4	0.01	0.14
		0.3	2.0	0.09	0.60
		0.4	2.5	0.16	1.00
		0.5	3.1	0.25	1.55
Sum	$\sum$	1.3	9.5	0.51	3.29
Mean	$\overline{\sum}$	0.26	1.9	0.102	0.658

$b = (0.26 \times 1.9 - 0.658)/(0.26 \times 0.26 - 0.102) = 4$, $a = 1.9 - 0.26 \times 4.8 \approx 0.65$ and the least-squares line is $y(x) = 0.65 + 4.8x$ (see Figure 4.6).

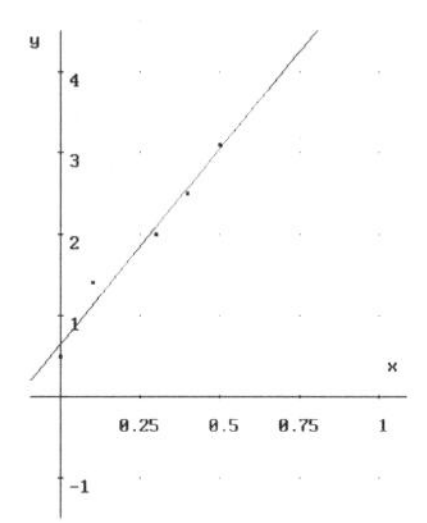

Figure 4.6 Linear least-squares fit.

DERIVE Experiment 4.3
In addition to interpolation, the *DERIVE* function `fit` can generate least-squares approximations, given $n+1$ data points. A polynomial p_m of degree m has $m+1$ coefficients. If $m = n$, the coefficients can be chosen so that p_n interpolates the data. If $m < n$, `fit` generates p_m as the least-squares estimate to the data. If $m = 1$ and $n > 1$, a *linear* least-squares approximation is computed.

To set up the data and the general form of the linear approximant for the data of Example 4.3, Author

```
d:=[[0,.9],[.2,1.9],[.4,2.8],[.6,3.3],[.8,4.2]]
p:=[x,a+bx]
```

The data can be plotted by highlighting the 5×2 data matrix and using `Plot Beside Plot` – make sure that `Options State` is `Discrete` – followed by `Algebra`. To generate the least-squares approximation, `Author` and `Simplify` `fit(p,d)` and `Plot` the resulting linear expression (see Figure 4.6).

4.2.1 Non-linear relations

Often the relation between two quantities will be non-linear.

Example 4.4 The data in Table 4.2 is taken from an experiment to measure the surface temperature of a hot body which is immersed in cold water maintained at a temperature of 0 °C. On plotting the data, connected by line segments for clarity (see Figure 4.7), a straight-line fit can be obtained but it is a very poor description of the underlying relationship between temperature T and time t.

Table 4.2 Newton's law of cooling.

Time t	0	1	2	4	6	7	9
Temperature T	100	56.80	38.60	13.50	4.78	2.90	1.16

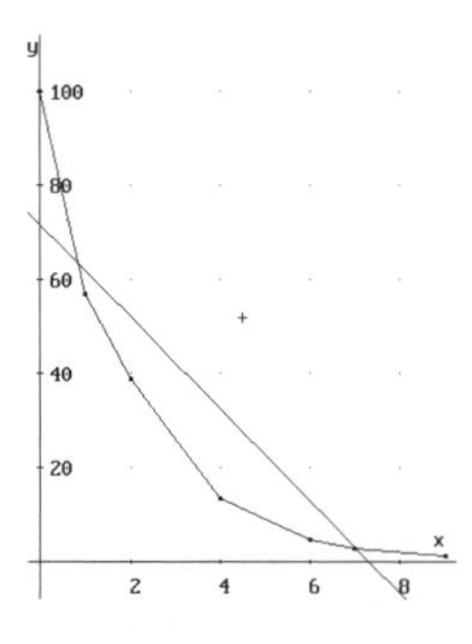

Figure 4.7 Newton's law of cooling and a linear least-squares fit.

A remedy is to seek an alternative approximant that is based upon the normal equations (4.15). For this example it is known that there is an exponential relation between T and t.

Let the variables x and y be related by the exponential form

$$y = ae^{bx}. \tag{4.17}$$

On taking natural logarithms,

$$\ln y = \ln a + bx, \tag{4.18}$$

and defining $A = \ln a$ and $Y = \ln y$, eqn (4.17) becomes

$$Y = A + bx. \qquad \textbf{(4.19)}$$

The significance of eqn (4.19) is that the modified data set $\{(x_i, Y_i)\}$ satisfies a linear relation which can be found using linear least squares, that is A and b can be found from the normal equations (4.15) based upon the data set $\{(x_i, Y_i)\}$. The parameter a is given by $a = e^A$.

Example 4.5 For Example 4.4, the modified data are appended to Table 4.2.

$x = t$	0	1	2	4	6	7	9
T	100	56.80	38.60	13.50	4.78	2.90	1.16
$Y = \ln T$	4.61	4.04	3.65	2.60	1.56	1.06	0.15

Applying linear least-squares to the data (x_i, Y_i):

	x_i	Y_i	x_i^2	$x_i Y_i$
	0	4.61	0	0
	1	4.04	1	4.04
	2	3.65	4	7.30
	4	2.60	16	10.40
	6	1.56	36	9.36
	7	1.06	49	7.42
	9	0.15	81	1.35
$\sum$	29	17.67	187	39.87
$\overline{\sum}$	4.14	2.52	25.71	5.70

gives the coefficients $A = 4.5899$ ($a = 98.4846$) and $b = -0.4986$. The curve is shown in Figure 4.8.

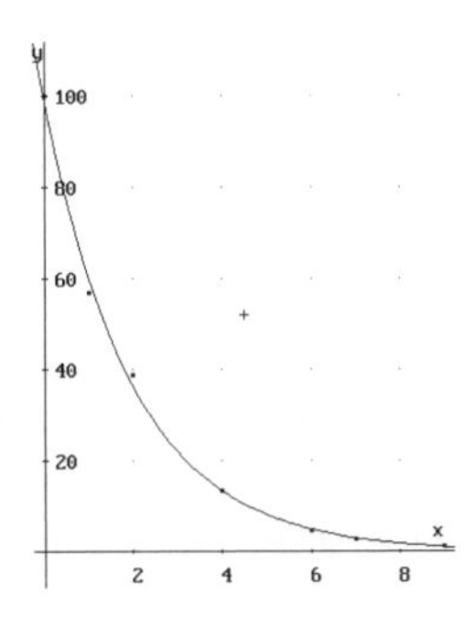

Figure 4.8 Exponential least-squares fit to the cooling data.

Example 4.5 illustrates the harmonious balance between mathematical 'acumen' and mathematical 'technique'. The 'technique' is least-squares, and it is clearly seen to apply to an exponential data distribution. The 'acumen' is in recognising that an exponential fit is appropriate, having inspected the data. No matter how proficient the technique, there is *always* a place for the mathematician's skill and experience.

DERIVE Experiment 4.4

In Experiment 4.3 the function `fit` was used for linear least-squares analysis. A little work is required to implement a non-linear analysis.

We shall base the illustration upon an exponential fit of the form (4.17). Let the data $\{x_i, y_i\}_{i=0}^n$ be stored in the $(n+1) \times 2$ data matrix `d`. The function

```
ld(d):=vector([element(d,i,1),ln(element(d,i,2))],
   i,1,dimension(d))
```

returns the data set $\{x_i, \ln(y_i)\}_{i=0}^n$. This data is approximated by $Y = A + bx$, whereby $y = e^Y = e^A e^{bx}$,

```
lfit(d):=exp(fit([x,a+bx],ld(d)))
```

Now define the data set of Example 4.5 and obtain an exponential least-squares approximation – `Author`

```
d:=[[0,100],[1,56.8],[2,38.6],[4,13.5],[6,4.78],
   [7,2.9],[9,1.16]]
```

and `Author` and `Simplify` `lfit(d)`. `Plot` the data and the exponential fit (it should look like Figure 4.8).

Summary

In this chapter two modifications of the basic interpolation concept were explored regarding the approximation of the data $\{(x_i, y_i)\}_{i=0}^n$. In Section 4.1 *piecewise* interpolation was introduced through piecewise linear interpolants of the form (4.1). The more sophisticated cubic spline (4.10) was developed to preserve first- and second-derivative continuity. The interpolation restriction was dropped in Section 4.2, where a linear approximation of the form $y = a + bx$ was sought for the data set by minimising point errors, leading to the least-squares method. In Section 4.2.1 the methodology was generalised to a non-linear form, with particular attention paid to exponential curves.

Exercises

4.1 Construct the linear spline approximation to x^2, $0 \le x \le 1$, using four equal intervals.

4.2 Find the natural cubic spline approximation to $\cos x$, $0 \le x \le \frac{\pi}{2}$, using two equal sub-intervals.

4.3 Obtain the natural cubic spline for x^4, $0 \le x \le 1$, on three equal intervals. Compare your answer with the cubic interpolating polynomial (using equispaced data).

4.4 Use the *DERIVE* functions given in Experiment 4.2 to compute and plot a natural cubic spline approximation for the following data:

t	0.0	0.1	0.2	0.3	0.4	0.5	0.6	0.7	0.8	0.9	1.0
F	0.0	0.5	3.35	3.3	1.65	1.6	1.6	1.6	1.6	0.6	0.0

Compare the graph with that of a single interpolating polynomial passing through the data set (use the *DERIVE* function `fit`).

4.5 For the Runge test function $f(x) = 1/(1 + 25x^2)$, $-1 \le x \le 1$, (a) determine polynomial interpolants of degree 1, 2, 3 and 4, each based upon equispaced nodes, and (b) construct the natural cubic spline using five equispaced nodes.

4.6 Find the natural cubic spline that matches the following data:

x	0	1	2
y	2	3	10

Given that the data originates from the cubic expression $2+x^3$, explain why the cubic spline does *not* recover the original expression (even though the spline is cubic and interpolates the data).

What end conditions would ensure that the cubic spline was precisely the originating cubic expression?

4.7 Find the linear least-squares approximation $y = a + bx$ to the data

x	0.0	0.1	0.2	0.3	0.4	0.5	0.7
$y = f(x)$	1.9	2.8	2.6	2.3	2.1	2.1	1.6

and write down the sum of the squares of the errors for this line.

4.8 (a) Show that the *linear* least-squares line passes through $(\overline{x}, \overline{f})$.
(b) Show that the *constant* least-squares line $y = a$ is defined by $a = \overline{f}$. Find the constant least-squares fit to the data of Exercise 4.7.

4.9 Reverse the rôles of x and y in Exercise 4.7 and find a least-squares fit of the form $x = a + by$ to the data set $\{(y_i, x_i)\}$: that is, minimise the sum of squares of the errors $x_i - (a + by_i)$. Compare the result with that of Exercise 4.7 and comment.

4.10 Fit a least-squares exponential curve $m = ae^{bt}$ to the following data, which shows the quantity of radioactive polonium-210 remaining at certain times given an initial amount of 5 mg.

Time t (days)	0	5	23	50	94	145	180	200
Mass m (mg)	5	4.88	4.46	3.90	3.14	2.44	2.05	1.88

Estimate the half-life of polonium-210 (the time for the amount of material to halve).

4.11 If the expression $y = a + bx + cx^2$ is a least-squares fit to the data $\{(x_i, f_i)\}_{i=0}^{n}$, show that the parameters a, b and c satisfy the equations

$$\begin{aligned}
(n+1)a + \sum_{i=0}^{n} x_i b + \sum_{i=0}^{n} x_i^2 c &= \sum_{i=0}^{n} f_i \\
\sum_{i=0}^{n} x_i a + \sum_{i=0}^{n} x_i^2 b + \sum_{i=0}^{n} x_i^3 c &= \sum_{i=0}^{n} x_i f_i \\
\sum_{i=0}^{n} x_i^2 a + \sum_{i=0}^{n} x_i^3 b + \sum_{i=0}^{n} x_i^4 c &= \sum_{i=0}^{n} x_i^2 f_i
\end{aligned} \tag{4.20}$$

Hence find the quadratic least-squares fit to the data of Exercise 4.7.

4.12 Determine an appropriate transformation to linearise the power law $y = ax^k$, where a and k are constants. Hence determine the power law relating the pressure to the flowrate from a pipe by applying a least-squares analysis to the following data.

Pressure (MPa)	**Flowrate (cm^3/s)**
70	7100
112	8900
175	11000
280	13500
420	17300

5 Iteration

The topic of this chapter is *sequences*. Sequences are a recurring theme in numerical analysis, as important as Taylor series, and arise from many of the numerical methods discussed in later chapters of this text. Here we concentrate on generating sequences using *iteration*, and look at some of their applications. Qualitative aspects of their behaviour are touched upon.

A primary application of sequences in numerical analysis is the solution of equations such as $f(x) = 0$. This large and important topic forms the subject of Chapter 6, and here we focus on the link with iteration.

5.1 Sequences, recurrence and iteration

Definition A *sequence* of numbers is a list of numbers separated by commas. A *term* is any single number appearing in a sequence.

Example 5.1 Several sequences are given in Table 5.1 and illustrated in Figure 5.1.

Table 5.1 Examples of sequences.

	Sequence
(a)	1, 3, 5, 7, 9, ...
(b)	$1, \frac{1}{2}, \frac{1}{4}, \frac{1}{8}, \frac{1}{16}, \ldots$
(c)	1, 2, 3, 2, 3, 2, 3, ...
(d)	1, 2.3, 7.7, 4.9, 3.3, 1.9, 3.6, 2.3, 6.3, ...
(e)	1, 1.6667, 1.9333, 1.9961, 2.000, 2.000, ...
(f)	1, 1, 1, 1, 1, ...

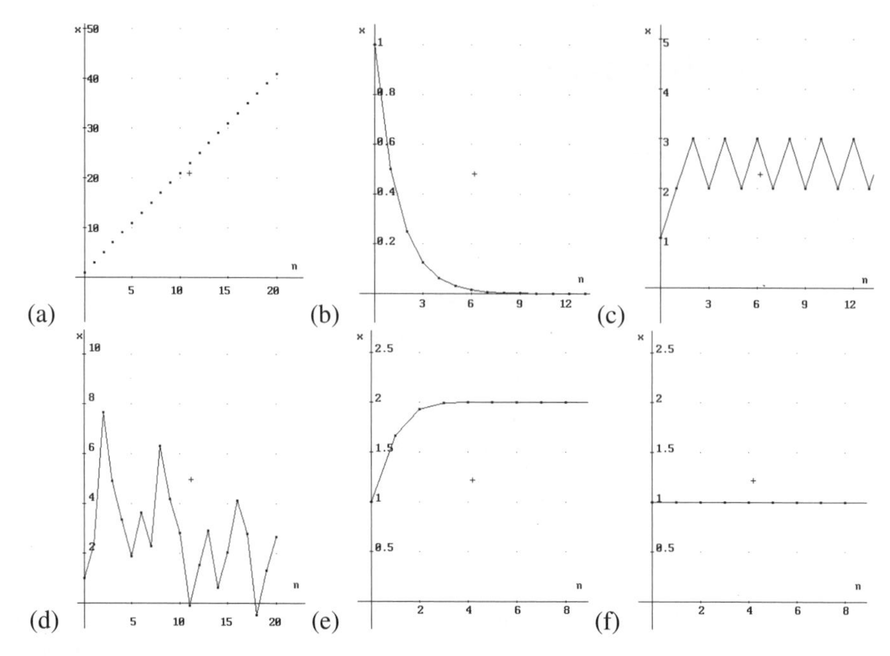

Figure 5.1 The sequences of Table 5.1: (a) increasing, (b) decreasing, (c) periodic, (d) chaotic, (e) limiting value, (f) constant.

There is a wide variety of behaviour in the sequences shown in Figure 5.1, and the remainder of this chapter addresses the following requirements:

Generation: *Is there a formula behind every sequence?*

Analysis: *Is it possible to predict the behaviour of a sequence?*

Application: *Can sequences be put to use?*

5.1.1 Generating formulae and notation

In this section a means of representing a sequence by a formula is developed. Consider the sequence

$$2, 5, 8, 11, 14, 17, \ldots \qquad \textbf{(5.1)}$$

What is the next term in the sequence ? I trust that your answer is 20, but why? In principle, three tasks have been mentally implemented:

(1) Pattern recognition.

(2) Formula construction (pattern representation).

(3) Application of the formula to the last known term to generate the next unseen term.

For sequence (5.1), the pattern is recognised as being 'a constant interval 3 between adjacent terms'. A formula might take the form

```
[next term] = [current term] + 3
```
(5.2)

and application to the last known term 17 generates the value $20 = 17 + 3$.

The power of a formula is that it may now be applied to the new term to further extend the sequence; 23 is the next term since $23 = 20 + 3$. In principle, the simple formula (5.2) represents the entire sequence, which contains an infinity of terms (it is always possible to extend the sequence no matter how many terms have already been generated):

$$26 = 23 + 3$$
$$29 = 26 + 3 \quad \text{etc.}$$

Definition *Iteration* is the repeated application of a formula.

However, the generating formula (5.2) is not specific enough, and it may lead to ambiguities if not used carefully. Consider the sequence

$$4, 7, 10, 13, 16, 19, \ldots \tag{5.3}$$

The 'pattern–formula' process leads to the generating formula

```
[next term] = [current term] + 3
```
(5.4)

and 'application' provides the first unseen term, $22 = 19 + 3$. All's well, or is it? Perhaps you can now see the ambiguity. Formulae (5.2) and (5.4) are identical, yet are related to two different sequences, eqns (5.1) and (5.3). What distinguishes sequence (5.1) from sequence (5.3)?

5.1.1.1 The first term

The first terms in eqns (5.1) and (5.3) are different, namely 2 and 4. To obtain a formula that uniquely identifies a sequence, the value of the *first term* in the sequence must be taken into account. However, formulae (5.2) and (5.4) provide no obvious way of incorporating the first term.

A rather cumbersome modified formula might take the form

```
if [current term] == [first term]
  [next term] = [first term] + 3
else
  [next term] = [current term] + 3
end
```
(5.5)

The logical expression `[current term]==[first term]` ensures that the `[first term]` in a sequence is treated separately. The sequence

$$32, 35, 38, 41, 44, 47, \ldots \tag{5.6}$$

which has yet a different first term, 32, can be generated and extended by applying the modified formula (5.5). This, however, is not the whole story. Consider the original formula (5.2), or (5.4), written in the form

```
[current term] = [next term] - 3
```
(5.7)

This shows that the `[current term]` may be computed from the `[next term]` – backwards iteration! Applying (5.7) to sequence (5.6), the next few terms working backwards are

$$
\begin{aligned}
29 &= 32 - 3 \\
26 &= 29 - 3 \\
23 &= 26 - 3 \\
20 &= 23 - 3 \\
17 &= 20 - 3 \\
14 &= 17 - 3 \quad \text{etc.}
\end{aligned}
$$

The later terms are precisely the same as those of the sequence (5.1), in reverse order. Essentially formulae (5.2) and (5.7) are the same, which implies that (5.1) and (5.6) may be parts of the same longer sequence!

This highlights another drawback of the present model for a formula – it cannot tell if two sequences are simply parts of a single longer sequence. A 'solution' to the generating formula, or equation, is needed. If two sequences form part of the same sequence, we might expect their 'solutions' to be identical. To progress, a more rigorous notation is required.

5.1.1.2 Indicial notation

When embarking on an investigation, mathematicians tend to represent 'everything' using *variables*. This permits generic rather than specific problems to be studied. To apply such a tack to sequences, two variables are required:

(1) The first variable denotes the *value* of a term, such as 2, 5, etc. in eqn (5.1).

(2) The second variable denotes the *location* of a term in the sequence. In eqn (5.1), 2 is the first term, 5 is the second term, etc.

Thus, a term has a value and a location.

Perhaps the most common variable is x: 'the unknown'. x will represent the value of a term. The location of a term is given as an integer n which appears as a subscript to x. Thus x_n represents the value of the term in position n. n is the *index*.

A quirk that mathematicians employ is to index the first term in a sequence with zero. Thus, x_0 represents the value of the first term, or *origin*, of a sequence and n represents the number of terms away from the origin, to the left ($n < 0$) or to the right ($n > 0$). For the sequence (5.1)

$$x_{-2} = -4, \quad x_{-1} = -1, \quad x_0 = 2, \quad x_1 = 5, \quad x_2 = 8, \quad x_3 = 11, \ldots$$

The generating formula (5.2), using the above notation, gives the equations

$$\begin{aligned} x_1 &= 5 = 2+3 = x_0+3 \\ x_2 &= 8 = 5+3 = x_1+3 \\ x_3 &= 11 = 8+3 = x_2+3 \\ x_4 &= 14 = 11+3 = x_3+3 \quad \text{etc.} \end{aligned}$$

Generalising, sequence (5.1) can be expressed by the formulae

$$x_0 = 2; \quad x_{n+1} = x_n + 3, \quad n = 0, 1, 2, \ldots \tag{5.8}$$

– a first term plus a generating formula that is repeatedly applied – *iteration*. The formula is a *recurrence relation*, or *difference equation*. x_n represents the `[current term]` in eqn (5.2) and x_{n+1} represents the `[next term]`, and the sequence is generated by simply writing down the recurrence relation (5.8) with values $n = 0, 1, 2, \ldots$ etc. Formula (5.8) is very specific, and precisely defines the sequence of which (5.1) represents the first six terms.

DERIVE Experiment 5.1

Now that a formal notation for recurrence relations is in place, it is a straightforward matter to 'program' *DERIVE* to execute an iterative process, based upon the function `iterates`. The specification is

```
iterates(g,x,x0,n)
```

- `g` right-hand side of recurrence relation
- `x` independent variable x
- `x0` first term x_0
- `n` number of iterations

To obtain the first five terms of the sequence generated by eqn (5.8), `Author` and `approX` `iterates(x+3,x,2,5)`. You can modify this simple expression to simulate many iterative processes – experiment!

To assist in identifying the behaviour of a sequence it is often useful to plot the *iterate* x_n against the index n. The following function sets up the required plotting data:

```
cit(g,x,x0,n):=[vector(i,i,0,n),iterates(g,x,x0,n)]`
```

To 'see' the sequence (5.8), `Author` and `approX` `cit(x+3,x,2,10)` and `Plot Beside` (for clarity, make sure that `Options State` is `Connected`) `Plot` the resulting data matrix.

5.1.2 Solutions

At the end of Section 5.1.1.1 the notion of a solution was loosely introduced. Here we attempt to be more precise about what is meant by the 'solution of a sequence' without being too concerned with how the solution is determined (see Section 10.2).

Definition The *solution of a sequence* whose general term x_n arises from a recurrence relation is an expression in n that gives the value of x_n.

For sequence (5.1) the following table can be constructed:

Index n	0	1	2	3	4	5	...
Value x_n	2	5	8	11	14	17	...

By inspection (and a little inspiration) the nth term in the sequence is seen to be given by the expression $3n + 2$, that is

$$x_n = 3n + 2, \quad n = 0, 1, 2, \ldots \tag{5.9}$$

Formula (5.9) is the 'solution of the sequence' (5.1) given by the recurrence relation (5.8). For example, if $n = 0$ then $x_0 = 3 \times 0 + 2 = 2$, the first term. If $n = 1$ then $x_1 = 3 \times 1 + 2 = 5$, if $n = 2$ then $x_2 = 3 \times 2 + 2 = 8$, etc.

How is (5.9) related to (5.8)? On the left-hand side of (5.8), replacing n by $n+1$ in eqn (5.9) yields $x_{n+1} = 3(n+1)+2 = 3n+5$. The right-hand side of (5.8) is x_n+3, that is $(3n + 2) + 3$, or $3n + 5$. In other words, if the solution $3n + 2$ is substituted into the relation (5.8), both sides of (5.8) take the *same* form. This is what 'solution' means – the recurrence relation is transformed into an *identity*.

Example 5.2 The sequence 1, 2, 1, 2, 1, 2, ... is generated by the recurrence relation

$$x_0 = 1; \quad x_{n+1} = 3 - x_n, \quad n = 0, 1, 2, \ldots \tag{5.10}$$

which has the solution

$$x_n = \frac{1}{2}[3 - (-1)^n], \quad n = 0, 1, 2, \ldots \tag{5.11}$$

To check the solution, noting that $(-1)^{n+1}$ equals $-(-1)^n$, the left-hand side of (5.10) becomes

$$x_{n+1} \to \frac{1}{2}\left[3 - (-1)^{n+1}\right] \to \frac{1}{2}[3 + (-1)^n],$$

which is identical to the right-hand side,

$$3 - x_n \to 3 - \frac{1}{2}[3 - (-1)^n] \to \frac{1}{2}[6 - 3 + (-1)^n] \to \frac{1}{2}[3 + (-1)^n].$$

Already it is difficult to 'spot' the solution from a table of values:

n	0	1	2	3	4	5	...
x_n	1	2	1	2	1	2	...

5.2 Classification of sequences

Recurrence relations can be classified by their properties, and here two particular characteristics are defined.

Definition The *order of a recurrence relation* equals the difference between the highest and lowest indices appearing in the relation.

Example 5.3

The relation $x_{n+1} = x_n + 3$ is a first-order recurrence relation since

```
[order] = [hi index] - [lo index] = n + 1 - n = 1.
```

Example 5.4

A well-known example of a second-order recurrence relation is that giving rise to the *Fibonacci sequence*

$$1, 1, 2, 3, 5, 8, 13, 21, \ldots$$

Can you spot the pattern? After the first two terms, each 'next' term in the sequence equals the sum of the two preceding terms,

```
[next term] = [current term] + [previous term].
```

Using the indicial notation with two first terms specified,

$$x_0 = 1, \quad x_1 = 1; \quad x_{n+2} = x_{n+1} + x_n, \quad n = 0, 1, 2, \ldots \tag{5.12}$$

The order of the relation is

```
[order] = [hi index] - [lo index] = n + 2 - n = 2.
```

The solution of the Fibonacci relation (5.12) is in no way available via tabulated values and/or inspection, as was solution (5.9). Its form is[1]

$$x_n = \frac{1}{\sqrt{5}}\left(\frac{1+\sqrt{5}}{2}\right)^{n+1} - \frac{1}{\sqrt{5}}\left(\frac{1-\sqrt{5}}{2}\right)^{n+1}, \quad n = 0, 1, 2, \ldots \tag{5.13}$$

The integer nature of the sequence is not obvious from the solution, which contains the irrational number $\sqrt{5}$ (see Experiment 5.2).

DERIVE Experiment 5.2

Author the expressions

```
s5:=sqrt(5)
r1(n):=((1+s5)/2)^(n+1)/s5
r2(n):=((1-s5)/2)^(n+1)/s5
```

[1] The case for a systematic method to develop solutions is becoming irresistible, and this topic is addressed in Section 10.2 on difference equations. See Exercise 5.2 for more on the Fibonacci sequence and its solution. The model (5.12) was proposed by Fibonacci to describe the population evolution of rabbits (see Example 10.14).

```
r(n):=r1(n)-r2(n)
w(n):=vector([i,r(i)],i,0,n)
```

`w` gives the month-by-month rabbit population for the first n months. `r1` and `r2` are the two components of the solution (5.13).

To reproduce the table in Example 5.4, `Author` and `approX` `w(6)`. So, how many rabbit pairs are there after two years? What is the behaviour of the components `r1` and `r2`?

Now, consider the following train of thought. Let x_n and y_n be the total number of rabbit pairs and the number of fertile pairs after n months, according to Example 5.4. We already use the form

```
[next total] = [current total] + [current fertile].
```

Further,

```
[next fertile] = [current total]
```

and, using indicial notation,

$$x_{n+1} = x_n + y_n, \qquad n = 0, 1, 2, \ldots$$
$$y_{n+1} = x_n, \qquad n = 0, 1, 2, \ldots$$

Thus, an alternative form for a *second-order* recurrence relation is *two coupled first-order* recurrence relations[2].

Further, formulae (5.9) and (5.11) which solve first-order recurrence relations each contain just one term involving n, whereas (5.13), which solves a second-order relation, contains two independent terms involving n.

A second useful property of recurrence relations is linearity.

Definition A recurrence relation is *linear* if terms such as x_n, x_{n+1}, etc. only appear in the relation (a) individually and (b) raised to the power of 1. Otherwise, the relation is non-linear.

Example 5.5 For the recurrence relations highlighted so far:

Recurrence relation	Order	Linear
$x_{n+1} = x_n + 3$	1	Yes
$x_{n+1} = 3 - x_n$	1	Yes
$x_{n+2} = x_{n+1} + x_n$	2	Yes

[2] Systems of first-order recurrence relations can be represented as matrix iterations. For this example $\boldsymbol{x}^{(n+1)} = \boldsymbol{A}\boldsymbol{x}^{(n)}$ where $\boldsymbol{x} = [x \;\; y]^{\mathrm{T}}$ and $\boldsymbol{A} = \begin{bmatrix} 1 & 1 \\ 1 & 0 \end{bmatrix}$.

but

Recurrence relation	Order	Linear
$x_{n+1} = x_n^2 - 5$	1	No
$x_{n+1} = x_n x_{n-1}$	2	No

5.3 Linear first-order recurrence relations

A linear *first-order* recurrence relation has the general form

$$x_0 = s; \quad x_{n+1} = ax_n + b, \quad n = 0, 1, 2, \dots \tag{5.14}$$

where a and b are constants. In Example 5.5, the first two relations have a and b given by

Recurrence relation	Order	Linear	a	b
$x_{n+1} = x_n + 3$	1	Yes	1	3
$x_{n+1} = 3 - x_n$	1	Yes	−1	3

For completeness, the solution of eqn (5.14) is[3]

$$x_n = x_0 a^n + b\left(\frac{1-a^n}{1-a}\right), \quad n = 0, 1, 2, \dots \tag{5.15}$$

5.4 Non-linear first-order recurrence relations

The behaviour expressed by linear first-order recurrence relations tends to be monotonic increasing/decreasing, with/without a limiting value, or 2-cyclic (periodic between two values). On the other hand, non-linear recurrence relations can exhibit all manner of exotic behaviour.

5.4.1 A card trick

To illustrate a non-linear first-order recurrence equation consider the following card trick[4].

[3] See Section 10.2 for details of the solution method.

[4] The following reference gives an excellent description of this problem and its generalizations: A.F. Bearon (1996). The dynamics of the 21-card trick, *Mathematical Spectrum*, **29**, 10–11.

> A magician fans a deck of 21 cards and asks the audience to memorise one card. The magician then deals the cards into three piles and asks the audience to indicate which pile the marked card is in. This process is completed two further times, at which point the magician produces the marked card!

If M_n represents the location of the marked card after n deals, where M_n lies between 1 and 21 inclusive, then successive locations are related by the first-order non-linear recurrence relation

$$M_{n+1} = 7 + \left\langle \frac{M_n}{3} \right\rangle . \tag{5.16}$$

The initial location of the card is between 1 and 21, $M_0 \in \{1, 2, \ldots, 21\}$ and $\langle x \rangle$ denotes the smallest integer k such that $x \leq k$. So how does the trick work? Mathematics provides the answer!

The model

First, what you were not told earlier!

> A magician fans a deck of 21 cards and asks the audience to memorise one card. The magician then deals the cards into three piles and asks the audience to indicate which pile the marked card is in. *On re-asssembling the deck, the pile containing the marked card is put in the centre.* This process is completed two further times, at which point the magician produces the marked card!

Assume that after n steps the deck is as follows:

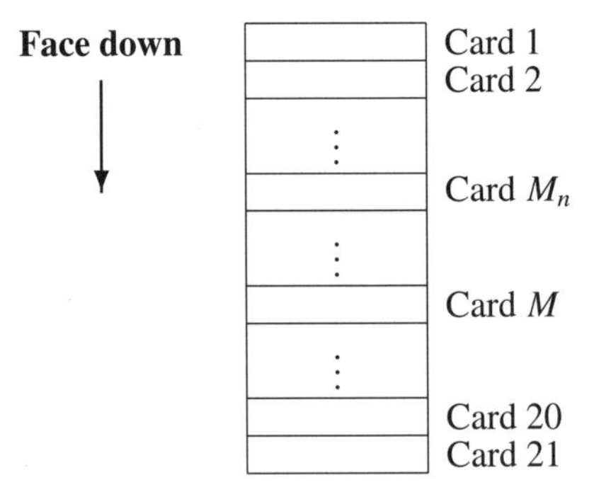

M denotes the middle card in the deck (position 11). On dealing the cards into 3 piles of 7 cards, the situation is

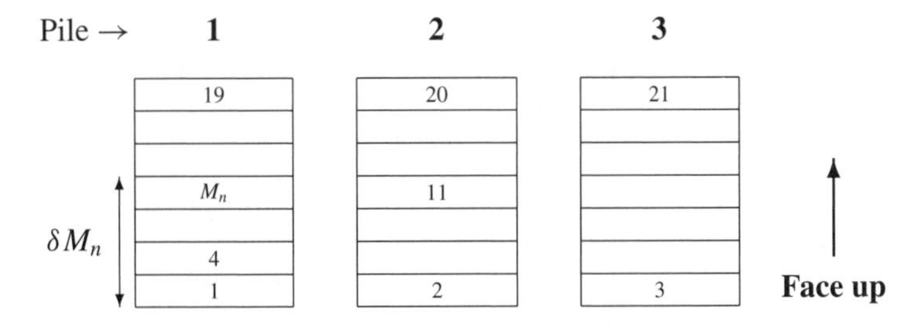

On reassembling the deck, such that *the pile containing the marked card is put in the centre*, the next location of the marked card, M_{n+1}, is determined by two values:

(1) 1 pile of 7 cards *before* (above) the pile with the marked card.

(2) The location of the marked card in its own pile, δM_n. If $M_n \leq 3$, the card is at the top of its pile, $\delta M_n = 1$. If $M_n \in \{4, 5, 6\}$, $\delta M_n = 2$, etc. So, δM_n equals the smallest integer above $M_n/3$, written $\langle M_n/3 \rangle$. The first few values of δM_n are

M_n	1	2	3	4	5	6	7	8	...
$M_n/3$	1/3	2/3	1	4/3	5/3	2	7/3	8/3	...
$\delta M_n = \langle M_n/3 \rangle$	1	1	1	2	2	2	3	3	...

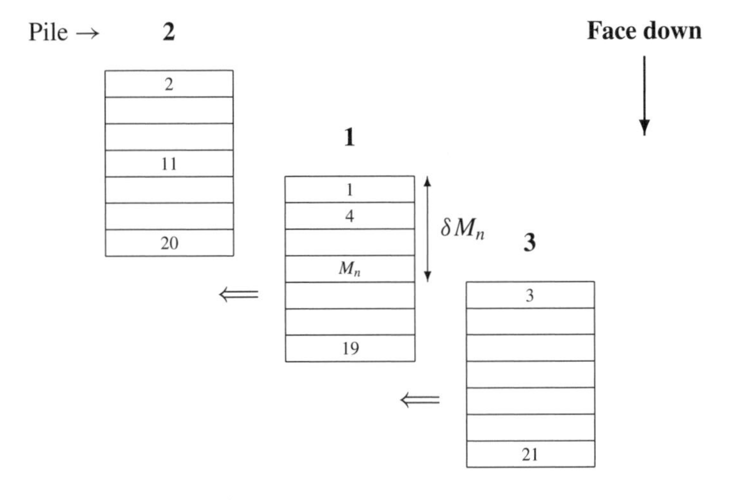

The next location of the marked card is given by eqn (5.16),

$$M_{n+1} = 7 + \delta M_n = 7 + \left\langle \frac{M_n}{3} \right\rangle .$$

The analysis

The location of the middle card is $M = 11$. If $M_n \leq 11$,

$$M_{n+1} = 7 + \langle M_n/3 \rangle \leq 7 + \langle M/3 \rangle = 7 + \langle 11/3 \rangle = 7 + 4 = 11.$$

In other words, $M_n \leq 11 \Rightarrow M_{n+1} \leq 11$. This analysis implies that if $M_n = 11$, then $M_{n+1} = 11$ – the card 'gets stuck' at the middle location. Further, $M_{n+1} > M_n$ (except for the case $M_n = 11$). This can be 'proven' by example. The table below shows M_{n+1} for each possible value of M_n in the range 1 to 11, using eqn (5.16):

M_n	1	2	3	4	5	6	7	8	9	10	11
M_{n+1}	8	8	8	9	9	9	10	10	10	11	11

The marked card works its way towards the middle position, where it stays! A similar analysis holds for $M_n \geq 11$ (the lower half of the deck).

DERIVE Experiment 5.3

DERIVE can play cards! Author

```
ceil(x):=if(mod(x),x,floor(x+1))
card(m0):=iterates(7+ceil(m/3),m,m0,3)
deck:=vector([vector(m0,i,1,4),card(m0)]',m0,1,21)
pack:=vector([vector(n,n,0,3),card(m0)]',m0,1,21)
```

ceil defines $\langle x \rangle$, card locates the marked card for three 'shuffles' starting at location m_0, and deck and pack repeat card for starting positions 1 to 21 (giving two different 'pictures').

Simplify deck and Plot Beside Plot – zoom out a few times until you have a reasonable picture (see Figure 5.2(a)). For each $M_0 \in \{1, \ldots, 21\}$ (horizontal axis), successive locations are plotted vertically. Simplify pack and Plot Plot to get Figure 5.2(b). For each M_0 (vertical axis) this picture shows the progression of the marked card (left to right). Note the tree-like structure of the diagram. What is the significance of this?

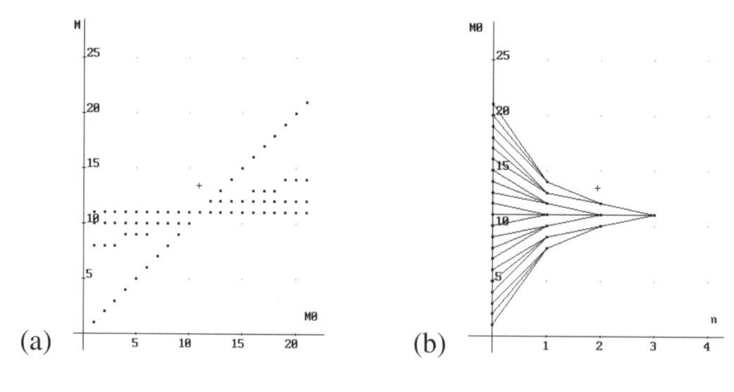

Figure 5.2 Motion of the marked card.

Generalise the $\mathcal{DERIVE}$ functions to decks having $2n + 1$ cards (odd number) and try to match the foregoing theory to this case. What happens if the deck has an even number of cards? Try it!

5.4.2 Further examples of recurrence relations

Three examples of non-linear first-order recurrence relations are presented. Iterate and observe!

5.4.2.1 Möbius sequences

Möbius sequences are generated by taking the ratio of two linear expressions

$$x_{n+1} = \frac{ax_n + b}{cx_n + d}, \qquad n = 0, 1, 2, \ldots \tag{5.17}$$

where a, b, c, and d are constants. If $c = 0$, eqn (5.17) reduces to the linear form (5.14), and if $ad - bc = 0$ then eqn (5.17) generates a constant sequence (see Exercise 5.5). The graphs in Figure 5.3 show several Möbius sequences. The terms in parentheses give the values of (a, b, c, d) used. In all cases $x_0 = 1$. Graphs (a)–(e) show limiting values or cyclic behaviour. Graph (f) indicates a quite different behaviour, vaguely cyclic, but differing amplitudes – is this chaos? None of the graphs indicate unbounded growth – is this a general property of Möbius sequences? Can you find values of a, b, c and d for which a Möbius sequence does exhibit 'unbounded' behaviour?

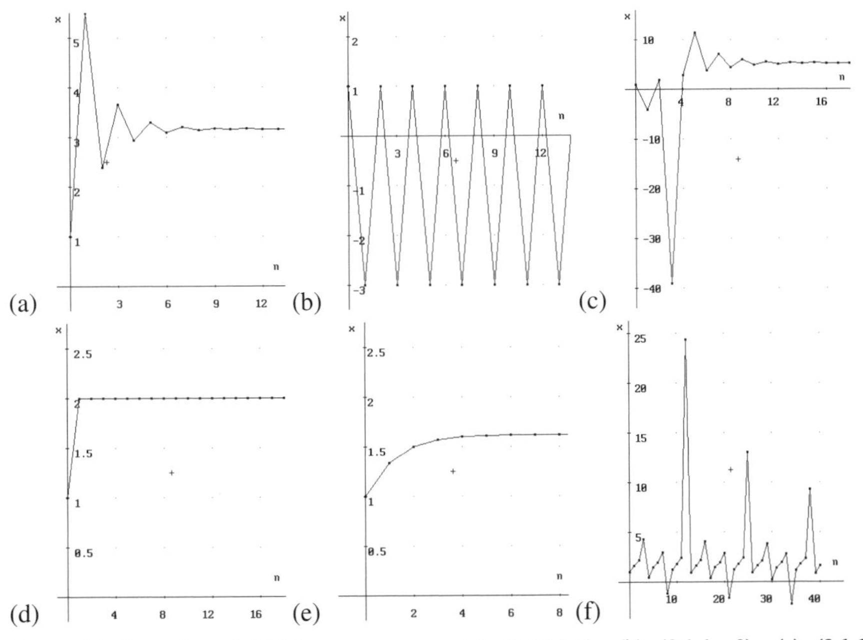

Figure 5.3 Examples of Möbius sequences: (a) (1,10,1,1), (b) (2,1,1,−2), (c) (3,1,1,−2), (d) (3,2,1.5,1), (e) (3,1,1,2), (f) (2,−7,2,−5).

5.4.2.2 Exponential sequences

An *exponential sequence* is generated by the recurrence relation

$$x_{n+1} = a^{x_n}, \quad n = 0, 1, 2, \ldots \tag{5.18}$$

given the value of a. For $x_0 = 0$, the graphs in Figure 5.4 illustrate the behaviour of eqn (5.18) for several values of a. The sequence generated by eqn (5.18) has attracted the minds of many mathematicians, not least being the great Euler himself[5]. For $a > e^{1/e} \approx 1.44466$ (see Figures 5.4(a)–(b)) the sequence increases without limit (see Exercise 6.15). What is the next term in graph (a) (the first 5 terms are 0, 1, 2, 4, 16)? In graph (c), $a = \sqrt{2} \approx 1.41421 < 1.44466$. For values of a in the interval 0 to 1, particle physicists have proposed the model as a description of the lengths of particle collections, or bunches, appearing in linear accelerators.

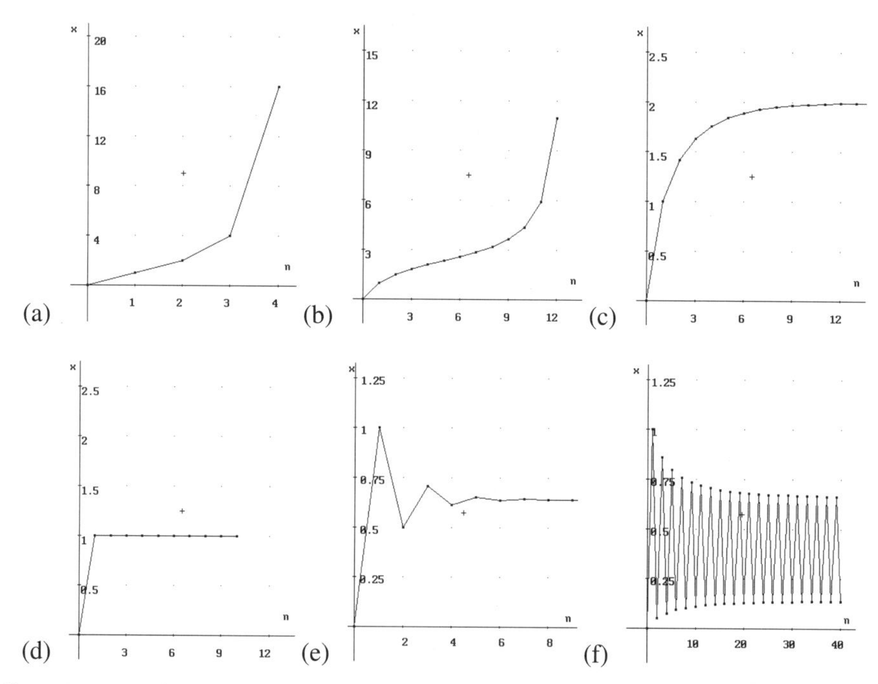

Figure 5.4 Examples of exponential sequences. (a) $a = 2$, (b) $a = 1.5$, (c) $a = \sqrt{2}$, (d) $a = 1$, (e) $a = 0.5$, (f) $a = 0.05$.

[5] Leonard Euler (1707–83), of Swiss origin, was probably the most prolific mathematician who lived and was said to 'write mathematics as effortlessly as most men breathe'. He studied the relation (5.18) around 1778. See p. 288 for more information on Euler.

5.4.2.3 Quadratic sequences

A *quadratic sequence* has the underlying relation $x_{n+1} \propto x_n^2$. One of the most enigmatic quadratic sequences is generated by the *logistic mapping*[6]

$$x_{n+1} = \lambda x_n(1 - x_n), \quad n = 0, 1, 2, \ldots \tag{5.19}$$

where $0 < \lambda < 4$. The range of behaviour of sequences generated by (5.19) is extraordinary! With $x_0 = 0.1$, Figure 5.5 tells a few of the stories. The behaviour includes monotonic decrease/increase to a limiting value, oscillatory movement to a limiting value, cyclic behaviour and random behaviour. For values of λ equal to 3.2, 3.5 and 3.55 (see Figure 5.6), the large-n behaviour of the sequence shows oscillation between two values ($\lambda = 3.2$), then four values ($\lambda = 3.5$), then eight values ($\lambda = 3.55$) to chaos ($\lambda = 3.8$; see Figure 5.5(f)).

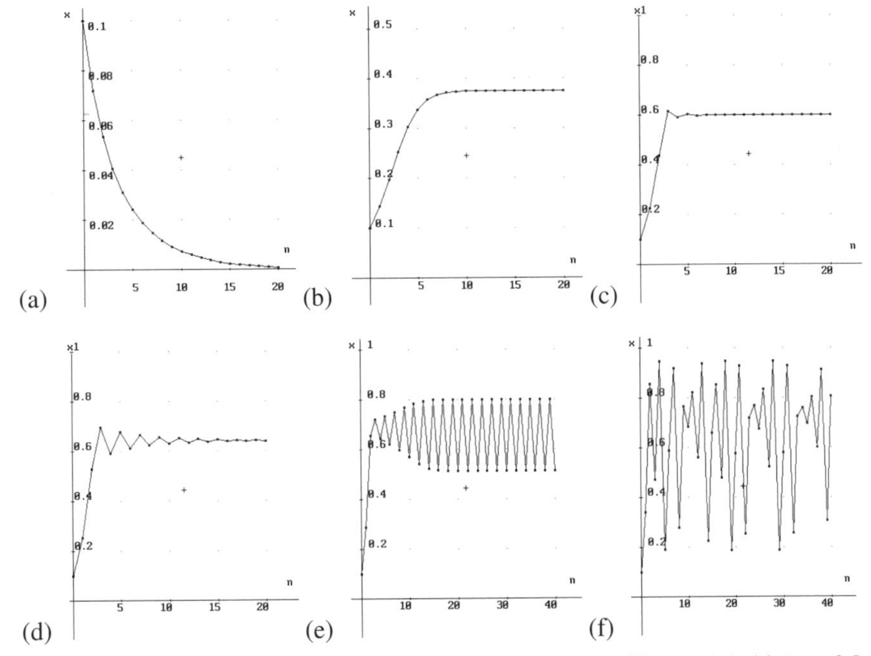

Figure 5.5 Behaviour of certain logistic sequences. (a) $\lambda = 0.8$, (b) $\lambda = 1.6$, (c) $\lambda = 2.5$, (d) $\lambda = 2.8$, (e) $\lambda = 3.2$, (f) $\lambda = 3.8$.

[6] In 1845, P. Verhulst proposed the logistic mapping as a description of population dynamics. Certain biological species increase when the population is small and decrease when the population is large (overcrowding, no food).

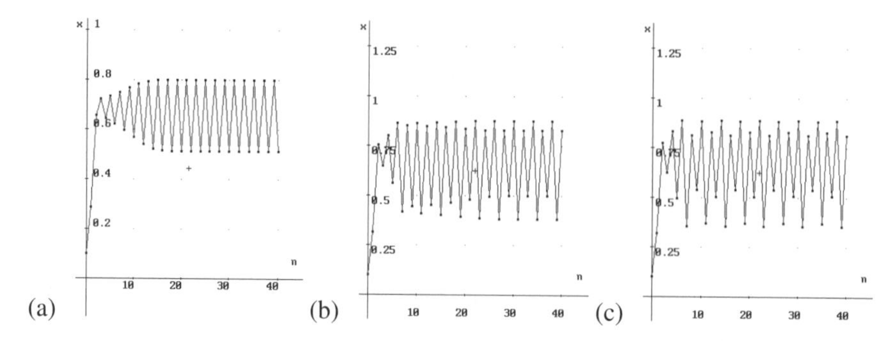

Figure 5.6 More logistic sequences. (a) $\lambda = 3.2$, (b) $\lambda = 3.5$, (c) $\lambda = 3.55$.

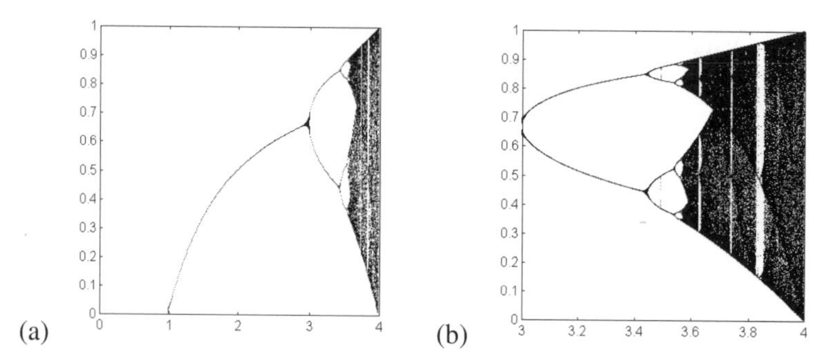

Figure 5.7 The logistic sequence 'at infinity'. (a) $0 < \lambda < 4$, (b) $3 < \lambda < 4$.

It is instructive to observe the behaviour as $n \to \infty$ for a given value of λ (see Figure 5.7(a)). For $0 < \lambda < 1$ the sequence approaches 0 and for $1 < \lambda < 3$ the sequence approaches a single positive limiting value (can you guess what this value might be?). For $3 < \lambda < 1 + \sqrt{6} \approx 3.4495$ the sequence has a 2-cycle limit, for $1 + \sqrt{6} < \lambda < 3.54$ (approximately) a 4-cycle limit, etc. Further refinement to the range $3 < \lambda < 4$ yields Figure 5.7(b).

DERIVE Experiment 5.4

The sequences discussed in this section are easily produced with the DERIVE function `iterates` (see Experiment 5.1). Author

```
nn(n):=vector(i,i,0,n)
mobi(a,b,c,d,x0,n):=[nn(n),iterates((ax+b)/(cx+d),x,x0,n)]'
expo(a,x0,n):=[nn(n),iterates(a^x,x,x0,n)]'
logi(l,x0,n):=[nn(n),iterates(xl(1-x),x,x0,n)]'
```

To obtain the first Möbius sequence (see Figure 5.3(a)), `Author` and `approX` `mobi(1,10,1,1,1,10)`, and `Plot` the resulting data matrix. You are encouraged to experiment with these functions to explore for yourself the wide range of behaviour encompassed within the recurrence relations discussed in this section.

5.5 General first-order recurrence form

All of the first-order recurrence relations encountered so far, both linear and non-linear, can be written in the general form

$$x_0 = s; \quad x_{n+1} = g(x_n), \quad n = 0, 1, 2, \ldots \tag{5.20}$$

Each 'next' term in a sequence is defined as a function g of the 'current' term. For linear relations $g(x) = ax + b$ and eqn (5.14) is recovered. Table 5.2 highlights those recurrence relations met thus far[7].

Table 5.2 Generating functions for first-order recurrence relations.

Description	$g(x)$
Linear relation (5.8)	$x + 3$
21-card trick	$7 + \langle x/3 \rangle$
Möbius sequence	$(ax + b)/(cx + d)$
Exponential sequence	a^x
Logistic mapping	$\lambda x(1 - x)$

5.6 Applications of recurrence relations

Recurrence relations are widely used to model *discrete dynamical systems*, that is systems in which certain variables take changing values at discrete 'instants' in time. Most realistic models are at least second order. Here a flavour of the diversity of applications is given[8].

5.6.1 Linear first-order relations

Compound interest

A well-used form of eqn (5.8) is the compound interest formula. If capital C is invested at a constant interest rate $r\%$, then what is the cumulative wealth at the end of each interest period?

Let w_n denote the wealth after n interest periods. At the end of each period, an amount equal to the interest rate times the 'current' wealth is added to the 'current' wealth to give the 'next' wealth. If $w_0 = C$, then after one period the wealth equals w_0 plus interest of $w_0 \times r/100$, that is $w_1 = w_0(1 + r/100)$. After the second period the wealth equals w_1 plus interest of $w_1 \times r/100$, that is $w_2 = w_1(1 + r/100)$, etc.

[7] [6] is an excellent text containing many iterative experiments and associated analyses.

[8] [4] contains a great many applications of iteration.

The process of accumulating wealth can be generalised as

```
[next wealth] = [current wealth] + [interest]
              = [current wealth] + [rate]*[current wealth]
              = (1 + [rate])*[current wealth]
```

In indicial notation, the sequence[9] $\{w_n\}$ is given by the recurrence relation

$$w_0 = C; \quad w_{n+1} = (1 + r/100)w_n, \quad n = 0, 1, 2, \ldots \tag{5.21}$$

Example 5.6 If $C = 1500$ (Euros) – the approximate value of a maintenance grant – then annually compounded interest, at a rate of 5.75% (UK base rate on 8 October 1996), generates the following results:

Year n	Interest	Wealth w_n
0	0	1500
1	86.25	1586.25
2	91.21	1677.46
3	96.45	1773.91

Is it worth saving your grant cheque? And remember, you can invest the second-year cheque for two years and the third-year cheque for one year (see Exercise 5.8).

Fish tank

The build-up of contaminants in a fish tank must be avoided to maintain healthy fish. If 2 L of clean water evaporates per week, and at the end of each week a further 5 L of contaminated water is removed and replaced with 7 L of tap water, how does the contaminant level in the fish tank vary from week to week?

Let c_n be the amount of contaminant, in kg, in the tank after n weeks and let q be the concentration of contaminants, in kg/L, in tap water. If V is the volume of the fish tank, in L, then $c_0 = Vq$. After one week:

(1) 2 L of clean water is lost, leaving a contaminant concentration of $c_0/(V-2)$ kg/L.

(2) 5 L of contaminated water is removed. The contaminant is reduced by $5c_0/(V-2)$ kg and the contaminant concentration is unaffected.

(3) 7 L of tap water is added which increases the contaminant by $7q$ kg.

[9] cf. eqn (5.8), $a = 1 + \frac{r}{100}$ and $b = 0$. The solution (5.15) is $w_n = C\left(1 + \frac{r}{100}\right)^n$.

Steps 2 and 3 result in a change in contamination equal to $\delta c_0 = -5c_0/(V-2) + 7q$. Thus

$$c_1 = c_0 + \delta c_0 = c_0 - \frac{5c_0}{V-2} + 7q = \left(\frac{V-7}{V-2}\right) c_0 + 7q.$$

The same argument applies to the situation after n weeks to arrive at the recurrence relation

$$c_{n+1} = \left(\frac{V-7}{V-2}\right) c_n + 7q, \quad n = 0, 1, 2, \ldots \tag{5.22}$$

This takes the form (5.14) with $a = (V-7)/(V-2)$ and $b = 7q$. Questions that naturally arise include (a) what is the limiting amount of contaminant ? and (b) what is the maximum limiting value ? (see Exercise 5.10).

5.6.2 Linear second- and higher-order relations

The economy

To forecast the national economy, the Treasury gives predictions of quantities such as GNP, M0 money supply, unemployment rate and interest rates. Typically the predictions cover intervals of one month and the nation is provided with 'snapshots' of its economic health. Industry and private investors use these 'snapshots' to guide their own business strategies.

For period n, the following indicators are used: I_n (national income), P_n (private investment), C_n (consumer spending) and G_n (government expenditure). Based upon observations, the following assumptions can be made:

(a) Total income equals the sum of all expenditure,
$I_n = P_n + C_n + G_n$.

(b) Current consumer spending depends upon previous income,
$C_n = aI_{n-1}$.

(c) Current investment depends upon changes in consumer spending,
$P_n = b(C_n - C_{n-1})$.

a and b are constants. Using (a), (b) and (c), a linear second-order recurrence relation can be derived for I_n,

$$I_{n+2} = a(1+b)I_{n+1} - abI_n + G_{n+2}, \quad n = 0, 1, 2, \ldots \tag{5.23}$$

Simplifying further, we might assume that G_n takes a constant value G. If $I_0 = cG$, $I_1 = dG$, $a = \frac{1}{2}$ and $b = 1$, then (for $c = 2$, $d = 3$) the solution is

$$I_n = 2\left[1 + \left(\frac{1}{\sqrt{2}}\right)^n \sin\frac{n\pi}{4}\right] G, \quad n = 0, 1, 2, \ldots \tag{5.24}$$

A 'picture' of the resulting solution (see Figure 5.8), normalised to G, makes interesting viewing! The solution for general a, b, c and d is difficult to obtain.

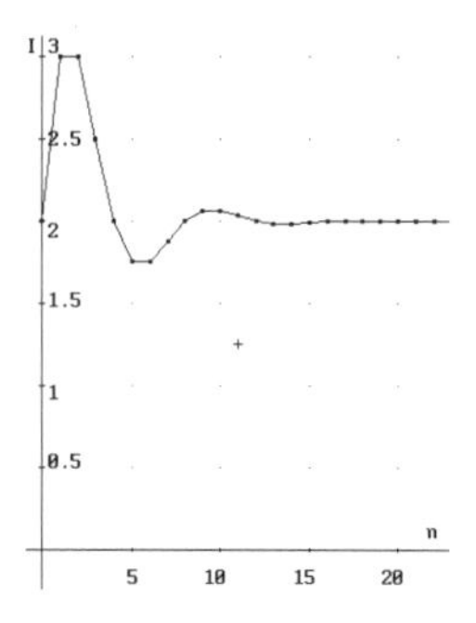

Figure 5.8 The national economy. Boom and bust?

In this situation it makes more sense to 'solve' the model by iterating with the recurrence relation (5.23) to generate the solution sequence. This permits the effect of a, b, c and d to be investigated (see Exercise 5.11).

Timber management

With the environment being a political 'hot potato', the management of natural resources is attracting significant attention. One example is the mangement of timber resources.

The trees in a Norwegian pine forest are classified into three age groups:

Group	Description	Age range
A:	up to 10 years	[0, 10]
B:	10+ to 30 years	(10, 30]
C:	30+ years	(30, ∞)

The assumptions are uniform age distribution, death by old age alone, and a time unit of 2 years during which one fifth of trees in A move to B, one tenth of trees in B move to C and one half of trees in C die of old age. Data gathered over time indicate that, on average, each A-tree seeds 5.2 new trees, each B-tree seeds 15 new trees and each C-tree seeds 2 new trees.

If A_n, B_n and C_n represent the tree populations in groups A, B and C, then how does the forest develop?

Group A: The next 2-year value, A_{n+1}, equals (a) the current number A_n plus (b) seeding rate times current number, plus (c) seeding rate times current number of B-trees, plus (d) seeding rate times current number of C-trees, minus (e) trees lost to B,

$$\begin{aligned} A_{n+1} &= A_n + 5.2A_n + 15B_n + 2C_n - 0.2A_n \\ &= 6A_n + 15B_n + 2C_n. \end{aligned} \quad \textbf{(5.25)}$$

Group B: The next 2-year value, B_{n+1}, equals (a) the current number, plus (b) trees gained from A, minus (c) trees lost to C,

$$\begin{aligned} B_{n+1} &= B_n + 0.2A_n - 0.1B_n \\ &= 0.2A_n + 0.9B_n. \end{aligned} \quad \textbf{(5.26)}$$

Group C: The next 2-year value, C_{n+1}, equals (a) the current number, plus (b) trees gained from B, minus (c) trees lost to old age

$$\begin{aligned} C_{n+1} &= C_n + 0.1B_n - 0.5C_n \\ &= 0.1B_n + 0.5C_n. \end{aligned} \tag{5.27}$$

In matrix form, eqns (5.25)–(5.27) can be written as

$$\boldsymbol{T}^{(n+1)} = \begin{bmatrix} 6 & 15 & 2 \\ 0.2 & 0.9 & 0 \\ 0 & 0.1 & 0.5 \end{bmatrix} \boldsymbol{T}^{(n)} \tag{5.28}$$

where $\boldsymbol{T}^{(n)} = [A_n \;\; B_n \;\; C_n]^{\mathrm{T}}$. To implement eqns (5.25)–(5.27), the initial values A_0, B_0 and C_0 are required.

Example 5.7 Let $A_0 = 10$, $B_0 = 20$ and $C_0 = 2$, in thousands (total population: 32 000 trees). After two years the scenario is

$$\begin{aligned} A_1 &= 6.0 \times 10 + 15.0 \times 20 + 2.0 \times 2 = 364 \\ B_1 &= 0.2 \times 10 + 0.9 \times 20 = 20 \\ C_1 &= 0.1 \times 20 + 0.5 \times 2 = 3 \end{aligned}$$

These equations can be written in matrix form as

$$\boldsymbol{T}^{(1)} = \begin{bmatrix} 6 & 15 & 2 \\ 0.2 & 0.9 & 0 \\ 0 & 0.1 & 0.5 \end{bmatrix} \begin{bmatrix} 10 \\ 20 \\ 2 \end{bmatrix} = \begin{bmatrix} 364 \\ 20 \\ 3 \end{bmatrix}.$$

After 4 years,

$$\boldsymbol{T}^{(2)} = \begin{bmatrix} 6 & 15 & 2 \\ 0.2 & 0.9 & 0 \\ 0 & 0.1 & 0.5 \end{bmatrix} \begin{bmatrix} 364 \\ 20 \\ 3 \end{bmatrix} = \begin{bmatrix} 2,490 \\ 90.8 \\ 3.5 \end{bmatrix}.$$

This model implies that Norwegian pines will 'take over the world', soon! Relevant questions might include 'how many B-trees per 2-year period can be felled ?', 'what strategy might be adopted for thinning A-trees ?', 'what is the effect of acid rain ?', 'what new planting schemes might be adopted ?', and 'what is the effect of animal damage on A-trees ?' (see Exercise 5.14).

DERIVE Experiment 5.5

Example 5.7 is a good illustration of matrix iteration (see eqn (5.28)). `Author`

```
x:=[[6,15,2],[.2,.9,0],[0,.1,.5]]
t:=[a,b,c]‘
forest(a,b,c,n):=iterates(x.t,t,[a,b,c],n)‘
```

To reproduce the values of Example 5.7, Author and approX the expression `forest(10,20,2,2)`.

If 30% of B-trees are felled then $0.3B_n$ must be subtracted from the right-hand side of eqn (5.26) with the resulting transfer matrix

$$\begin{bmatrix} 6 & 15 & 2 \\ 0.2 & 0.6 & 0 \\ 0 & 0.1 & 0.5 \end{bmatrix}$$

Edit the second row of `x` to read `[.2,.6,0]` and re-Author and approX `forest(10,20,2,2)` to see the effect.

Summary

The material in this chapter has covered some basic ground concerning the concept of *iteration* and *recurrence relations*. In Section 5.1 the term *sequence* was introduced. The tools for defining sequences (generating formula and first term) were discussed and then formalised with an *indicial notation* in Section 5.1.1.2 to yield the standard representation of a recurrence relation (difference equation). In Section 5.1.2, solutions were touched upon. Useful terms for classifying sequences (see Section 5.2) were followed by discussions on linear and non-linear first-order relations (see Sections 5.3 and 5.4). The general first-order form was presented in Section 5.5 and the chapter concluded with some illustrative applications of recurrence relations (see Section 5.6).

Exercises

5.1 Using the 'pattern–formula–application' approach, determine the next five terms in each of the following sequences:

(a) 0, 1, 3, 7, 15

(b) 1, 2, 4, 7, 11

(c) 0, 1, 3, 6, 10

(d) $1, \frac{1}{2}, \frac{1}{3}, \frac{1}{4}$

(e) $1, \frac{1}{4}, \frac{1}{9}, \frac{1}{16}$

(f) 10, 6, 13, 1, 13, 10, 10 (obscure!)

(g) 1, 1, 2, 3, 5, 8

5.2 Use eqn (5.12) to obtain the first 14 Fibonacci numbers. Use your calculator to obtain the ratios of successive terms, x_{k+1}/x_k, $k = 1, \ldots, 11$. What do you notice?

Lucas sequence: What is the next distinct pair of starting values x_0 and x_1 for a Fibonacci-like sequence? Determine a few values of the sequence and investigate the ratio of successive terms. Comment!

In a 'tribonacci' sequence each term equals the sum of the previous *three* terms. Write down the recurrence relation to generate this sequence and for $x_0 = x_1 = x_2 = 1$

compute the next few terms in the sequence. Again investigate the ratio of successive terms.

5.3 Show that the expression for x_n given by eqn (5.15) is a solution of the difference equation (5.14). [*Hint*: Substitute the expression (5.15) into the RHS of eqn (5.14).]

5.4 For the card trick of Section 5.4.1, complete a table of moves for M_n until all columns show 11. What is the maximum number of moves required for the marked card to arrive at the middle location?

Repeat the analysis for the case $M_n \geq 11$ (the bottom half of the deck). That is (a) show that the marked card must end up at the middle location and (b) complete a table of all possible moves.

5.5 If $ad - bc = 0$ show that the Möbius sequence is constant.

5.6 Generate the first few terms of the exponential sequence using various values of x_0.

5.7 Using the pictures shown in Section 5.4.2.3 (see Figure 5.7), estimate the value of λ such that the logistic mapping iterates to a 3-cycle. Use $\mathcal{DERIVE}$ to implement your answer.

5.8 Apply the compound interest relation (5.21) to your grant cheque (1500 Euros) over a three-year period. Remember to include the second and final year cheques and assume an interest rate of 6%. Repeat the calculation if the interest is compounded monthly!

5.9 Compound interest: if interest is compounded m times per year where r is the annual percentage interest rate and C is the initial capital, the accrued wealth after n interest periods is

$$w_{n+1} = \left(1 + \frac{r}{100m}\right) w_n.$$

Apply this relation for a few steps (by hand) to deduce that

$$w_n = C\left(1 + \frac{r}{100m}\right)^n.$$

Consider a one-year period with $r = 10\%$ and $C = 20\,000$ Euros. What is the trend in accumulated wealth after one year ($m = n$) as m increases? Is there a limiting value to this trend? If so, what formula can you obtain for the limiting case of wealth accumulation? (You may find $\mathcal{DERIVE}$ useful.)

5.10 Fish tank: for the fish tank model discussed in the text, (a) determine the limiting contaminant concentration (if $V \geq 8$) and (b) the maximum limiting value. [*Hint*: Apply the recurrence relation to obtain c_1, c_2 and c_3 in terms of n, V and q, and propose a general form for c_n – geometric series!]

Suppose instead that at the end of each week, just the 2 L of evaporated water was replaced with tap water, and every fourth week a further 20 L of contaiminated water is removed and replaced with tap water. Develop the difference equation and find the limiting contaminant concentration.

5.11 Economy: iterate eqn (5.23) for the case $a = \frac{3}{4}$ (spend, spend, spend!), $b = \frac{1}{5}$, $c = 2$, $d = 3$.

5.12 Supply and demand: s_n and p_n are the supply and price of a commodity in year n. For positive constants a, b and c, $p_n = a - bs_n$ and $s_{n+1} = cp_n$. Interpret these recurrence relations in terms of 'supply s and demand (price) p'. Obtain the recurrence relation satisfied by p_n and determine what restrictions on the constants ensure that the price attains a limiting value after a sufficiently long time.

5.13 Pollution: the outflow of Lake Erie (USA) flows into Lake Ontario and each year about 38% of Erie's water and 13% of Ontario's water is replaced by fresh water (e.g. rain and rivers).

E_n and O_n denote the amount of pollution in the two lakes in year n. New laws introduced at year 0 prevent any further pollution entering the lakes. Obtain the recurrence relations for E_n and O_n. Show that Erie's pollution is reduced by about 90% after 5 years. If $E_0 = 3O_0$, show that a similar reduction in Ontario's pollution level takes about 29 years (see Exercise 10.6).

5.14 Timber management: what is the maximum percentage of B-trees that can be felled without leading to a decreasing population of B-trees? How is this value affected if indigenous animals destroy 15% of A-trees per 2-year period?

6 The equation $f(x) = 0$

An important application of iteration and recurrence relations concerns the problem of finding the zero(s) of a function f. If f is a function of one variable x, the problem is concisely written as

Find all x such that $f(x) = 0$.

Definition A *root* of the equation $f(x) = 0$ is a value of x which when substituted into the equation gives the identity $0 = 0$. The value x is a *zero* of the function f.

Example 6.1 A sphere of radius r and density ρ is floating in water of unit density (see Figure 6.1). It is required to determine the depth h of the sphere that is below the waterline. Equating the mass of the sphere, $\frac{4}{3}\pi r^3 \rho$, to the displaced water mass[1], $\frac{1}{3}\pi h^2(3r - h)$, leads to the equation $4r^3\rho = h^2(3r - h)$. Defining the ratio $x = h/r$, this equation can be written as $x^3 - 3x^2 + 4\rho = 0$. The left-hand side is a *cubic* polynomial which

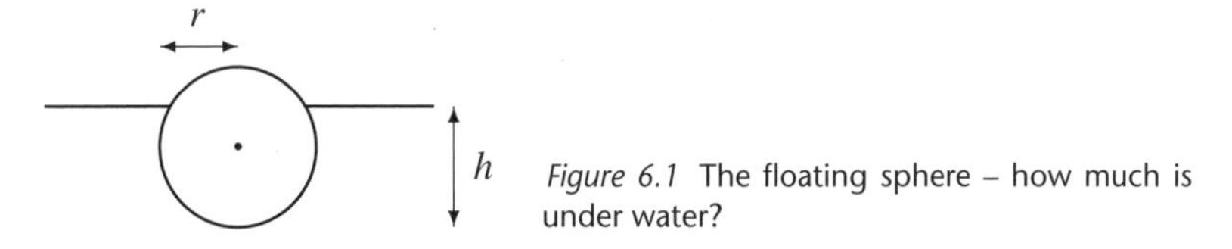

Figure 6.1 The floating sphere – how much is under water?

[1] Archimedes' principle: a body partially submerged in a fluid is buoyed up by a force equal to the weight of the displaced fluid.

has three zeros[2]. If $\rho = 0.6$, to what depth does the sphere sink as a fraction of its radius?

The type of problem described in Example 6.1 can be visualised as finding the values of x for which the graph of $y = f(x)$ touches the x-axis (see Figure 6.2). The discussion here treats functions of a single variable, with no zeros, one zero, an infinity of zeros, or any combination of multiple zeros, repeated zeros and complex zeros (see Figure 6.3).

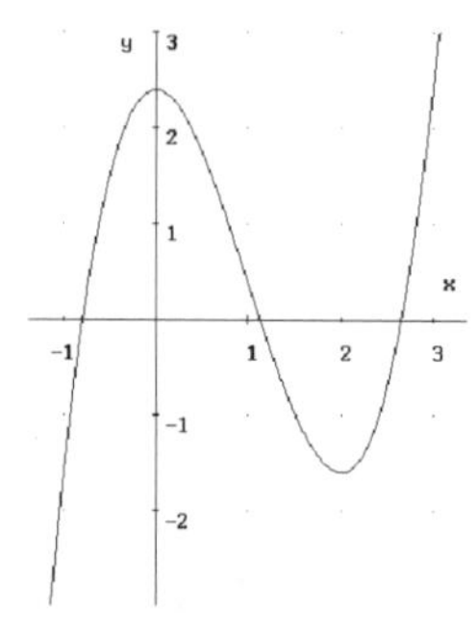

Figure 6.2 The floating sphere 'solution' curve, $y = x^3 - 3x^2 + 2.4$.

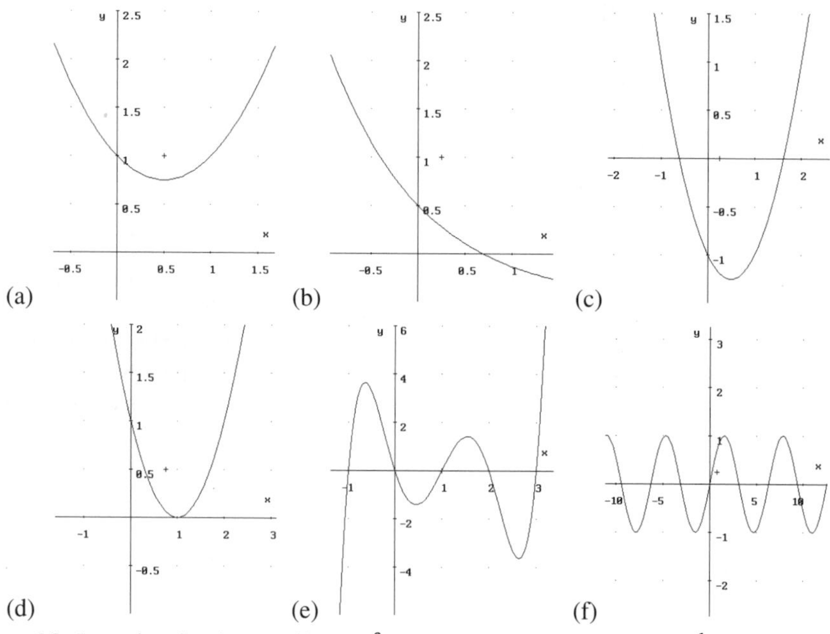

Figure 6.3 Examples of real roots. (a) $y = x^2 - x + 1$, no roots. (b) $y = e^{-x} - \frac{1}{2}$, one root. (c) $y = x^2 - x - 1$, two roots. (d) $y = x^2 - 2x + 1$, double root. (e) $y = x^5 - 5x^4 + 5x^3 + 5x^2 - 6x$, multiple (five) roots. (f) $y = \sin x$, infinity of roots.

---

[2] First fundamental theorem of algebra: a polynomial of degree n has n zeros.

The need for *numerical methods* arises since equations of the form $f(x) = 0$ rarely provide *closed-form* expressions for the roots. A well-known equation having a closed-form expression for its roots is the *quadratic equation*

$$ax^2 + bx + c = 0. \qquad \textbf{(6.1)}$$

The roots of eqn (6.1) are defined explicitly by the formula

$$x = \frac{-b \pm \sqrt{b^2 - 4ac}}{2a}, \quad a \neq 0, \qquad \textbf{(6.2)}$$

and simply require the substitution of values for a, b and c into eqn (6.2).

Example 6.2 For the quadratic equation $x^2 - x - 1 = 0$ (see Figure 6.3(c)), $a = 1$, $b = -1$ and $c = -1$, and formula (6.2) gives the roots as

$$x = \frac{1 \pm \sqrt{1+4}}{2} = \frac{1 \pm \sqrt{5}}{2}.$$

Root-finding *recurrence relations* are usually first-order and non-linear. The next 15 terms of the sequence from the logistic mapping $x_{n+1} = \lambda x_n(1 - x_n)$, with $\lambda = 1.6$ and $x_0 = 0.1$, are (see Figure 6.4(a))

$$\begin{gathered}0.1440,\ 0.1972,\ 0.2533,\ 0.3026,\ 0.3377,\\ 0.3578,\ 0.3677,\ 0.3720,\ 0.3738,\ 0.3745,\\ 0.3748,\ 0.3749,\ 0.3750,\ 0.3750,\ 0.3750\end{gathered} \qquad \textbf{(6.3)}$$

with limiting value 0.375. Dropping subscripts from the logistic mapping gives the equation $x = \lambda x(1 - x)$ which can be arranged to the familiar quadratic form $\lambda x^2 + (1 - \lambda)x = 0$. Application of formula (6.2) yields the roots $x = 0$ and $x = 1 - 1/\lambda$. If $\lambda = 1.6$ the second root is $x = 0.6/1.6 = 0.375$ – the limiting value of the sequence (6.3). Thus:

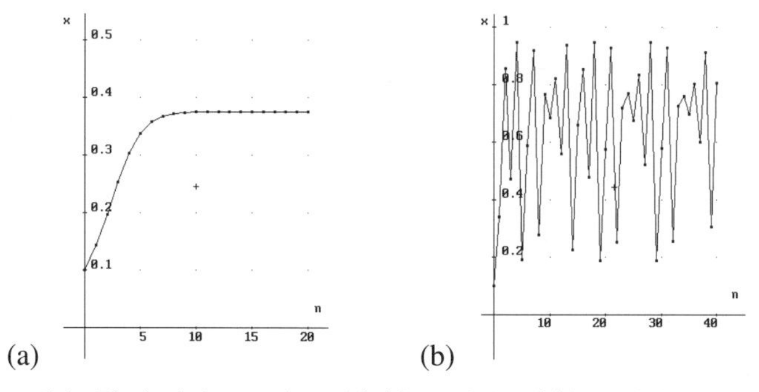

(a) (b)

Figure 6.4 The logistic mapping with (a) $\lambda = 1.6$ and (b) $\lambda = 3.8$.

I Recurrence relations *can find* roots of certain equations

For $\lambda = 3.8$ the non-zero root is $x = 2.8/3.8 \simeq 0.736842$. If $x_0 = 0.1$ the sequence generated by $x_{n+1} = 3.8x_n(1 - x_n)$ is

$$\begin{aligned}&0.3420,\ 0.8551,\ 0.4707,\ 0.9467,\ 0.1916,\\&\quad 0.5886,\ 0.9202,\ 0.2790,\ 0.7645,\ 0.6842,\\&\qquad 0.8211,\ 0.5583,\ 0.9371,\ 0.2240,\ 0.6606\end{aligned}$$

This sequence is 'going nowhere' (see Figure 6.4(b)). Thus:

II Recurrence relations *cannot always find* the roots of equations

In this chapter we (a) systematise the development of recurrence relations capable of finding the root(s) of the equation $f(x) = 0$, and (b) develop analyses capable of distinguishing between cases I and II.

6.1 Locating a root

To develop a numerical method for finding the root(s) of $f(x) = 0$, it is useful to first determine an interval $a \le x \le b$ that brackets (contains) *at least one* root. We need to ask:

what properties must a function 'f of x' satisfy on the interval $a \le x \le b$ to guarantee at least one root in this interval?

Example 6.3 A graph of the polynomial $x^3 - 3x^2 + 2x$ is shown in Figure 6.5. The curve cuts the x-axis at $x = 0$, $x = 1$ and $x = 2$, indicating that the cubic equation $x^3 - 3x^2 + 2x = 0$ has *three* solutions (roots).

As x increases from -0.5, there is a *sign change* in the value of $f(x)$ as x 'passes through' the root at $x = 0$. A similar change occurs as x 'passes through' the values $x = 1$ and $x = 2$.

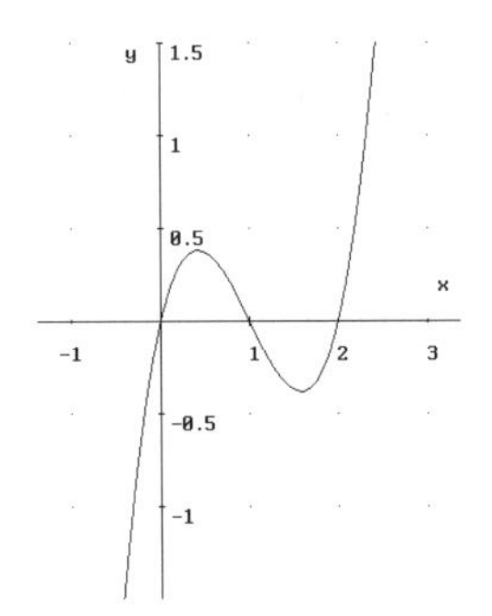

Figure 6.5 The graph of $y = x^3 - 3x^2 + 2x$, $0.5 \le x \le 2.5$.

It appears that $f(x) = 0$ has *at least one* root on the interval $[a, b]$ if $f(x)$ changes sign[3] on $[a, b]$, that is $f(a)f(b) < 0$ – the *sign property*.

Example 6.4 For the function defined by $f(x) = 1/(1 - x)$, any interval $[a, b]$ with $a < 1$ and $b > 1$ contains the point $x = 1$. Further, $f(a) > 0$ and $f(b) < 0$, and $f(x)$ satisfies the sign property. However, $f(x)$ is *not* zero at $x = 1$; the function f is undefined at $x = 1$ (see Figure 6.6)

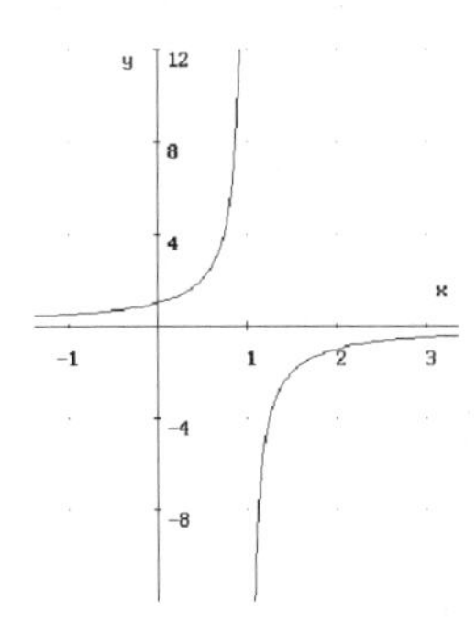

Figure 6.6 The graph of $y = 1/(1 - x)$, $-1 \le x \le 3$.

Continuity is required.

Definition $f(x) = 0$ has *at least one* root on the interval $a \le x \le b$ if

(a) $f(x)$ is *continuous* on (a, b) and

(b) $f(a)f(b) < 0$, that is $f(x)$ *changes sign* on $[a, b]$.

This definition[4] facilitates the location of one or more roots.

6.2 Refining the location by bisection

Once a root α has been located in an interval $[a, b]$ the next step is to improve (refine) the estimate. If $[\mathcal{A}]$ the mid-point $c = (a + b)/2$ is taken as an estimate to α, the question then arises:

can a smaller interval $[a_1, b_1]$ be found that contains the root?

[3] This condition implies an *odd* number of roots in $[a, b]$.

[4] If the sign property is not satisfied, this does not mean that no roots exist. It simply means that an even number of roots may exist, which includes zero. For example, for the cubic polynomial $x^3 - 3x^2 + 2x$, the product $f(-0.5)f(1.5)$ is positive (no apparent change of sign), yet two roots exist within the interval $[-0.5, 1.5]$.

If the root α lies in the interval $[a, b]$, it must lie in one of the sub-intervals $[a, c]$ or $[c, b]$, since $[a, b] = [a, c] \cup [c, b]$. The length of both $[a, c]$ and $[c, b]$ is one-half the length of $[a, b]$ – *smaller* intervals. So, which of $[a, c]$ and $[c, b]$ contains α? $[\mathcal{B}]$ This is answered by investigating the sign of $f(a)f(c)$. If $f(a)f(c) < 0$ then $\alpha \in [a, c]$ since a sign change occurs. Otherwise the sign change must occur on $[c, b]$, and $\alpha \in [c, b]$.

The *bisection* at $[\mathcal{A}]$ is repeated on the new bracketing interval $[a, c]$ or $[c, b]$. Steps $[\mathcal{A}]$ and $[\mathcal{B}]$ are repeated until the bracketing interval is sufficiently small. This process of interval halving is the *bisection method.*

Example 6.5 For the function defined by $f(x) = \sin x$, $f(1) = 0.84$, $f(5) = -0.96$ and there is a *sign change* on $[1, 5]$. Since f is *continuous* on any finite interval, *at least one* root exists on $[1, 5]$ (see Figure 6.7(a)). The mid-point is $x = 3$ and $f(3) = 0.14$, which is positive. Since $f(5) < 0$, the new bracketing interval is $[3, 5]$. Table 6.1 summarises several steps of the bisection method (see Figure 6.7(b)).

Table 6.1 A bisection example.

Interval	Sign		Mid-point	
$[a, b]$	$f(a)$	$f(b)$	c	$f(c)$
[1, 5]	+	−	3	0.14
[3, 5]	+	−	4	−0.76
[3, 4]	+	−	3.5	−0.35
[3, 3.5]	+	−	3.25	−0.11
[3, 3.25]	+	−	3.125	0.02
[3.125, 3.25]	+	−	3.1875	−0.05

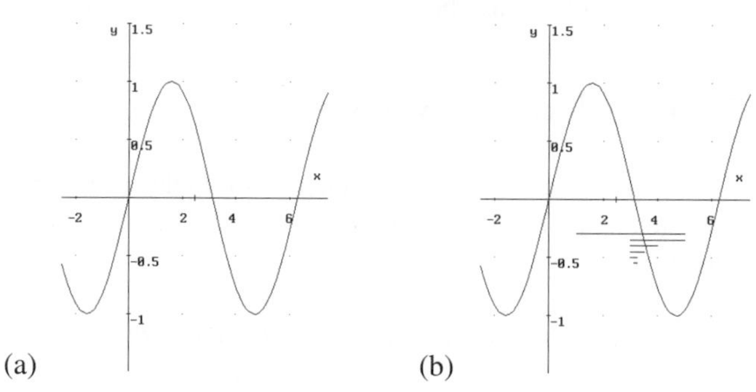

Figure 6.7 (a) Graph of $y = \sin x$. (b) Bisection interval halving.

The next step is to develop an algorithm that can be implemented on a computer. The bracketing interval at any stage is denoted by $[x_0, x_1]$, and y_0 and y_1 denote the values of $f(x_0)$ and $f(x_1)$.

Algorithm 6.1 Bisection method
Function: Root of $f(x) = 0$

Line	Description	
1	**Problem**	`input` a, b, n_{max}
2	Root exists?	`if` $f(a)f(b) \geq 0$ `then stop`
3	Initialise	$x_0 = a,\ y_0 = f(a),\ x_1 = b,\ y_1 = f(b),\ c = \frac{a+b}{2}$
4		`for` $n = 1$ `to` n_{max}
5	Root in $[a, c]$	`if` $y_0 f(c) < 0$ `then`
6		$x_1 = c,\ y_1 = f(c)$
7	Root in $[c, b]$	`else`
8		$x_0 = c,\ y_0 = f(c)$
9		`end`
10	Bisect	$c = (x_0 + x_1)/2$
11		`end`
12	**Solution**	`output` c

Line 2 checks for a root in $[a, b]$, line 3 initialises the variables (interval $[x_0, x_1]$ and function values) and line 5 tests for a sign change on $[x_0, c]$. If this is true, the right-hand point x_1 is 'moved' to $x_1 = c$ and y_1 is redefined as $f(c)$. If line 5 is false, the left-hand point x_0 is 'moved' to $x_0 = c$. The redefined interval $[x_0, x_1]$ is then bisected (line 10) and the process is repeated n_{max} times.

DERIVE Experiment 6.1

Author the following functions to set up *DERIVE* to simulate the bisection method.

```
v(f,x,x0):=lim(f,x,x0)
bis(f,x,a,b,n):=iterates(if(v(f,x,l)v(f,x,c)<0,
    [l,(l+c)/2,c],[c,(c+r)/2,r]),[l,c,r],[a,(a+b)/2,b],n)
```

`v` evaluates the expression $f(x)$ at $x = x_0$. `l`, `c` and `r` denote the variables x_0, c and x_1.

To produce Example 6.5, `Author bis(sinx,x,1,5,5)` and `approX` – the columns in the resulting matrix give values of x_0, c and x_1.

6.2.1 Error analysis

Numerical methods require an associated error analysis since the very use of an 'approximation' implies the introduction and propagation of errors. For the bisection method the analysis is quite straightforward.

The aim is to identify the maximum error after n bisections (iterations) if the mid-point c is taken to be the latest estimate of α. The *worst case* occurs when α

equals either x_0 or x_1 (the root lies at one end point of the bracketing interval) and the error equals one-half of the interval size,

$$|\alpha - c| \leq \frac{|x_1 - x_0|}{2}.$$

At each bisection the interval is halved, and after n bisections the interval size equals 2^{-n} times the *initial* interval size,

$$|x_1 - x_0| = \frac{|b - a|}{2^n}.$$

The error after n bisections is therefore bounded by

$$|\alpha - c| \leq \frac{|b - a|}{2^{n+1}}. \qquad \textbf{(6.4)}$$

Example 6.6 That eqn (6.4) does define an *upper bound* for the error is illustrated in Table 6.2 which extends Table 6.1 to include the observed error and the maximum error (6.4). The exact value of the root is π (3.1416 to 4 decimal places).

Table 6.2 Monitoring the error of bisection.

		Sign		**Mid-point**			
n	**Interval** $[a, b]$	$f(a)$	$f(b)$	c	$f(c)$	**Obs. error**	**Max. error**
0	[1, 5]	+	−	3	0.14	0.1416	2
1	[3, 5]	+	−	4	−0.76	−0.8584	1
2	[3, 4]	+	−	3.5	−0.35	−0.3584	0.5
3	[3, 3.5]	+	−	3.25	−0.11	−0.1084	0.25
4	[3, 3.25]	+	−	3.125	0.02	0.0166	0.125
5	[3.125, 3.25]	+	−	3.1875	−0.05	−0.0459	0.0625

Note that the *observed* error does not decrease monotonically. For example, the initial estimate 3 is better than the next estimate 4. Figure 6.7(b) shows the evolution of the bracketing interval as the bisection method progresses.

DERIVE Experiment 6.2

The function `bis` in *Experiment 6.1* can be modified to include the maximum error at each iteration. Author

```
ac(a,b):=append(a',b')'
bism(f,x,a,b,n):=ac(bis(f,x,a,b,n),
   vector(abs(b-a)/2^i,i,1,n+1)')
```

`ac(a,b)` appends the columns of matrix `b` to matrix `a`.

For Example 6.6, Author `bism(sinx,x,1,5,5)` and `approX` – the columns in the output matrix give x_0, c, x_1 and $|x_1 - x_0|/2$.

6.2.2 Tolerance

How many bisections are needed to achieve a specified accuracy?

If the value of α is required to an accuracy of k decimal places, $|\alpha - c|$ must not exceed $\frac{1}{2} \times 10^{-k}$ (see Section 2.3.2 on rounding errors). From eqn (6.4)

$$|\alpha - c| \le \frac{|b-a|}{2^{n+1}} \le \frac{1}{2} \times 10^{-k} \Rightarrow n = \tilde{n} \ge \frac{\log|b-a| + k \log 10}{\log 2}. \quad \mathbf{(6.5)}$$

After $\tilde{n}$ bisections the error in the estimate c will not exceed $\frac{1}{2} \times 10^{-k}$. $\tilde{n}$ must be an integer so the process is completed once[5] $n = \langle \tilde{n} \rangle$. This is a *termination condition.* The bisection algorithm is modified:

Algorithm 6.2 Bisection method (modified)
Function: Root of $f(x) = 0$

Line	Description	
1	**Problem**	`input` a, b, k
3a	Max. iters.	$\tilde{n} = \langle [\log|b-a| + k \log 10]/ \log 2 \rangle$
4		`for` $n = 1$ `to` $\tilde{n}$

Thus, a *predetermined* number of iterations are performed to achieve a specified accuracy (tolerance).

Example 6.7 The equation $x^3 + 4x^2 - 10 = 0$ has at least one root ($x = 1.3652$) in the interval $[1, 2]$ (since $f(1) = -5 < 0$, $f(2) = 14 > 0$ and $f(1)f(2) < 0$, where $f(x) =$

Table 6.3 More bisections!

n	Interval $[a, b]$	Sign $f(a)$	$f(b)$	Mid-point c	$f(c)$	Obs. error	Max. error
0	[1, 2]	−	+	1.5	2.38	−0.1348	0.5
1	[1, 1.5]	−	+	1.25	−1.80	0.1152	0.25
2	[1.25, 1.5]	−	+	1.375	0.16	−0.0098	0.125
3	[1.25, 1.375]	−	+	1.3125	−0.85	0.0527	0.0625
4	[1.3125, 1.375]	−	+	1.3438	−0.35	0.0215	0.0313
5	[1.3438, 1.375]	−	+	1.3594	−0.10	0.0059	0.0156
6	[1.3594, 1.375]	−	+	1.3672	0.03	−0.0020	0.0078
7	[1.3594, 1.3672]	−	+	1.3633	−0.03	0.0019	0.0039

[5] The notation $\langle\ \rangle$ means 'the smallest integer upper bound', sometimes called ceil( ). Thus, $\langle 3 \rangle = 3$ whereas $\langle 5.1 \rangle = 6$.

$x^3 + 4x^2 - 10$). For 2 decimal places of accuracy:

$$n \geq \frac{\log|b-a| + k\log 10}{\log 2} = \frac{\log 1 + 2\log 10}{\log 2} = 6.64 \text{ iterations.}$$

In other words, $\tilde{n} = 7$ iterations ensures 2 decimal places of accuracy (see Table 6.3). Note that the estimate after 2 iterations, $c = 1.375$, is a better approximation than the next two iterations!

Summary *The bisection method is (a) simple – interval halving, (b) has a straightforward error analysis, and (c) is not very efficient – it does not account for values of $f(x)$.*

6.2.3 Parallel bisection

A brute force remedy for the inefficiency of the bisection method is to use a parallel computing architecture, such as a transputer array. This consists of many identical processors linked by hard-wired communication channels, such as the 'loop' configuration of Figure 6.8.

A typical *sequential machine* has a single processor and processes data as shown in Figure 6.9. For the bisection method, the original interval $[a, b]$ is repeatedly halved and the processor interrogates the sign of $f(a)f(c)$.

For p processors working in *parallel* (see Figure 6.10), $[\mathcal{A}]$ the interval $[x_0, x_1]$ is divided into $p+1$ sub-intervals, $[x_0, c_1], [c_1, c_2], \ldots, [c_{p-1}, c_p], [c_p, x_1]$, of size $(x_1 - x_0)/(p+1)$. $[\mathcal{B}]$ Processor 1 interrogates $[x_0, c_1]$, processor 2 interrogates $[c_1, c_2], \ldots$, and processor p interrogates $[c_{p-1}, c_p]$. The interval containing a sign change, say $[c_{j-1}, c_j]$ (processor j), is taken as the new bracketing interval for step $[\mathcal{A}]$. If no processor detects a sign change then the root lies in $[c_p, x_1]$.

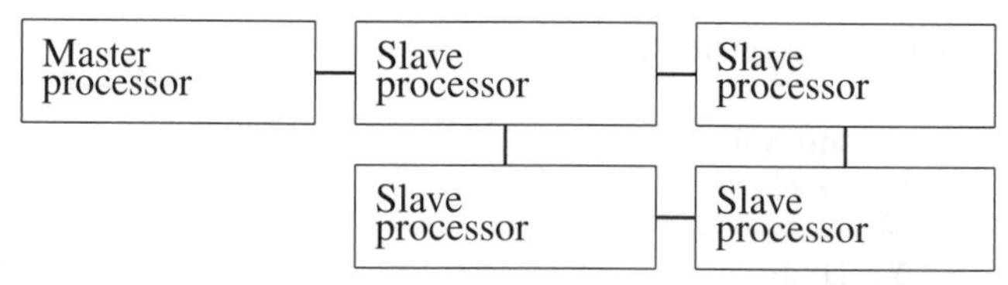

Figure 6.8 A typical parallel 'wiring' diagram. Each processor communicates directly with two adjacent processors.

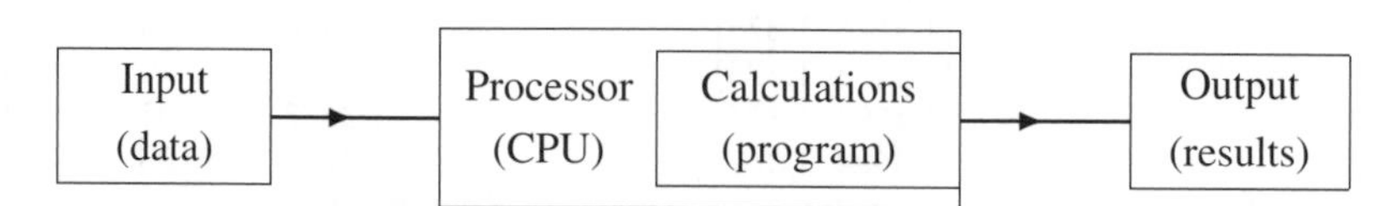

Figure 6.9 Schematic diagram of sequential data processing.

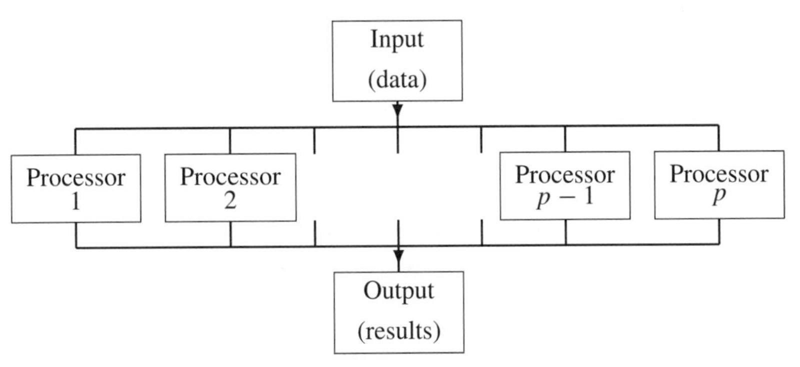

Figure 6.10 Schematic diagram of parallel data processing.

At each step the interval size is reduced by a factor $1/(p+1)$ and after m iterations $|x_1 - x_0| = |b - a|/(p+1)^m$. If the mid-point c of the interval $[x_0, x_1]$ is used to estimate α then k decimal places of accuracy require

$$|\alpha - c| \leq \frac{|b-a|}{2(p+1)^m} \leq \frac{1}{2} \times 10^{-k} \Rightarrow m \geq \frac{\log|b-a| + k\log 10}{\log(p+1)}. \tag{6.6}$$

The *speed-up* S_p of parallel code using p processors is defined by the ratio

$$S_p = \frac{T_1}{T_p} = \frac{\text{Time for sequential code}}{\text{Time for parallel code}}. \tag{6.7}$$

In the case of the bisection method[6],

$$S_p = \frac{n}{m} = \frac{\frac{\log|b-a|+k\log 10}{\log 2}}{\frac{\log|b-a|+k\log 10}{\log(p+1)}} = \frac{\log(p+1)}{\log 2}. \tag{6.8}$$

Example 6.8 Repeating Example 6.7 with three processors, the bisection table is shown in Table 6.4.

Table 6.4 Parallel bisection results.

n	x_0	x_1	Error	c_1	c_2	c_3	$f(c_1)$	$f(c_2)$	$f(c_3)$
0	1	2	0.5	1.25	1.5	1.75	−1.80	2.38	7.61
1	1.25	1.5	0.125	1.3125	1.375	1.4375	−0.85	0.16	1.24
2	1.3125	1.375	0.0313	1.3281	1.3438	1.3594	−0.60	−0.35	−0.10
3	1.3594	1.375	0.0078		1.3672				

[6] The speed-up given here ignores *communication time*, that is the time spent by the processors passing information 'to and fro'.

6.2.4 Termination conditions

We have seen that the number of iterations $\tilde{n}$ for the bisection method to give k decimal places of accuracy can be obtained from the inequality

$$|\alpha - c| \leq \frac{1}{2} \times 10^{-k}$$

(see eqn (6.5)). For many numerical methods the determination of $\tilde{n}$ is not possible and two further termination conditions should be included:

(1) Stop if the *magnitude* of $f(c)$ is 'close to' zero, $|f(c)| \leq \frac{1}{2} \times 10^{-k}$.

(2) Stop once a specified maximum number of iterations have been performed, $n = n_{\max}$. This avoids iterating forever!

6.3 First-order fixed-point iteration

The bisection method (see Section 6.2) is simple, has a clear graphical interpretation (the interval size is 'seen' converging to zero) and has an uncomplicated error analysis. However, it is not an efficient method.

We now explore a class of numerical methods for solving the equation $f(x) = 0$ which, given a *first term* x_0, generate a *sequence* $\{x_n\}$ such that

(1) x_{n+1} is obtained from x_n (*first-order recurrence relation*), and

(2) $x_n \to \alpha$ as $n \to \infty$, where α is a root of the equation $f(x) = 0$.

The requirements are (a) to generate a sequence based on the equation $f(x) = 0$ and (b) to ensure that $x_n \to \alpha$ as $n \to \infty$ (convergence).

6.3.1 Fixed points

It is first necessary to develop the notion of a *fixed point*.

Example 6.9 Let f be defined by $f(x) = 2x^2 - x$. The two roots of the equation $f(x) = 0$ are $x = 0$ and $x = 0.5$ (see Figure 6.11(a)). Consider the sequence generated by the recurrence relation

$$x_0 = 0.1; \quad x_{n+1} = 2x_n(1 - x_n), \quad n = 0, 1, \ldots \tag{6.9}$$

n	0	1	2	3	4	5	6
x_n	0.100	0.180	0.295	0.416	0.486	0.500	0.500

The sequence approaches the root $x = 0.5$ where it 'gets stuck'. On dropping subscripts from eqn (6.9), the equation $x = 2x(1 - x)$ is obtained (an arrangement of $2x^2 - x = 0$). In Figure 6.11(b) it is seen that the graphs of $y = x$ and $y = 2x(1 - x)$ intersect at $x = 0.5$. Thus, when the right-hand side of eqn (6.9) is

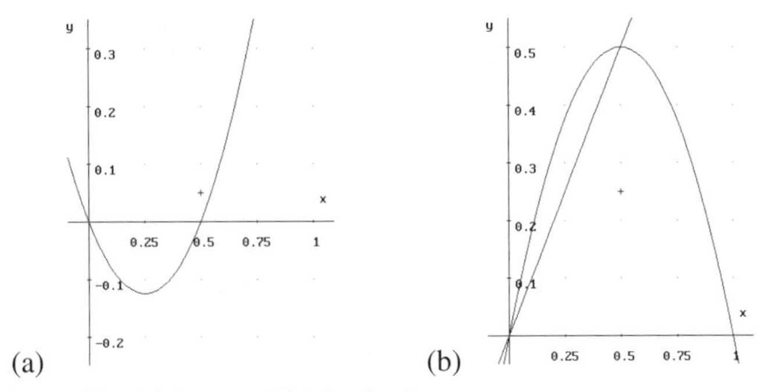

Figure 6.11 (a) Roots and (b) fixed points.

evaluated at $x_n = 0.5$, the value $x_{n+1} = 0.5$ is obtained. All subsequent terms in the sequence, x_{n+2}, x_{n+3}, etc., will equal the fixed value 0.5.

Definition The value $x = \alpha$ is a *fixed point* of the function[7] g if $\alpha = g(\alpha)$.

Example 6.9 suggests that if a sequence generated by a recurrence relation of the form $x_{n+1} = g(x_n)$ has a *limiting value* α, then α is a fixed point of the function g. Such a sequence is said to be *convergent*.

6.3.2 Fixed-point iteration

The problem at hand is to find a value of x such that the equation $f(x) = 0$ is satisfied. To utilise recurrence the equation must be written in the form $x = g(x)$. This can always be accomplished by adding x to each side,

$$f(x) = 0 \Rightarrow x + f(x) = x \Rightarrow x = x - f(x) \Rightarrow x = g(x).$$

There are an *infinity* of ways of arranging $f(x) = 0$ in the form $x = g(x)$.

Example 6.10 For the equation $f(x) = 0$ where $f(x) = x^3 + 4x^2 - 10$,

$$x = \sqrt[3]{10 - 4x^2} = g_1(x) \tag{6.10}$$

$$x = \frac{1}{2}\sqrt{10 - x^3} = g_2(x) \tag{6.11}$$

$$x = \sqrt{\frac{10}{4 + x}} = g_3(x) \tag{6.12}$$

are just three possible arrangements in the form $x = g(x)$.

[7] In the text the statements 'fixed point of a function g' and 'fixed point of an equation $x = g(x)$' are taken to have the same meaning.

We have seen that if α is a zero of f, $f(\alpha) = 0$, then α is a fixed point of g, $\alpha = g(\alpha)$. If the recurrence relation $x_{n+1} = g(x_n)$ generates a sequence with limit α, then α will be a fixed point of g and hence a zero of f. The systematic iterative approach we require to determine a zero of f is

$\mathcal{A}$ write $f(x) = 0$ in the form $x = g(x)$,

$\mathcal{B}$ define the first-order recurrence relation $x_{n+1} = g(x_n)$, and

$\mathcal{C}$ given x_0, use the recurrence relation to generate the sequence $\{x_n\}$.

Example 6.11 Step $\mathcal{A}$ applied to $x^3 + 4x^2 - 10 = 0$ gives eqns (6.10)–(6.12). Applying step $\mathcal{B}$ to each equation yields the following recurrence relations:

$$x_{n+1} = \sqrt[3]{10 - 4x_n^2} = g_1(x_n) \tag{6.13}$$

$$x_{n+1} = \frac{1}{2}\sqrt{10 - x_n^3} = g_2(x_n) \tag{6.14}$$

$$x_{n+1} = \sqrt{\frac{10}{4 + x_n}} = g_3(x_n) \tag{6.15}$$

Step $\mathcal{C}$ with $x_0 = 1.5$ gives the sequences shown in Table 6.5.

Table 6.5 Examples of fixed-point iteration.

n	$g_1(x_{n-1})$	$g_2(x_{n-1})$	$g_3(x_{n-1})$
1	1	1.2870	1.3484
2	1.8171	1.4025	1.3674
3	−1.4748	1.3455	1.3650
4	1.0914	1.3752	1.3653
5	1.7364	1.3601	1.3652
6	−1.2725	1.3678	1.3652
7	1.5215	1.3639	
8		1.3659	
⋮	Not	⋮	
⋮	convergent	⋮	
14		1.3652	

The sequences show a range of behaviour. The first does not approach a limiting value, the second attains a limiting value after 14 iterations, and the third after just 5 iterations. But, *remember*, all three equations (6.10)–(6.12) have the *same* fixed point.

DERIVE Experiment 6.3

It is a simple matter to set up a fixed-point iteration process in *DERIVE* – just Author

```
fixpt(g,x,x0,n):=iterates(g,x,x0,n)
```

To reproduce the third sequence of Example 6.11, Author and approX the expression `fixpt(sqrt(10/(4+x)),x,1.5,6)`.

What properties does g require near a fixed point α for α to be the limiting value of a sequence generated by $x_{n+1} = g(x_n)$?

Example 6.12 The three forms of g given in eqns (6.10)–(6.12) are shown in Figure 6.12(a). All three curves intersect the line $y = x$ at the *same* point – they all have the same *fixed point*. The difference is in the slope of each curve at the fixed-point (root), shown in Figure 6.12(b).

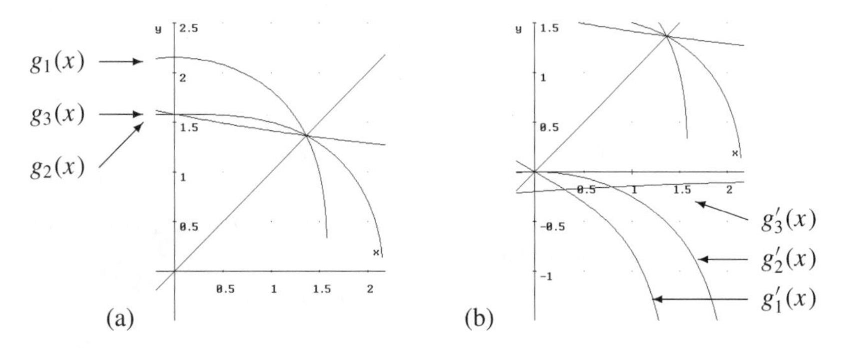

Figure 6.12 Convergence property. (a) $g(x)$ and (b) $g'(x)$.

6.3.3 Error analysis I – mean-value theorem

To identify the property of $g'(x)$ at $x = \alpha$ that generates a convergent sequence with limit α, a simple error analysis is required.

Let e_n represent the error after n iterations, $e_n = x_n - \alpha$. The value of e_n is not needed but simply its behaviour as n increases. Specifically, under what conditions does e_n approach 0 as $n \to \infty$? This gives rise to the concept of *convergence*. Subtracting $\alpha = g(\alpha)$ from the relation $x_{n+1} = g(x_n)$ yields

$$x_{n+1} - \alpha = g(x_n) - g(\alpha).$$

Using the *mean-value theorem*[8], we obtain[9]

$$x_{n+1} - \alpha = (x_n - \alpha)g'(\xi), \quad x_n < \xi < \alpha,$$

and with the definition of the error e_n,

$$e_{n+1} = e_n g'(\xi), \quad x_n < \xi < \alpha. \tag{6.16}$$

The error e_{n+1} is less in magnitude than e_n if $|g'(\alpha)| < 1$. That is, if the slope of g is sufficiently small 'close to' $x = \alpha$ then the sequence $\{x_n\}$ generated by $x_{n+1} = g(x_n)$ will converge to α.

Example 6.13 For Example 6.10 the derivatives of the functions g_1, g_2 and g_3 are shown in Figure 6.12(b). In a *neighbourhood* of $x = 1.365$ the slope of g_1 is greater than 1 in magnitude and the associated sequence is not convergent. The slopes of g_2 and g_3 are both less than 1 and the associated sequences do converge. The slope of g_3 is less than that of g_2 and the sequence generated by g_3 converges more quickly.

Summary *(fixed-point convergence)*

- *Convergence of the sequence $\{x_n\}$ generated by $x_{n+1} = g(x_n)$ occurs if $|g'(x)| < 1$ in a neighbourhood of α.*
- *Smaller $|g'(x)|$ implies faster convergence.*

6.3.4 Cobweb diagrams and fixed-point classification

The concept of fixed-point convergence can be visualised with *cobweb diagrams*. These are pictures that display the progress of the sequence $\{x_n\}$ as it approaches (or does not) a root α.

Example 6.14 Figure 6.13 shows monotonic convergence, monotonic divergence, and oscillatory convergence and divergence.

---

[8] Mean-value theorem: For a function f that is continuous on the interval $[a, b]$ there exists a $\xi \in (a, b)$ such that the slope of f at ξ, $f'(\xi)$, equals the slope of the chord drawn from $f(a)$ to $f(b)$, $[f(b) - f(a)]/(b - a)$.

[9] This assumes that $x_n < \alpha$. If $x_n > \alpha$ then $x_n < \xi < \alpha$ is replaced with $\alpha < \xi < x_n$.

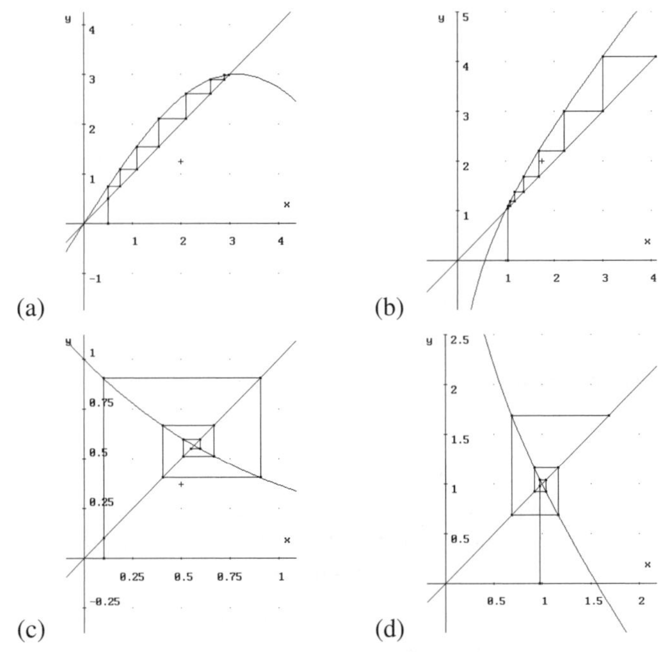

Figure 6.13 Cobweb diagrams: (a) $x = 3\sin\frac{x}{2}$, $x_0 = 0.5$. (b) $x = \ln x + x$, $x_0 = 1.05$. (c) $x = e^{-x}$, $x_0 = 0.1$. (d) $x = 2 - x - \ln x$, $x_0 = 0.98$.

Table 6.6 Classification of fixed points.

Class		Condition	Sequence
Attracting	if	$\lvert g'(\alpha)\rvert < 1$	Convergent
Indifferent	if	$\lvert g'(\alpha)\rvert = 1$	Uncertain behaviour
Repelling	if	$\lvert g'(\alpha)\rvert > 1$	Divergent

The cobweb nature of Figure 6.13 is clear! If the *slope* of g is *positive* then *monotonic* convergence or divergence results. If the slope of g is *negative* then *oscillatory* convergence or divergence[10] results. Table 6.6 lists the three types of fixed point, classified according to the local behaviour of g.

Example 6.15 The equation $x^2 - 11x + 10 = 0$ has two roots, $x = 1$ and $x = 10$. Arranging the equation as $x = (x^2 + 10)/11$ provides the basis for a recurrence relation. For x_0

[10] It should be pointed out that 'divergence' might mean divergence or it might mean convergence to a different attracting fixed point (root).

$= 9.95$, Figure 6.14(a) indicates apparent divergence. On 'zooming out' Figure 6.14(b) shows convergence to the smaller root.

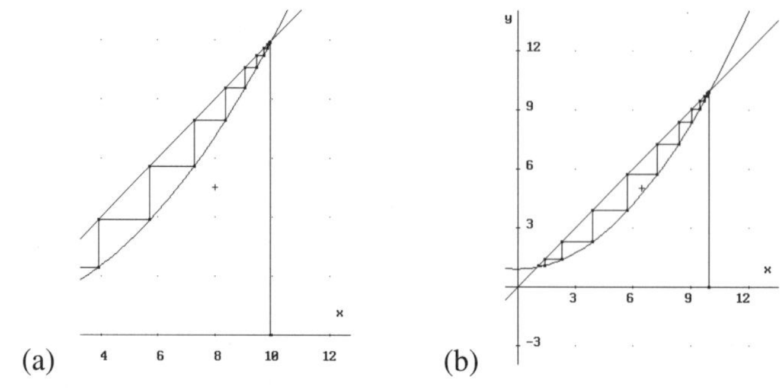

Figure 6.14 (a) Apparent divergence and (b) alternative convergence.

The question is how to 'get to' repelling fixed points. The requirement is known (a g-slope of magnitude less than 1 at α) but not how to achieve it, except by accident! This issue is addressed in Sections 6.3.6–6.3.8.

DERIVE Experiment 6.4

The cobweb diagrams of Examples 6.14 and 6.15 can be helpful in understanding the nature of the recurrence relation $x_{n+1} = g(x_n)$. The following function will produce cobwebs[11].

```
cobweb(g,x,x0,n):=[g,x,iterate(append(v,lim([[x,g],[g,g]],
   x,v sub dimension(v) sub 2)),v,[[x0,0],[x0,x0]],n)]
```

Note that `iterate` is used (this is the same as `iterates` but only the last iterate is returned). Figure 6.13(a) is produced by plotting the result of `Author cobweb(3sin(x/2),x,.5,6) approX`, making sure that `Options State` is `Connected`.

6.3.5 Initial guess

Fixed-point methods require a good initial value x_0. A simple solution is to find an interval $[a, b]$ on which the original function f satisfies the sign property and then use the mid-point $x_0 = (a + b)/2$ as the initial value.

Example 6.16 One of Kepler's laws of planetary motion states that for a planet P executing an elliptical orbit about a sun S, with S at the focus of the ellipse, the line SP sweeps out area at a constant rate.

---

[11] See A. Martinez, R. Minano and F. Rincon (1997). Teaching some numerical methods with DERIVE, in [11] pp. 175–89.

From a knowledge of ellipses, *Kepler's equation* can be derived[12]:

$$2\pi p = \theta - e \sin\theta$$

where p is the fraction of the planet's year elapsed between A and P, e is the eccentricity of the orbit and the angle θ is the eccentric anomaly of the planet (see Figure 6.15). Given p and e, the problem is to determine the location θ of the planet. A natural arrangement of this equation,

$$\theta = 2\pi p + e \sin\theta,$$

provides the basis for a first-order non-linear recurrence relation

$$\theta_{n+1} = 2\pi p + e \sin\theta_n.$$

What is a suitable[13] interval $[a, b]$ from which to derive the *initial value* θ_0? With $p = 0.25$ and $\theta_0 = 1$, Table 6.7 shows the first 10 iterations for several planets of the Solar System.

Table 6.7 Results for Kepler's law.

	x_n				
n	**Circle** $e = 0$	**Earth** $e = 0.017$	**Mercury** $e = 0.206$	**Pluto** $e = 0.249$	**Comet** $e = 0.9$
0	1.0000	1.0000	1.0000	1.0000	1.0000
1	1.5708	1.5851	1.7441	1.7803	2.3281
2	1.5708	1.5878	1.7737	1.8144	2.2248
3		1.5878	1.7726	1.8124	2.2851
4			1.7726	1.8126	2.2508
5				1.8126	2.2706
6					2.2593
7					2.2658
8					2.2620
9					2.2642
10					2.2630
Deg.	90	90.97	101.56	103.85	129.66

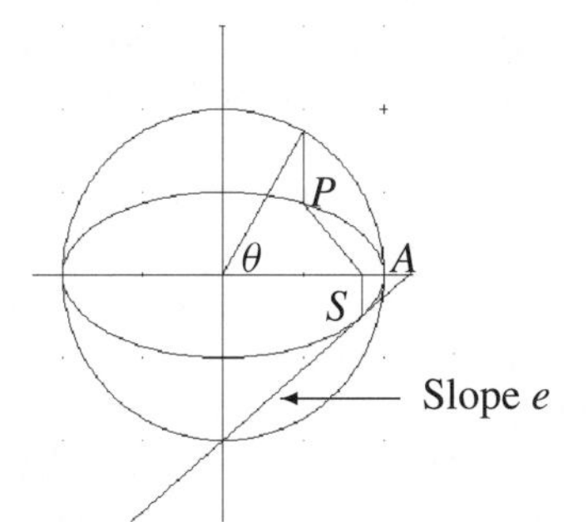

Figure 6.15 Kepler's law.

[12] See D.E. Rutherford (1964). *Classical Mechanics*, Oliver & Boyd, Edinburgh.

[13] θ_0 must lie in the interval $[0, 2\pi]$.

Observations *e affects the convergence rate.* $g(\theta) = 2\pi p + e\sin\theta$, $g'(\theta) = e\cos\theta$ *and smaller e implies smaller* $g'(\theta)$ *– faster convergence. For* $e = 0$ *the update formula is* $\theta_{n+1} = 2\pi p$ *– a constant sequence – and just one iteration is required.*

JOHANNES KEPLER
(1571–1630)
German astronomer and physicist

LIFE

Kepler led a fascinating but ill-starred life. His alcoholic father made him work in a tavern as a child, later withdrawing him from elementary school and hiring him out as a field labourer where the boy contracted smallpox (crippling his hands and impairing his eyesight). In later years, Kepler's first wife and several children died, his mother was accused of witchcraft, and being a Protestant he was subjected to persecution by Catholic authorities. He was often impoverished, eking out a living as an astrologer and prognosticator. It was his mother who left an indelible mark on the six-year-old Kepler by showing him the comet of 1577 (he later prepared her defence against the witchcraft charges). Kepler became acquainted with the work of Copernicus while at the University of Tübingen, where he received his master's degree (1591). He continued as a theological student, but at the urging of the university officials he accepted a position as a mathematician and teacher in Graz. He was expelled from the city when it came under Catholic control, and in 1600 moved on to Prague where he assisted at the observatory of the Danish astronomer Tycho Brahe.

WORK

Kepler gave three mathematical laws of planetary motion. He published his *Mysterium Cosmographicum* (1596) which argues for the truth of the Copernican system by giving a mathematical explanation of its structure (in terms of regular polyhedra). When Tycho Brahe died (1601) Kepler went on to use his observations to calculate planetary orbits to unprecedented accuracy. He showed that a planet moves round the Sun in an elliptical orbit which has the Sun in one of its two foci (it was Newton who discovered the laws of gravitation that explained the reason for elliptical orbits). He also showed that a line joining the planet to the Sun sweeps out equal areas in equal times as the planet describes its orbit. These laws were first formulated for the planet Mars and published in *Astronomia Nova* (1609). The calculations for the Area Law involved a numerical technique that (with hindsight) resembles integral calculus. Kepler's third law – that the squares of the periods of planets are proportional to the cubes of the mean radii of their orbits – appeared in *Harmonice Mundi* (1619).

Kepler proved that sight was by the reception of light rays in the eye (1604), and wrote on the optics of the telescope, introducing the design using two convex lenses (1611). He also wrote on the new star of 1604 (called 'Kepler's supernova') (1606) and sent Galileo an enthusiastic endorsement of his telescopic discoveries. Kepler's *Epitome Astronomiae Copernicanae* (1618 to 1621) became a widely used textbook. His *Rudolphine Tables* (1627), based on Tycho's observations and Kepler's laws, proved to be accurate over a long time scale. Their success did much to gain general acceptance for heliocentric astronomy.

Kepler also did important work on polyhedra and on logarithms.

6.3.6 Accelerated convergence I – Aitken's method

From the fixed-point processes seen so far it is clear that the choice of g is instrumental in defining the convergence properties of the sequence $\{x_n\}$ generated by $x_{n+1} = g(x_n)$. We now investigate how to accelerate the convergence of $\{x_n\}$.

The first approach uses the sequence $\{x_n\}$ itself and simply modifies the terms x_n to improve convergence[14].

ALEXANDER CRAIG AITKEN
(1895–1967)
New Zealand mathematician

LIFE

Aitken left the Otago Boys' High School in Dunedin (1913) having won a scholarship to Otago University where he studied languages and mathematics. His university career was interrupted by World War I; he enlisted in 1915 and served in Gallipoli, Egypt and France, being wounded at the battle of the Somme. His war experiences were to haunt him for the rest of his life. In 1918 he returned to university, graduating in 1920 with First Class Honours in French and Latin but only Second Class in mathematics. Aitken became a school teacher at Otago Boys' High School. His mathematical genius bubbled under the surface and, encouraged by the professor of mathematics at Otago University, Aitken went to Scotland in 1923 and studied for a PhD at Edinburgh under Whittaker. His PhD thesis was considered so outstanding that he was awarded a DSc. In 1925 he was appointed to Edinburgh where he spent the rest of his life. After holding lecturing posts in actuarial mathematics, in statistics, and in mathematical economics, he became a Reader in statistics in 1936, the year he was elected a Fellow of the Royal Society. In 1946 he was appointed to Whittaker's chair. Aitken had an incredible memory (he knew π to 2000 places) and could instantly multiply, divide and take roots of large numbers.

---

[14] Aitken suggested the approach in 1926.

WORK

Aitken's mathematical work was in statistics, numerical analysis and algebra. He introduced the idea of accelerating the convergence of a numerical method, and a method of progressive linear interpolation. In algebra he contributed to the theory of determinants. He saw clearly how invariant theory came under the theory of groups, but never followed through his ideas. Aitken wrote several books: *The Theory of Canonical Matrices* (1932) was written jointly with Turnbull. With Rutherford he was editor of a series of the University Mathematical Texts and himself wrote *Determinants and Matrices* (1939) and *Statistical Mathematics* (1939).

Using the *mean-value theorem*, it has been established that successive error terms $e_n = x_n - \alpha$ satisfy the relation

$$x_{n+1} - \alpha = (x_n - \alpha)g'(\xi_n), \quad x_n < \xi_n < x_{n+1},$$

that is $e_{n+1} = e_n g'(\xi_n)$. Applying the formula to the *previous* iteration

$$x_n - \alpha = (x_{n-1} - \alpha)g'(\xi_{n-1}), \quad x_{n-1} < \xi_{n-1} < x_n.$$

If g' were constant then these two equations could be used to eliminate g',

$$\frac{x_{n+1} - \alpha}{x_n - \alpha} = \frac{x_n - \alpha}{x_{n-1} - \alpha},$$

from which the root α may be found. This is unlikely to be the case. However, if g' *is* a 'slowly varying' function then we might reasonably assume that a number $x^*_{n+1} \simeq \alpha$ exists such that

$$\frac{x_{n+1} - x^*_{n+1}}{x_n - x^*_{n+1}} = \frac{x_n - x^*_{n+1}}{x_{n-1} - x^*_{n+1}}.$$

Some simple manipulation yields the result[15]

$$x^*_{n+1} = x_{n+1} - \frac{(x_{n+1} - x_n)^2}{x_{n+1} - 2x_n + x_{n-1}}. \tag{6.17}$$

This scheme is called *Aitken acceleration* and we can often expect the sequence $\{x^*_n\}$ to converge more rapidly than the original sequence $\{x_n\}$.

Example 6.17 In Example 6.11 fixed-point iteration was used to find the real root of the equation $x^3 + 4x^2 - 10 = 0$, specifically $x_{n+1} = \frac{1}{2}\sqrt{10 - x_n^3}$. The process took 14 iterations to converge to 4 decimal places; $x_{14} = 1.3652$. Table 6.8 shows the associated Aitken sequence $\{x^*_n\}$ obtained by applying eqn (6.17). Convergence to 4 decimal places is now achieved after just 6 iterations.

---

15 Equation (6.17) can be written as $x^*_{n+1} = x_{n+1} - (\Delta x_n)^2/\Delta^2 x_{n-1}$, and is sometimes referred to as Aitken's delta-squared method.

Table 6.8 Aitken acceleration.

n	x_n	x_n^*
0	1.5	–
1	1.2870	–
2	1.4025	1.3619
3	1.3455	1.3643
4	1.3752	1.3650
5	1.3601	1.3652
6	1.3678	1.3652

DERIVE Experiment 6.5

To simulate Aitken acceleration, Author the following expressions:

```
dim(x):=dimension(x)
fixpt(g,x,x0,n):=iterates(g,x,x0,n)
el(x,i):=element(x,i)
d1(x,i):=el(x,i)-el(x,i-1)
d2(x,i):=el(x,i)-2el(x,i-1)+el(x,i-2)
accel(x):=vector(el(x,i)-d1(x,i)^2/d2(x,i),i,3,dim(x))
aitken(x):=[x,append([0,0],accel(x))]
```

x is the sequence $x_0, \ldots, x_n$.

To execute Example 6.17 Author and approX the expression

```
aitken(fixpt(sqrt(10-x^3)/2,x,1.5,6))
```

or Author `z:=fixpt(sqrt(10-x^3)/2,x,1.5,6)`, then Author approX `aitken(z)`.

6.3.7 Accelerated convergence II – the 'g' factor

A second approach to accelerating convergence is based upon the choice of the function g. Earlier, several ideas were established regarding the convergence of a sequence $\{x_n\}$, generated by $x_{n+1} = g(x_n)$, to a fixed point α of g, that (a) $x_n \to \alpha$ if $|g'(\alpha)| < 1$ and (b) convergence is *accelerated* for smaller values of $|g'(\alpha)|$. This suggests the following questions:

- *Can g always be chosen such that $|g'(\alpha)| < 1$?*
- *Can the value of $|g'(\alpha)|$ be optimised (minimised)?*

To answer the first question let us add the quantity λx to each side of the equation

$x = g(x)$, where $\lambda \neq -1$, to obtain

$$x + \lambda x = g(x) + \lambda x.$$

The equation can be arranged to give

$$(1 + \lambda)x = g(x) + \lambda x \quad \Rightarrow \quad x = \frac{g(x) + \lambda x}{1 + \lambda} = G(x), \tag{6.18}$$

which has the same root(s) as $x = g(x)$. Clearly λ cannot equal -1.

Example 6.18 The quadratic equation $x^2 - 3x + 2 = 0$ has two real *roots*, $x = 1$ and $x = 2$. In the form $x = g(x)$, for example $x = (x^2 + 2)/3$, then $g(1) = (1 + 2)/3 = 1$, $g(2) = (4 + 2)/3 = 2$, and $x = 1$ and $x = 2$ are *fixed points* of the expression $(x^2 + 2)/3$.

If, for example, $7x$ is added to both sides and then arranged to the form (6.18) we obtain

$$x = \frac{\frac{x^2+2}{3} + 7x}{8} = \frac{x^2 + 21x + 2}{24}.$$

$G(x) = (x^2 + 21x + 2)/24$, $G(1) = 1$ and $G(2) = 2$. Thus, $x = 1$ and $x = 2$ are fixed points of G, and hence roots of $x = G(x)$.

To find a fixed point α of G, the relation $x_{n+1} = G(x_n)$ could be used to generate a sequence $\{x_n\}$ with limiting value α. To be efficient $G'(\alpha)$ should be small, ideally close to zero. Differentiating G,

$$G(x) = \frac{g(x) + \lambda x}{1 + \lambda} \quad \Rightarrow \quad G'(x) = \frac{g'(x) + \lambda}{1 + \lambda}$$

shows that $G'(\alpha)$ *equals* zero if $g'(\alpha) + \lambda = 0$, that is if $\lambda = -g'(\alpha)$. In practice the value of α is unknown, so this 'best' value of λ cannot be obtained. Typically, if a root is sought in the vicinity of x_0, then λ is taken as $-g'(x_0)$. The better the initial guess x_0, the better the value of λ.

Example 6.19 Consider again the equation $x^3 + 4x^2 - 10 = 0$. The acceleration scheme is applied to all three iterative processes of Example 6.11, $x_{n+1} = g_i(x_n)$, $i = 1, 2, 3$, where $\lambda = -g_i'(1.5)$.

Compare the convergence of the sequences in Table 6.9 (11, 5 and 3 iterations) with the original sequences in Table 6.5 (divergence, 14 and 6 iterations). In both the originally convergent cases, the iteration count has been halved. What is perhaps most astounding is the fact that the *divergent process* $x_{n+1} = g_1(x_n)$ has formed the basis of the *convergent process* $x_{n+1} = G_1(x_n)$ (see Figure 6.16).

Table 6.9 Accelerated convergence.

λ / n	4.0000	0.6556	0.1226
	x_n		
0	1.5000	1.5000	1.5000
1	1.4000	1.3713	1.3650
2	1.3785	1.3657	1.3652
3	1.3705	1.3653	1.3652
4	1.3674	1.3652	
5	1.3661	1.3652	
6	1.3656		
7	1.3654		
8	1.3653		
9	1.3653		
10	1.3652		
11	1.3652		

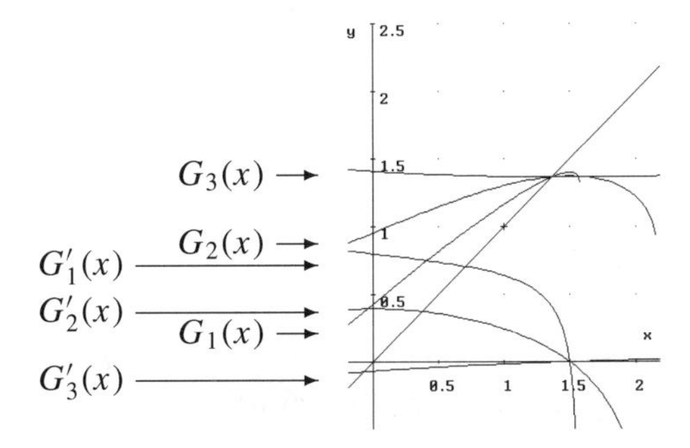

Figure 6.16 Accelerated convergence.

DERIVE Experiment 6.6

To implement the 'g' factor, Author

```
l(g,x,x0):=-lim(dif(g,x,x0))
fixla(g,x,x0,n,l):=iterates((g+lx)/(1+l),x,x0,n)
```

For the second sequence in Table 6.9, $g(x) = g_2(x) = \sqrt{(10-x^3)}/2$, Author

```
g:=sqrt(10-x^3)/2
z:=l(g,x,1.5)
```

and Author and approX `fixla(g,x,1.5,5,z)`.

6.3.8 Accelerated convergence III – optimised

Let the sequence $\{x_n\}$ generated by the *modified* relation $x_{n+1} = G(x_n)$ converge to a limiting value equal to the fixed point α of the equation $x = g(x)$. It is reasonable to assume that each term x_{n+1} will be an improved estimate to α when compared with the estimate x_n.

To *further* accelerate the sequence we might update the value of λ *at each step* using $\lambda = -g'(x_n)$ in place of the fixed value $\lambda = -g'(x_0)$ (x_n should be closer to α than x_0 for a convergent sequence). The effect of this modification is now investigated and concerns the question of optimising g'.

First a recurrence relation needs to be developed,

$$x_{n+1} = G(x_n) = \frac{g(x_n) + \lambda x_n}{1 + \lambda} = \frac{g(x_n) - g'(x_n)x_n}{1 - g'(x_n)}. \tag{6.19}$$

The function g is related to the function f appearing in the original equation $f(x) = 0$. Adding x to each side yields $x + f(x) = x$ from which $x = x - f(x)$. Hence $g(x) = x - f(x)$, $g'(x) = 1 - f'(x)$ and eqn (6.19) becomes

$$\begin{aligned} x_{n+1} &= \frac{x_n - f(x_n) - [1 - f'(x_n)]x_n}{1 - 1 + f'(x_n)} \\ &= x_n - \frac{f(x_n)}{f'(x_n)}. \end{aligned} \tag{6.20}$$

Example 6.20 For the equation $f(x) = 0$ with $f(x) = x^3 + 4x^2 - 10$, $f'(x) = 3x^2 + 8x$. The recurrence relation (6.20) becomes (see Table 6.10).

$$x_{n+1} = x_n - \frac{f(x_n)}{f'(x_n)} = x_n - \frac{x_n^3 + 4x_n^2 - 10}{3x_n^2 + 8x_n} = \frac{2x_n^3 + 4x_n^2 + 10}{3x_n^2 + 8x_n}.$$

Table 6.10 Optimally accelerated convergence.

n	0	1	2	3	4
x_n	1.5000	1.3733	1.3653	1.3652	1.3652

This is a fast method, and no 'fancy' λs are needed. Why? We shall return to answer this shortly. But first...

6.4 Newton–Raphson iteration

JOSEPH RAPHSON
(1648–1715)
English mathematician

LIFE

Raphson attended Jesus College Cambridge and graduated with an MA in 1692. He was elected to the Royal Society in 1689, the year before he graduated, based on the strength of his book *Analysis aequationum universalis* (1690). In 1702 Raphson published a mathematical dictionary.

WORK

Raphson's book *Analysis aequationum universalis* (1690) contained the Newton method for approximating the roots of an equation. In *Method of Fluxions*, Newton describes the same method and, as an example, finds the root of $x^3 - 2x - 5 = 0$ lying between 2 and 3. Although written in 1671 it was not published until 1736, so Raphson published the result nearly 50 years before Newton.

If you are familiar with any numerical method for finding roots of $f(x) = 0$, this is probably it! *Newton–Raphson iteration* can be generated by graphical means (see Figure 6.17).

Let x_n estimate the root α. The estimate x_{n+1} is obtained by finding the point of intersection of the tangent to f at $x = x_n$ with the x-axis. From elementary geometry the slope of the tangent line equals 'y divided by x',

$$f'(x_n) = \tan\theta = \frac{f(x_n) - 0}{x_n - x_{n+1}}.$$

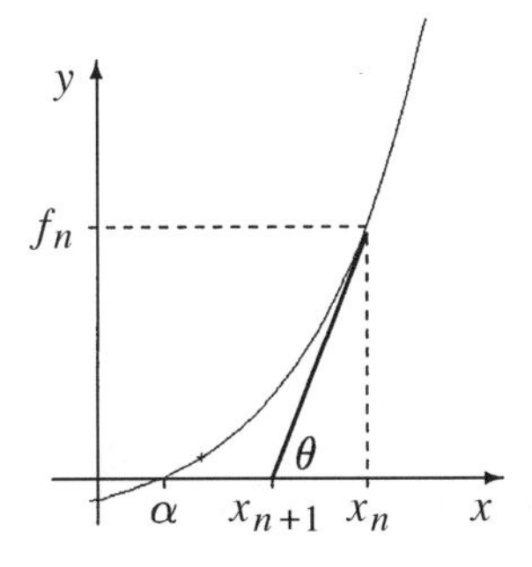

Figure 6.17 Newton–Raphson iteration.

Elementary manipulation yields the equation

$$x_{n+1} = x_n - \frac{f(x_n)}{f'(x_n)}. \quad \textbf{(6.21)}$$

This is the *Newton–Raphson* formula and is identical to the optimised formula (6.20). The question is *'why does it work so well?'*.

Written as a general relation $x_{n+1} = g(x_n)$, it is not difficult to see that eqn (6.21) uses a function g defined by $g(x) = x - f(x)/f'(x)$. Remember that the derivative of g is the key descriptor of convergence properties. Here

$$\begin{aligned} g'(x) &= 1 - \frac{f'(x)f'(x) - f(x)f''(x)}{[f'(x)]^2} \\ &= 1 - \frac{f'(x)f'(x)}{[f'(x)]^2} + \frac{f(x)f''(x)}{[f'(x)]^2} \\ &= \frac{f(x)f''(x)}{[f'(x)]^2}. \end{aligned}$$

At a root $x = \alpha$ the value $f(\alpha)$ is zero by definition. Hence, for the Newton–Raphson process (6.21), $g'(\alpha) = 0$ – 'optimal' convergence! However, from the previous error analysis, this appears to imply that $e_{n+1} = e_n g'(\alpha) = 0$. A refined error analysis is required.

6.4.1 Error analysis II – Taylor series

The mean-value theorem used in Section 6.3.3 gives the leading (most significant) term in the error analysis of fixed-point methods. To delve deeper we return to the original error equation

$$x_{n+1} - \alpha = g(x_n) - g(\alpha)$$

where $\alpha = g(\alpha)$ (α is a fixed point of g) and the error is $e_n = x_n - \alpha$. Consequently, $x_n = \alpha + e_n$ and the above equation becomes

$$e_{n+1} = g(\alpha + e_n) - g(\alpha).$$

Using Taylor's theorem,

$$\begin{aligned} e_{n+1} &= g(\alpha) + e_n g'(\alpha) + \frac{1}{2}e_n^2 g''(\alpha) + \cdots - g(\alpha) \\ &= e_n g'(\alpha) + \frac{1}{2}e_n^2 g''(\alpha) + \cdots \end{aligned} \quad \textbf{(6.22)}$$

It is assumed that e_n is sufficiently small that $e_n \gg e_n^2 \gg e_n^3$ etc. For the relation $x_{n+1} = g(x_n)$, $e_{n+1} \propto e_n$ (since $g'(\alpha) \neq 0$). Such a method is said to be *first-order*. For Newton–Raphson iteration, $g'(\alpha) = 0$, so $e_{n+1} \propto e_n^2$. The method is *second-order* – each error is proportional to the square of the previous error. This explains the rapid convergence associated with Newton–Raphson iteration. However, ...

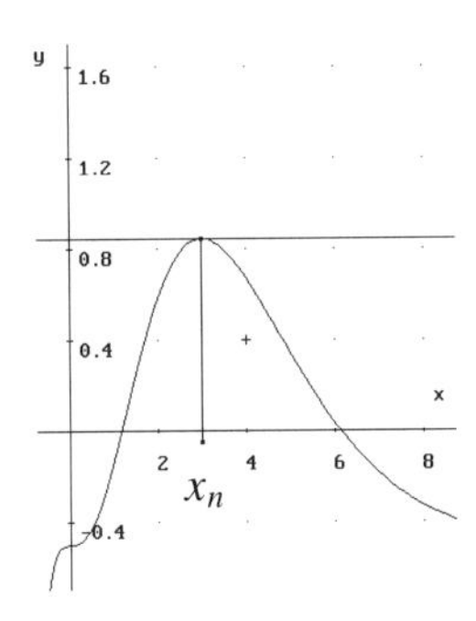

Figure 6.18 'Bad' Newton–Raphson I – undefined iterate.

6.4.2 Bad behaviour

While Newton–Raphson iteration works in most cases, there are several pitfalls. Here three of the most common difficulties are noted.

(1) *Iterate undefined*: if the slope of f is zero at $x = x_n$ then $f(x_n)/f'(x_n)$ is undefined – the tangent line is *parallel* to the x-axis and never intersects it. Consequently, x_{n+1} is *not* defined (see Figure 6.18).

(2) *Oscillation*: while a root may exist, the sequence $\{x_n\}$ 'gets stuck' between two (or more) values, that is $x_{n+2} = x_n$.

Example 6.21 For the equation $f(x) = 0$ with $f(x) = x^3 - 2x + 2$, $f'(x) = 3x^2 - 2$ and $x_0 = 0$, the Newton–Raphson sequence

$$x_{n+1} = x_n - \frac{x_n^3 - 2x_n + 2}{3x_n^2 - 2} = \frac{2x_n^3 - 2}{3x_n^2 - 2}$$

oscillates between 0 and 1. Why? A graph of $y = x^3 - 2x + 2$ shows the existence of a root, at $x = -1.76929$ (see Figure 6.19). However, $f(0) = 2$, the slope of f is -2, and the tangent line intersects the x-axis at $x = 1$. Further, $f(1) = 1$, the slope of f is 1, and the tangent line passes through the point $x = 0$.

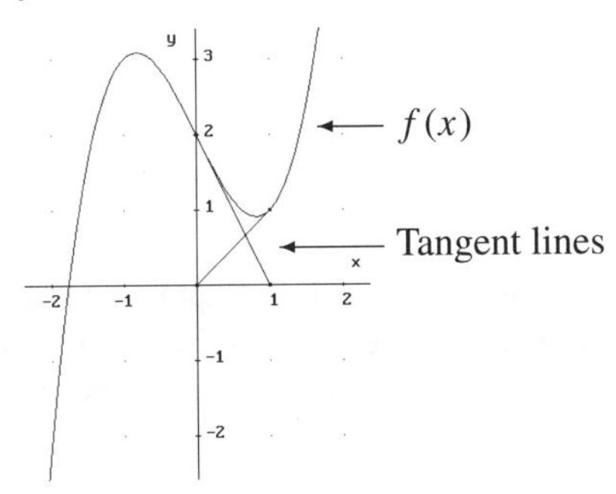

Figure 6.19 'Bad' Newton–Raphson II – oscillation.

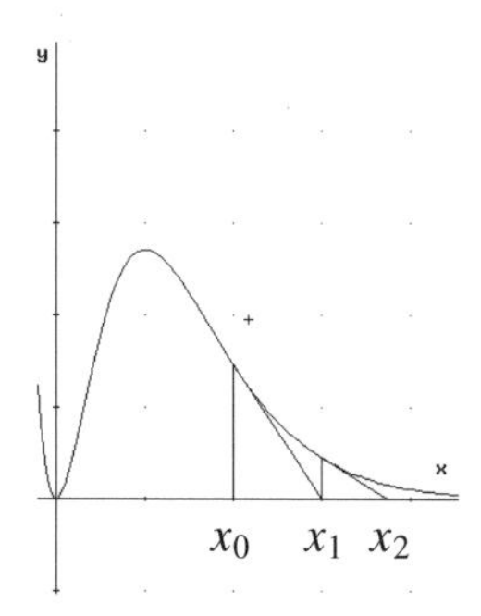

Figure 6.20 'Bad' Newton–Raphson III – divergence.

(3) *Divergence*: if x_0 is not 'sufficiently close' to the required root α then the sequence $\{x_n\}$ may diverge from α (see Figure 6.20).

Example 6.22 The equation $xe^{-x} = 0$ has one real root at $x = 0$. Starting the iterative process at $x_0 = 0.5$ produces a rapidly converging sequence of estimates to the root (see Table 6.11). If $x_0 = 2$, the sequence $\{x_n\}$ 'disappears to infinity'. This introduces the idea of *basins of attraction* and relates the behaviour of Newton–Raphson to the initial guess x_0. The topic involves *Julia sets* and *fractal maps* (e.g. the *Mandlebrot set*).

Table 6.11 'Bad' Newton–Raphson III – divergence.

n	x_n	
0	0.5	2
1	−0.5	4
2	−0.1667	5.3333
3	−0.0238	6.5641
4	−0.0006	7.7438
5	−0.0000	8.8921

6.4.3 Square root process

Perhaps the most famous application of Newton–Raphson iteration is that of determining the square root of a positive number c (the solution of the quadratic equation $x^2 - c = 0$). Here $f(x) = x^2 - c$, $f'(x) = 2x$ and the recurrence relation (6.21) becomes

$$x_{n+1} = x_n - \frac{f(x_n)}{f'(x_n)} = x_n - \frac{x_n^2 - c}{2x_n} = \frac{x_n^2 + c}{2x_n}. \qquad \textbf{(6.23)}$$

The derivative of the right-hand side is

$$g'(x) = \frac{\mathrm{d}}{\mathrm{d}x}\left(\frac{x^2+c}{2x}\right) = \frac{x^2-c}{2x^2}$$

and equals zero at the root $x = \sqrt{c}$. This implies that the relation (6.23) is second order. Alternatively, using the error definition $e_n = x_n - \alpha$,

$$e_{n+1} = x_{n+1} - \alpha = \frac{x_n^2+c}{2x_n} - \alpha = \frac{(x_n-\alpha)^2}{2x_n} = \frac{e_n^2}{2x_n}.$$

Thus, $e_{n+1} \propto e_n^2$.

> *DERIVE Experiment 6.7*
> To define the Newton–Raphson process, Author
>
> ```
> nr(f,x,x0,n):=iterates(x-f/dif(f,x),x,x0,n)
> ```
>
> Table 6.10 is given by Author `nr(x^3+4x^2-10,x,1.5,4)` approX – the square root process can be implemented with `nr(x^2-c,x,x0,n)` or by `nrs(c,x0,n):= nr(x^2-c,x,x0,n)`. The square root of 5 is then obtained with, for example, Author `nrs(5,1,4)` approX to give the sequence 1, 3, 2.33333, 2.23809, 2.23606 .

6.4.4 Secant method

A drawback of Newton–Raphson iteration is the need to determine the derivative of the function f (often this is practically impossible). An estimate to $f'(x_n)$ can be obtained from the derivative formula

$$f'(x_n) \approx f_n' = \frac{f(x_n) - f(x_{n-1})}{x_n - x_{n-1}},$$

which gives the slope of a chord from $(x_{n-1}, f(x_{n-1}))$ to $(x_n, f(x_n))$, and the Newton–Raphson formula (6.21) is modified to

$$x_{n+1} = x_n - \frac{(x_n - x_{n-1})f(x_n)}{f(x_n) - f(x_{n-1})}, \quad n = 1, 2, \ldots \tag{6.24}$$

This is the *secant method* (note that *two initial values*, x_0 and x_1, are used).

Example 6.23 Consider the now familiar equation $x^3 + 4x^2 - 10 = 0$. If $x_0 = 1.5$ and x_1 is the 'Newton–Raphson' value $x_1 = 1.3733$, the secant method converges just as fast as Newton–Raphson. As the sequence $\{x_n\}$ converges to α, so successive x_n get closer together and the derivative approximation $[f(x_n) - f(x_{n-1})]/(x_n - x_{n-1})$ approaches its limiting value $f'(x_n)$.

n	0	1	2	3	4
x_n	1.5000	1.3733	1.3657	1.3652	1.3652

Of course, one should be aware of the potential loss of significance as $x_n \to \alpha$. In this case the derivative approximation approaches $\frac{0}{0}$.

DERIVE Experiment 6.8

The secant method generates a second-order difference equation since x_{n+1} depends upon the previous two iterates. The *DERIVE* implementation is slightly more tricky. `Author`

```
d(f,x,a,b):=(b-a)lim(f,x,b)/(lim(f,x,b)-lim(f,x,a))
secant(f,x,x0,x1,n):=
   iterates([b,b-d(f,x,a,b)],[a,b],[x0,x1],n-1)
```

The table in Example 6.23 is obtained from `Author`

```
secant(x^3+4x^2-10,x,1.5,1.3733,4)
approX
```

6.5 Logistic mapping revisited

With several analysis tools now in place we take a second look at sequences generated by the *logistic mapping* $x_{n+1} = \lambda x_n(1 - x_n)$.

The recurrence relation derives from the equation $x = \lambda x(1 - x)$. This is a quadratic equation, $\lambda x^2 + (1 - \lambda)x = 0$, with *two* roots $x = 0$ and $x = 1 - 1/\lambda$. Convergence of the sequence $\{x_n\}$ is governed by the derivative of $g(x) = \lambda x(1-x)$, that is $g'(x) = \lambda(1 - 2x)$.

At $x = 0$, the value $g'(0) = \lambda$ lies in the interval $[-1, 1]$ if $\lambda \in [-1, 1]$, that is convergence to $x = 0$ occurs if $\lambda \in [0, 1]$ (see Figure 6.21(a), $0 < \lambda < 1$).

At $x = 1 - 1/\lambda$, $g'(x) = \lambda(1 - 2(1 - 1/\lambda)) = 2 - \lambda$ which lies in the interval $[-1, 1]$ if $\lambda \in [1, 3]$. That is, convergence to $x = 1 - 1/\lambda$ occurs if $\lambda \in [1, 3]$ (see Figure 6.21(a), $1 < \lambda < 3$). This root exists for *all* $\lambda > 0$.

For $\lambda > 3$ the graph of x_∞.v.λ *bifurcates* at $\lambda = 3$ and the sequence $\{x_n\}$ approaches a limiting *2-cycle*, until $\lambda \approx 3.45$ (see Figure 5.7(b)).

What is the significance of the limiting values in the 2-cycle?

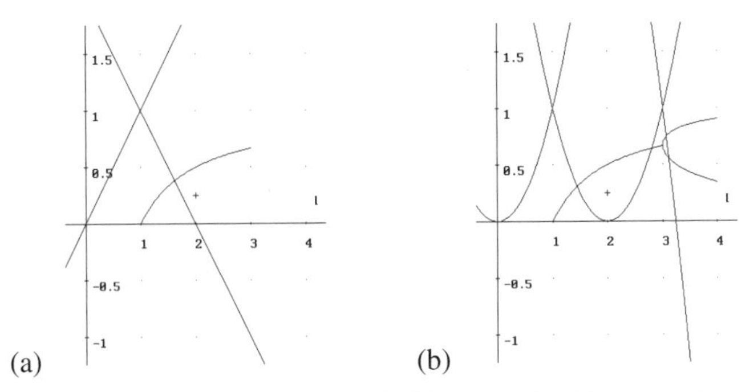

Figure 6.21 Logistic mapping (analysis). (a) $0 < \lambda < 3$. (b) $0 < \lambda < 4$.

A 2-cycle implies that $x_{n+2} = x_n$. Here, $x_{n+2} = g(x_{n+1}) = g(g(x_n))$, that is $x_{n+2} = G(x_n)$ where $G(x) = g(g(x))$. Dropping subscripts, the originating equation for the 2-cycle is

$$\begin{aligned} x = G(x) = g(g(x)) &= g(\lambda x(1-x)) \\ &= \lambda^2 x(1-x)(1-\lambda x(1-x)). \end{aligned}$$

This is a quartic equation which can be written as

$$[\lambda x^2 + (1-\lambda)x]\lambda[\lambda^2 x^2 - \lambda(1+\lambda)x + 1 + \lambda] = 0. \tag{6.25}$$

The first term in eqn (6.25) is the original quadratic equation, so two of the four roots of the quartic are $x_1 = 0$ and $x_2 = 1 - 1/\lambda$. The second pair of roots are ($\lambda \neq 0$)

$$x_{3,4} = \frac{1 + \lambda \pm \sqrt{\lambda^2 - 2\lambda - 3}}{2\lambda}.$$

Convergence of the sequence $\{x_n\}$ generated by $x_{n+2} = G(x_n)$ is governed by the derivative $G'(x) = \lambda^2(1-2x)(2\lambda x^2 - 2\lambda x + 1)$. After some effort the following expressions are obtained for $G'(x)$ at the four roots.

$$\begin{aligned} x &= x_1, \quad G'(x_1) = \lambda^2, \\ x &= x_2, \quad G'(x_2) = (2-\lambda)^2, \\ x &= x_3, \quad G'(x_3) = -\lambda^2 + 2\lambda + 4, \\ x &= x_4, \quad G'(x_4) = -\lambda^2 + 2\lambda + 4. \end{aligned}$$

For $\lambda \in (0,1)$, $|G'(x_1)| < 1$ and $\{x_n\}$ converges to the root $x_1 = 0$. For $\lambda \in (1,3)$, $|G'(x_2)| < 1$ and $\{x_n\}$ converges to the root $x_2 = 1 - 1/\lambda$. For $\lambda > 3$, $|G'(x_3)| < 1$ and $|G'(x_4)| < 1$, and $\{x_n\}$ converges to the root $x_{3,4} = [1+\lambda \pm \sqrt{\lambda^2 - 2\lambda - 3})]/2\lambda$. At a certain value of $\lambda > 3$, the value of $G'(x_3)$ (or $G'(x_4)$) drops below -1. At this point, given by $\lambda = 1 + \sqrt{6}$, the sequence no longer converges to x_3 (or x_4) – the effect is a further bifurcation to a *4-cycle, ad infinitum*.

Table 6.12 Newton–Raphson applied to the logistic mapping.

	n							
λ	0	1	2	3	4	5	6	7
0.5	0.2	0.029	0.001	0				
	1.2	0.424	0.097	0.008	0			
1	0.2	0.100	0.050	0.025	0.013	0.006	0.003	0.002
	1.2	0.600	0.300	0.150	0.075	0.038	0.019	0.009
1.5	0.2	0.600	0.415	0.347	0.334	0.333	0.333	
	1.2	0.697	0.458	0.360	0.335	0.333	0.333	
2	0.2	−0.400	−0.123	−0.020	0			
	1.2	0.758	0.566	0.507	0.500	0.500		
2.5	0.2	−0.200	−0.040	−0.002	0			
	1.2	0.800	0.640	0.602	0.600	0.600		
3	0.2	−0.150	−0.023	−0.001	0			
	1.2	0.831	0.694	0.667	0.667			
3.5	0.2	−0.127	−0.017	0				
	1.2	0.854	0.734	0.715	0.714	0.714		
4	0.2	−0.114	−0.013	0				
	1.2	0.873	0.765	0.750	0.750	0.750		

Newton–Raphson iteration has no difficulty in finding the root for any value of λ. Taking $f(x) = x - \lambda x(1-x)$, then $f'(x) = 1 - \lambda(1-2x)$ and the Newton–Raphson method produces the recurrence relation

$$x_{n+1} = x_n - \frac{x_n - \lambda x_n(1-x_n)}{1-\lambda(1-2x_n)} = \frac{\lambda x_n^2}{1-\lambda(1-2x_n)}.$$

Table 6.12 shows a selection of typical sequences, starting at $x_0 = 0.2$ and $x_0 = 1.2$ for a range of values of λ.

6.6 Deflation for multiple roots

Throughout this chapter many equations of the form $f(x) = 0$ have had *multiple roots*, that is several values of x for which the graph of $y = f(x)$ touches the x-axis. Further, having chosen the value x_0 to start an iterative solution method, we have little control over which root may be determined. In this section we briefly discuss *deflation* as a means of systematically finding multiple roots.

Example 6.24 The quartic polynomial $f(x) = (8x^4 - 36x^3 + 38x^2 + 9x - 10)/8$ has four zeros at $x = -0.5, 0.5, 2$ and 2.5 ($y = f(x)$ in Figure 6.22). Since $x = -0.5$ is a root, $x + 0.5$ is a *factor* of the function f, that is $f(x) = (x + 0.5)g(x)$ where g is a

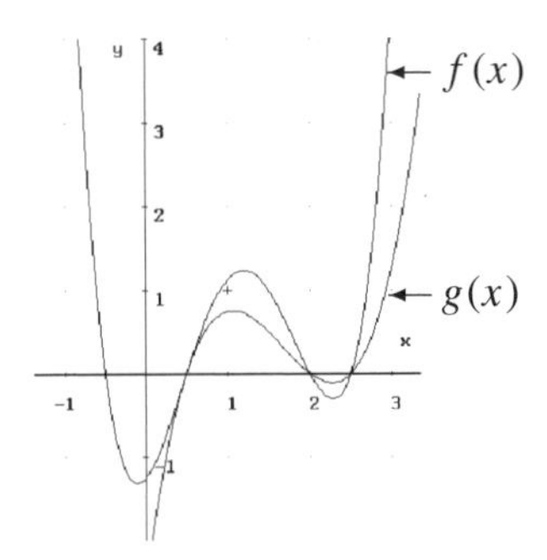

Figure 6.22 Extracting multiple roots.

function having three zeros equal to the remaining three zeros of f ($y = g(x)$ in Figure 6.22). In fact $g(x) = (4x^3 - 20x^2 + 29x - 10)/4$.

Having found the root $x = -0.5$ (by some numerical method), subsequent roots of $f(x) = 0$ can be found by solving $g(x) = 0$ – a problem that does not contain the first root (unless it is a *repeated root*).

The idea of Example 6.24 (which will be expanded for polynomials in Section 6.7) is equally applicable to non-polynomial equations.

Example 6.25 The equation $f(x) = 0$, where $f(x) = x^2e^{-x} - 0.5$, has *three* roots at $x = -0.539835$, $x = 1.48796$ and $x = 2.61786$ (see Figure 6.23(a)). Defining the function g by $g(x) = f(x)/(x - 1.48796)$ results in a new function having zeros equal to the first and third roots of $f(x) = 0$ – the middle root has been 'factored out' (see Figure 6.23(b)).

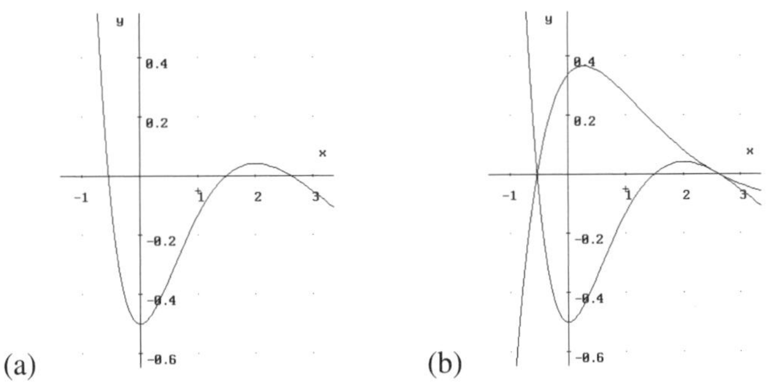

Figure 6.23 (a) Multiple roots of $x^2e^{-x} - \frac{1}{2}$. (b) Deflated function $g(x)$.

The process of successive factoring is called *deflation*. Given a function f with n zeros, $\alpha_1, \ldots, \alpha_n$, successive zeros are found by the following process.

(1) Set $f_1(x) = f(x)$ and find a zero, say α_1.

(2) For values of i from 2 to n, set $f_i(x) = f_{i-1}(x)/(x - \alpha_{i-1})$ and find a zero, say α_i.

Example 6.26 Use Newton–Raphson iteration to find the three roots of the function defined in Example 6.25; $f(x) = x^2 e^{-x} - \frac{1}{2}$, $f'(x) = x(2-x)e^{-x}$ and eqn (6.21) becomes

$$x_{n+1} = x_n - \frac{f(x_n)}{f'(x_n)} = \frac{2x_n^2(1 - x_n) + e^{x_n}}{2x_n(2 - x_n)}.$$

With $x_0 = 1$ the following sequence is obtained:

n	0	1	2	3	4	5
x_n	1	1.3591	1.4730	1.4877	1.4880	1.4880

The next step is to find a root of $f_2(x) = f(x)/(x - 1.4880)$. Again using $x_0 = 1$, we obtain

n	0	1	2	3	4	5
x_n	1	2.3601	2.5955	2.6177	2.6179	2.6179

Finally, the root of $f_3(x) = f(x)/[(x - 1.4880)(x - 2.6179)]$ is sought (using $x_0 = 1$), to generate

n	0	1	2	3
x_n	1	9.5356	13.4071	19.0849

The divergent nature of the sequence is more easily understood on inspecting the 'shape' of the final function f_3 (see Figure 6.24). It is simply a case of choosing an appropriate value for x_0 – a good choice is $x_0 = -1$, leading to the solution sequence

n	0	1	2	3	4	5
x_n	−1	−0.6664	−0.5517	−0.5399	−0.5398	−0.5398

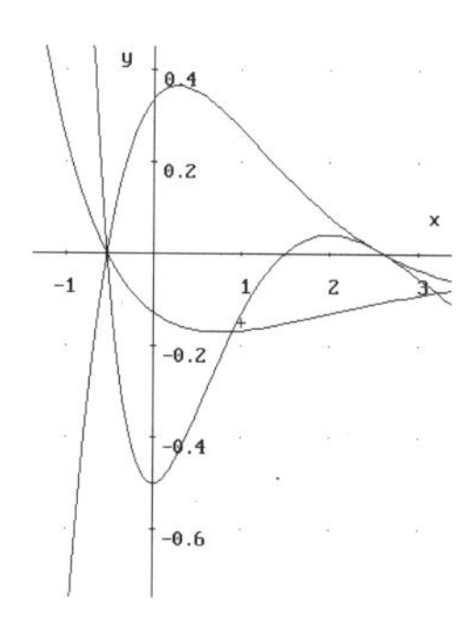

Figure 6.24 Twice deflated function $f_3(x)$.

DERIVE Experiment 6.9

To automate deflation 'robustly' is not easy. The difficulty lies in the automatic selection of x_0 for each root (as seen in Example 6.26, a poor choice can lead to divergence). With these cautionary words, the following expressions will generate a *DERIVE* implemention of deflation with Newton–Raphson iteration as the basic root finder (don't be surprised if the function fails in certain instances!). `Author`

```
e1(x,i):=element(x,i)
r(f,x,x0,n):=iterate(x-f/dif(f,x),x,x0,n)
deflate(f,x,x0,m,n):=e1(iterates([r(v/(x-s),x,s,n),
   v/(x-s)],[s,v],[r(f,x,x0,n),f],m-1)',1)
```

`f` is the original function (in terms of `x`), `x0` is the initial guess for the first zero, `m` is the number of zeros required and `n` is the maximum number of iterations performed per zero (to avoid iterating forever). Each subsequent iterative process is started at the value of the preceding root, `s`. The use of `e1` simply extracts the column of `s` values. Note also the use of `iterate` and `iterates`.

To obtain the three roots of the function defined in Example 6.26, `Author u:= x^2exp-x-1/2` and `Author approX deflate(u,x,-1,3,10)` to get the vector [−0.5398, 1.4880, 2.6179]. The fact that the roots are found in ascending order is coincidence. Figure 6.25 shows the original and deflated functions for this particular root ordering (compare with Figures 6.23 and 6.24 in which the root extraction order was different). Try different values of x_0 – what do you observe?

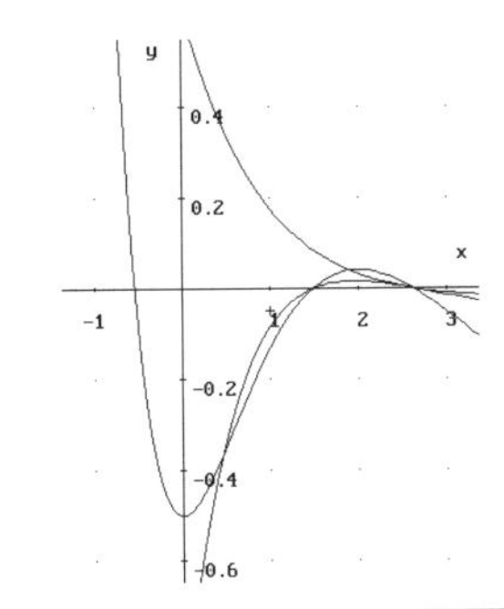

Figure 6.25 f and its deflations.

6.7 Polynomial roots

In this section the solution of $f(x) = 0$ is sought where f is a *polynomial of degree* n, say p_n. No new methods are introduced and the discussion will focus upon using the special form of polynomials in order (a) to improve the efficiency of Newton–Raphson iteration and (b) to find all roots of the equation $p_n(x) = 0$ using deflation (see Section 6.6) where

$$p_n(x) = \sum_{k=0}^{n} a_k x^k. \tag{6.26}$$

If α_1 is a zero of $p_n(x)$ then $p_n(\alpha_1) = 0$ and the monomial $(x - \alpha_1)$ is a factor of $p_n(x)$,

$$p_n(x) = (x - \alpha_1)q_{n-1}(x)$$

where q_{n-1} is a polynomial of degree $n - 1$. If α_2 is a zero of $q_{n-1}(x)$ then $q_{n-1}(\alpha_2) = 0$ and $(x - \alpha_2)$ is a factor of $q_{n-1}(x)$,

$$\begin{aligned} p_n(x) &= (x - \alpha_1)q_{n-1}(x) \\ &= (x - \alpha_1)(x - \alpha_2)r_{n-2}(x) \end{aligned}$$

where $r_{n-2} \in P_{n-2}$. Clearly, α_2 is also a root of $p_n(x) = 0$. After n steps of the factorisation process

$$p_n(x) = (x - \alpha_1)(x - \alpha_2)\dots(x - \alpha_{n-2})(x - \alpha_{n-1})(x - \alpha_n).$$

This rather informal presentation shows that a polynomial of degree n has exactly n zeros. These zeros may be real, complex or repeated.

Example 6.27 The polynomial $p_4(x) = x^4 - 8x^3 + 23x^2 - 30x + 18$ can be factorised as $p_4(x) = (x-3)^2(x^2 - 2x + 2)$. Thus, p_4 has two real zeros (a repeated zero of multiplicity 2) $\alpha_1 = \alpha_2 = 3$ and a pair of complex conjugate zeros, $\alpha_{3,4} = 1 \pm \mathrm{i}$, that are the roots of $x^2 - 2x + 2 = 0$.

DERIVE Experiment 6.10

It is worth noting that *DERIVE* has an intrinsic `factor` function with the specification (given here for a function of one variable)

```
factor(f,how,x)
  f     object function f(x)
  how   type of factorisation
  x     independent variable x
```

`how` can be `Trivial`, `Squarefree`, `Rational`, `raDical` or `Complex`.

To rework Example 6.27, `Author p:=x^4-8x^3+23x^2-30x+18`. Now `Author` and `Simplify factor(p,Rational,x)`. To extract the complex roots explicitly, `Author` and `Simplify factor(p,Complex,x)`.

Finding the roots of the polynomial equation $p_n(x) = 0$ by analytical methods often proves to be quite difficult (unless p_n has a special form). The obvious numerical method is Newton–Raphson iteration,

$$x_{j+1} = x_j - \frac{p_n(x_j)}{p_n'(x_j)}, \quad j = 0, 1, \ldots \tag{6.27}$$

The issues in using eqn (6.27) are (a) how to evaluate $p_n(x_j)$ and $p_n'(x_j)$ efficiently and (b) how to use eqn (6.27) to find all n zeros of p_n.

The first issue is motivated by the need to make full use of the rapid (quadratic) convergence offered by the Newton–Raphson process – there is little point in taking few iterations to converge if the computational effort for each iteration is significant. The nested multiplication algorithm (see Algorithm 3.2 with x replaced by x_j) is an efficient algorithm to evaluate the polynomial $p_n(x_j)$, requiring just $2n$ flops.

To compute $p_n'(x_j)$ efficiently we shall require the intermediate values appearing in the nested routine. A suitably modified algorithm is

Algorithm 6.3 Nested polynomial (modified)
Function: Power form evaluation

List	Description	
1	**Problem**	`input` $n, a_0, \ldots, a_n, x$
2	Initialise	$b_n := a_n$
3		`for` $k = n - 1$ `to` 0 `step` -1
4		$b_k := a_k + b_{k+1} * x_j$
5		`end`
6	**Solution**	`output` $b_0, \ldots, b_n$ $(p_n(x) = b_0)$

b_k stores the 'value' of p_n after $n-k$ cycles of the `for` loop. The coefficient b_0 equals $p_n(x_j)$. To see this, expanding the polynomial (6.26) gives

$$p_n(x) = a_n x^n + a_{n-1}x^{n-1} + \cdots + a_2 x^2 + a_1 x + a_0.$$

Dividing $p_n(x)$ by the monomial $(x - z)$ gives a polynomial of degree $n - 1$,

$$q_{n-1}(x) = b_n x^{n-1} + b_{n-1}x^{n-2} + \cdots + b_2 x + b_1$$

with *remainder* b_0. We have

$$p_n(x) = (x - z)q_{n-1}(x) + b_0, \tag{6.28}$$

that is

$$a_n x^n + a_{n-1}x^{n-1} + \cdots + a_2 x^2 + a_1 x + a_0$$
$$= (x - z)(b_n x^{n-1} + b_{n-1}x^{n-2} + \cdots + b_2 x + b_1) + b_0.$$

Comparing coefficients of both sides:

$$\begin{array}{llll}
O(x^n) & : a_n = b_n & \Rightarrow b_n & = a_n \\
O(x^{n-1}) & : a_{n-1} = b_{n-1} - zb_n & \Rightarrow b_{n-1} & = a_{n-1} + zb_n \\
O(x^{n-2}) & : a_{n-2} = b_{n-2} - zb_{n-1} & \Rightarrow b_{n-2} & = a_{n-2} + zb_{n-1} \\
 & & \vdots & \\
O(x^2) & : a_2 = b_2 - zb_3 & \Rightarrow b_2 & = a_2 + zb_3 \\
O(x) & : a_1 = b_1 - zb_2 & \Rightarrow b_1 & = a_1 + zb_2 \\
O(1) & : a_0 = b_0 - zb_1 & \Rightarrow b_0 & = a_0 + zb_1
\end{array}$$

from which $b_n = a_n$ and $b_k = a_k + zb_{k+1}$ for $k = n - 1, \ldots, 0$. This establishes the nested multiplication algorithm.

Replacing x by z in eqn (6.28) gives $p_n(z) = (z - z)q_{n-1}(z) + b_0 = b_0$. This establishes the fact that the final coefficient b_0 in the nested multiplication algorithm equals the polynomial value at $x = z$. Thus, the algorithm is an efficient $O(n)$ method for polynomial evaluation.

The next stage is to evaluate the derivative $p_n'(z)$ efficiently. Differentiating eqn (6.28) with respect to x gives

$$p_n'(x) = q_{n-1}(x) + (x - z)q_{n-1}'(x). \qquad \textbf{(6.29)}$$

At $x = z$, $p_n'(z) = q_{n-1}(z)$. Thus, if $q_{n-1}(z)$ can be determined then the value of the first derivative of p_n at $x = z$ is known. To evaluate $q_{n-1}(z)$ efficiently, nested multiplication can be used. Dividing $q_{n-1}(x)$ by the monomial $(x - z)$ gives a polynomial of degree $n - 2$

$$r_{n-2}(x) = d_n x^{n-2} + d_{n-1}x^{n-3} + \cdots + d_3 x + d_2,$$

with remainder $d_1 = q_{n-1}(z) = p_n'(z)$. The d_k are defined by $d_n = b_n$ and $d_k = d_k + d_{k+1}x_j$, $k = n - 1, \ldots, 1$, requiring $2(n - 1)$ flops.

Example 6.28 Find a root of $p_4(x) = x^4 - 10x^3 + 35x^2 - 50x + 24$ near $x = 0.5$. The solution of this problem suggests a tabular format for hand calculations (in practice a computer program would be implemented). Coefficient values are rounded to two decimal places (and carried forward to subsequent calculations).

$x_0 = 0.5$	k	4	3	2	1	0	
	a_k	1	−10	35	−50	24	
	b_k	1	−9.5	30.25	−34.88	**6.56**	$= p_4(0.5)$
	d_k	1	−9	25.75	**−22.01**		$= p'_4(0.5)$

$$x_1 = x_0 - \frac{p_4(0.5)}{p'_4(0.5)} = 0.5 - \frac{6.56}{-22.01} = 0.80$$

$x_1 = 0.80$	k	4	3	2	1	0	
	b_k	1	−9.2	27.64	−27.89	**1.69**	$= p_4(0.80)$
	d_k	1	−8.4	20.92	**−11.15**		$= p'_4(0.80)$

$$x_2 = x_1 - \frac{p_4(0.80)}{p'_4(0.80)} = 0.80 - \frac{1.69}{-11.15} = 0.95$$

$x_2 = 0.95$	k	4	3	2	1	0	
	b_k	1	−9.05	26.40	−24.92	**0.33**	$= p_4(0.95)$
	d_k	1	−8.10	18.71	**−7.15**		$= p'_4(0.95)$

$$x_3 = x_2 - \frac{p_4(0.95)}{p'_4(0.95)} = 0.95 - \frac{0.33}{-7.15} = 1.00$$

$x_3 = 1.00$	k	4	3	2	1	0	
	b_k	1	−9	26	−24	**0**	$= p_4(1.00)$

$p_4(x) = 0$ at $x = 1.00$, so 1 is a root of the equation $p_4(x) = 0$.

Given the coefficients a_k, the computational tableau determines the b_k (from $b_k = a_k + x_j b_{k+1}$) and the d_k (from $d_k = b_k + x_j d_{k+1}$). The values b_0 and d_1 give the polynomial value and its derivative at $x = x_j$ for use in the Newton–Raphson formula.

6.7.1 Deflation for polynomial roots

Deflation can be used to find additional roots of the polynomial equation $p_n(x) = 0$. Once the first root has been found, say $x = \alpha_1$, then $b_0 = p_n(\alpha_1) = 0$ and

$$p_n(x) = (x - \alpha_1) q_{n-1}(x) \qquad \textbf{(6.30)}$$

(from eqn (6.28)), where the coefficients of q_{n-1} are $b_n, \ldots, b_1$. The method can be applied to find a zero of the deflated polynomial q_{n-1}, which by virtue of eqn (6.30) will also be a zero of p_n.

Example 6.29 To find a second root of $p_4(x) = 0$, where p_4 is defined in Example 6.28, the method is applied to the deflated polynomial equation $q_3(x) = 0$ where $q_3(x) = x^3 - 9x^2 + 26x - 24$ (see the last line of b_k values in Example 6.28). If $x_0 = 0.5$,

$x_0 = 0.5$	k	3	2	1	0	
	a_k	1	−9	26	− 24	
	b_k	1	−8.5	21.75	**−13.13**	$= q_3(0.5)$
	d_k	1	−8	**17.75**		$= q_3'(0.5)$

$$x_1 = x_0 - \frac{q_3(0.5)}{q_3'(0.5)} = 0.5 - \frac{-13.13}{17.75} = 1.24$$

$x_1 = 1.24$	k	3	2	1	0	
	b_k	1	−7.76	16.38	**−3.69**	$= q_3(1.24)$
	d_k	1	− 6.52	**8.30**		$= q_3'(1.24)$

$$x_2 = x_1 - \frac{q_3(1.24)}{q_3'(1.24)} = 1.24 - \frac{-3.69}{8.30} = 1.68$$

$x_2 = 1.68$	k	3	2	1	0	
	b_k	1	−7.32	13.70	**−0.98**	$= q_3(1.68)$
	d_k	1	−5.64	**4.22**		$= q_3'(1.68)$

$$x_3 = x_2 - \frac{q_3(1.68)}{q_3'(1.68)} = 1.68 - \frac{-0.98}{4.22} = 1.91$$

$x_3 = 1.91$	k	3	2	1	0	
	b_k	1	−7.09	12.46	**−0.20**	$= q_3(1.91)$
	d_k	1	−5.18	**2.57**		$= q_3'(1.91)$

$$x_4 = x_3 - \frac{q_3(1.91)}{q_3'(1.91)} = 1.91 - \frac{-0.20}{2.57} = 1.99$$

$x_4 = 1.99$	k	3	2	1	0	
	b_k	1	−7.01	12.05	**−0.02**	$= q_3(1.99)$
	d_k	1	−5.02	**2.06**		$= q_3'Z(1.99)$

$$x_5 = x_4 - \frac{q_3(1.99)}{q_3'(1.99)} = 1.99 - \frac{-0.02}{2.06} = 2.00$$

$x_5 = 2.00$	k	3	2	1	0	
	b_k	1	−7	12	**0**	$= q_3(2.00)$

$q_3(x) = 0$ at $x = 2.00$, so 2 is a root of the equation $q_3(x) = 0$, and is also a root of $p_4(x) = 0$.

Observation *Even though the same initial guess, $x_0 = 0.5$, is used, the method does not converge to the root $x = 1$ since this is* not *a zero of the deflated polynomial q_3 (convergence is slower since 0.5 is not such a good initial guess for the second root). In this fashion, as roots of $p_n(x) = 0$ are found they are factored out to give successive deflated polynomials.*

Summary

This chapter dealt with the development of systematic (iterative) methods for solving the algebraic equation $f(x) = 0$, starting with basic definitions (roots and zeros) and motivating the need for numerical methods, with recurrence as a possible methodology.

Section 6.1 dealt with the location of a root within a bracketing interval $[a, b]$, which was then refined (reduced) by the bisection method (see Section 6.2). If the mid-point c of the latest interval is taken as an estimate of the root, Section 6.2.1 developed an error bound (6.4). The method always converges. In Section 6.2.2, accuracy (tolerance) was specified in terms of decimal places. The inefficiency of the method was addressed via parallel processing (see Section 6.2.3) and a measure of the speed-up was given (6.8). Additional termination conditions were specified in Section 6.2.4.

Section 6.3 concerned (first-order) fixed-point methods. The notion of a fixed point was described (see Section 6.3.1), followed by fixed-point iteration (see Section 6.3.2). An error analysis based upon the mean-value theorem (see Section 6.3.3) led to a condition for convergence, $|g'(\alpha)| < 1$. Visualisation using cobweb diagrams and a discussion of the initial value were given in Sections 6.3.4 and 6.3.5. The acceleration (improvement) of convergence was tackled by Aitken's method (see Section 6.3.6) and choice of g (see Section 6.3.7), and was then optimised in Section 6.3.8.

Optimal convergence was shown to occur in the guise of Newton–Raphson iteration (see Section 6.4). An improved error analysis based on Taylor's theorem indicated the second-order nature of the method (see Section 6.4.1) and certain failure situations were described in Section 6.4.2. Its application to finding the square root of a positive number was developed in Section 6.4.3, and the avoidance of derivative calculations using the secant method was described in Section 6.4.4. The analyses developed in this chapter were applied to the logistic mapping in Section 6.5.

Section 6.6 dealt with the determination of multiple roots using deflation, and Section 6.7 described the special case of finding the roots of a polynomial equation.

Exercises

6.1 Show that the equation $1 - 2x + \cos x = 0$ has a root in $[0, 1]$. How many iterations of the bisection method are required to estimate this root to an accuracy of $\frac{1}{2} \times 10^{-3}$?

6.2 Use the bisection method to find the positive root of the equation $x^2 + x - 1 = 0$ in the interval $[0, 1]$, to an accuracy of one decimal place.

6.3 A goat is tethered by a rope of length x to a pole on the boundary of a circular field of unit radius (see Figure 6.26). How long should the rope be so that the goat can graze over 25% of the area of the field?

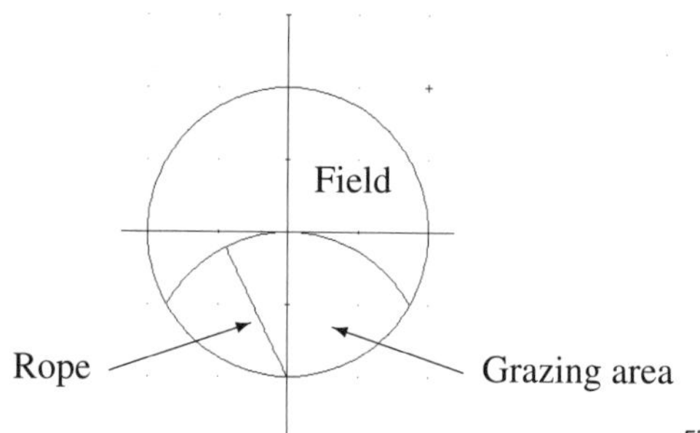

Figure 6.26 The 'tethered goat' problem.

The area of the field is π and the grazing area can be shown to equal $(x^2 - 2)\cos^{-1}(x/2) - x\sqrt{1 - x^2/4} + \pi$. Equating this to $\pi/4$ results in the equation $f(x) = 0$ where

$$f(x) = x - 2\cos\left[\frac{2x\sqrt{4 - x^2} - 3\pi}{4(x^2 - 2)}\right].$$

Solve this problem using bisection to an accuracy of at least 0.001.

6.4 A parallel bisection method is implemented on a p-processor array. Show that the speed-up S_p over the single-processor implementation behaves approximately like $\log_2(p + 1)$. Tabulate S_p for $p = 1, 2, \ldots, 10$ and comment on the values you obtain. Suggest why these speed-up values are unlikely to be attained in practice.

6.5 Ordinary annuity equation: the money required to pay off a fixed-term mortgage is described by the *ordinary annuity equation*

$$C = \frac{P}{i}[1-(1+i)^{-n}].$$

C is the borrowed capital, P is the (monthly) payment, i is the interest rate per payment period (expressed as a fraction of 100; i.e. an interest rate of 4.5% $\rightarrow$ $i = 0.045$) and n is the number of payment periods.

A borrower can afford a monthly payment of £450. On a 25-year mortgage of £ 50 000 what is the maximum annual interest rate that the borrower can afford?

6.6 Annuity due equation: the accumulated money in a savings account is described by the *annuity due equation*

$$S = \frac{P}{i}[(1+i)^{n}-1].$$

S is the account balance, P is the amount of a regular deposit, i is the interest rate per payment period (see the previous question) and n is the number of deposit periods. The aim is to have an account balance of £2000 at the end of a three-year degree programme given a monthly deposit of £30. What is the minimum annual interest rate at which this amount can be deposited if the interest is compounded quarterly?

6.7 *Regula falsi*: a simple modification to the bisection method is obtained by replacing the mid-point $c = \frac{a+b}{2}$ by the point at which the line interpolating $(a, f(a))$ and $(b, f(b))$ meets the x-axis. Use linear interpolation (see eqn (3.12)) to show that this line is given by

$$y = f(a) + \left(\frac{f(b)-f(a)}{b-a}\right)(x-a).$$

The estimate to α is taken as the root of this equation when y equals zero. Show that

$$c = a - f(a)\left(\frac{b-a}{f(b)-f(a)}\right).$$

The next interval is found by using the sign property on $[a, c]$ and $[c, b]$.

Apply the method to the problem of Exercise 6.1. How many iterations are required to estimate α to an accuracy of $\frac{1}{2} \times 10^{-3}$?

6.8 Compare the amount of calculation required to find the root of the equation $2^x - 5x + 1 = 0$ to four decimal places of accuracy by using

(a) the bisection method with $a = 0$ and $b = 1$, and
(b) the fixed-point method $x_0 = 0.1$, $x_{n+1} = (2^{x_n} + 1)/5$, $n \geq 0$.

6.9 Choose a constant λ so that the iterative process

$$x_{n+1} = \frac{1}{1+\lambda}\left(\lambda x_n + \frac{2^{x_n}+1}{5}\right)$$

gives a rapidly convergent method for finding the root of $2^x - 5x + 1 = 0$ near $x = 0.1$. Calculate the first few iterates, beginning with $x_0 = 0.1$.

6.10 Apply Aitken acceleration to the method of Exercise 6.8(b) for finding a root of $2^x - 5x + 1 = 0$. Start with $x_0 = 0.1$.

6.11 Apply Aitken acceleration to the 'comet' iterations of Example 6.16 (Kepler's law).

6.12 The value of the resistance of a certain resistor contained within an electrical circuit varies with current i as $R(i) = 50 + i^{\frac{2}{3}}$. With Ohm's law, $V = iR$, and using an iterative process of the form $i_{n+1} = g(i_n)$, determine the current flowing in a circuit for which $V = 7$ volts.

6.13 Obtain a quadratic equation associated with the iterative method

$$x_{n+1} = \frac{bx_n^2 + 2cx_n}{c - ax_n^2}.$$

Show that if $c \neq 0$ and $ax^2 \neq c$ at a root then the method is exactly second order. What happens if $c = 0$?

6.14 Show that the equation $x^p - c = 0$ can be written $x = c/x^{p-1}$, and that the associated iterative process $x_{n+1} = c/x_n^{p-1}$ will not be successful in finding the pth root of the positive number c where $p \in \{2, 3, 4, \ldots\}$.

6.15 Show that $x = a^x$, $a > 0$, only has a root for $0 < a \leq e^{1/e}$. For $a = e^{1/e}$, find the root analytically and by using the exponential recurrence relation (5.18). Interpret the result graphically.

6.16 This exercise concerns the Newton–Raphson process (6.21).

(a) If $x_0 = 0.1$ then apply the formula (6.21) to find the root of $2^x - 5x + 1 = 0$. How many iterations are required in order to obtain four decimal places of accuracy?
(b) Derive an iterative process for finding the pth root of a positive number c by applying formula (6.21) to the equation $x^p - c = 0$.
(c) What happens in Newton's square root process if the first guess x_0 is negative? Prove it!
(d) Apply (6.21) to the equation $\frac{1}{x} - c = 0$, $c \neq 0$. What is the descriptive word for the limiting value? Any comments?
(e) On applying (6.21) to find the first positive root of $\sin x = 0$ what value of x_n propels the sequence $\{x_n\}$ to a 2-cycle (i.e. $x_{n+2} = x_n$)? Why can this situation never arise? [*Hint*: Show that the 'critical' value of x_n satisfies the equation $\tan x = 2x$.]
(f) Determine the recurrence relation resulting from an application of (6.21) to the

logistic equation $x = \lambda x(1-x)$. You may assume that $f(x) = x - \lambda x(1-x)$. Implement the recurrence relation for values of λ ranging from 0.5 to 4 in steps of 0.5, starting each sequence with the value $x_0 = 0.5$ – comments!

(g) Finally, write down the linear Taylor series expansion, p_1, of a function 'f of x' about the point $x = x_0$. Find the root of the equation $p_1(x) = 0$. What have you found?

6.17 Use the secant method to find the area lying between the perimeter of a circle and a chord subtending an angle of θ radians at the centre of the circle – the formula is

$$A = \frac{R^2[\theta - \sin\theta]}{2}$$

where R is the radius of the circle. If $R = 5$ and the area equals 1.2, find the angle θ.

6.18 Find three roots of the equation $\sin x = 0$ using Newton–Raphson iteration coupled with deflation. For each root use the initial guess $x_0 = 0.5$. What three roots do you find?

6.19 Continue the deflation process used in Example 6.29 to find the remaining two zeros of the quartic polynomial $p_4(x) = x^4 - 10x^3 + 35x^2 - 50x + 24$.

6.20 Root-finding problems frequently appear within industrial, commercial and scientific contexts. Here are two instances.

(a) For a melting material, heated on its surface, the location of the liquid–solid interface from the heated surface is given by $s(t) = 2\lambda\sqrt{t}$ where λ is the root of $\alpha\sqrt{\pi}\lambda e^{\lambda^2}\text{erf}(\lambda) = 1$.
$\alpha = L/(c\Delta T)$ where L is latent heat of fusion, c is specific heat and ΔT is the difference in temperature between the applied surface temperature and the material's melting temperature. For copper ($L/c = 550$) and medium density polymer ($L/c = 120$), find the depth of molten material at a time $t = 2$ if $\Delta T = 250$.

(b) The radius of a stable ice cap can be crudely modelled by the equation $\int_0^R \cos(r^2)\,\mathrm{d}r - \frac{R}{2} = 0$, with the cap's base radius R given in units of 10 000 m. Approximate the integral by Simpson's rule using two intervals and solve the resulting algebraic equation for R.

6.21 The roots of the polynomial

$$p_{20}(x) = \prod_{i=1}^{20}(x - i)$$

are given by the set $x \in \{1, 2, 3, 4, 5, \ldots, 19, 20\}$. If the coefficient of x^{19} (originally equal to -210) is modified by 2^{-23}, then find the roots of the resulting polynomial. This example illustrates the ill-conditioned nature of the polynomial problem.

7 Numerical differentiation

A rocket is launched from Earth with the aim of reaching an orbital state. From lift-off the telemetry system sends information on the distance travelled which is sampled by a ground receiver at regular intervals of time. For the safe orbit of the rocket, certain *speeds* and *accelerations* are required, where speed is the rate of change of distance, or *first derivative* of distance with respect to time, and acceleration is the rate of change of speed. The problem of estimating derivative values using *discrete* data of the type shown in Table 7.1 is addressed in this chapter.

Table 7.1 Rocket telemetry data.

Time t (s)	0	4	8	12	16	20
Distance s (km)	0	0.8	3.5	8.4	15.9	27.0

7.1 Preliminaries

Differentiation is a fundamental operation of calculus. Let f be a function of a single real variable x. Graphically, the first derivative of f with respect to x, written $\mathrm{d}f/\mathrm{d}x$ or f', at $x = x_0$ is equal to the *slope* of the curve $y = f(x)$ at $x = x_0$. In other words, $f'(x_0)$ gives the gradient of a straight line – a *tangent* – which touches the curve $y = f(x)$ at the point $(x_0, f(x_0))$ (see Figure 7.1).

Physically, a derivative may represent a rate of change.

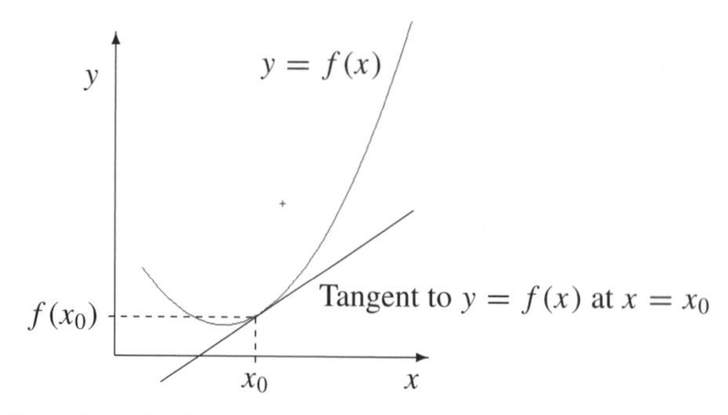

Figure 7.1 The first derivative as a tangent to a curve.

Example 7.1 A ball thrown vertically up in the air with an initial speed v_0 attains a height h at time t given by

$$h(t) = v_0 t - \tfrac{1}{2} g t^2$$

where g is the downward acceleration due to gravity. The instantaneous speed of the ball, v, equals the rate of change of height with respect to time

$$v(t) = \frac{\mathrm{d}h}{\mathrm{d}t} = v_0 - gt.$$

Visually $v(t)$ is the slope (first derivative) of the graph of $y = h(t)$ at time t. For $g = 9.8$ and $v_0 = 20$, Figure 7.2 shows the nature of the curves. When the ball is at its zenith, the speed is zero.

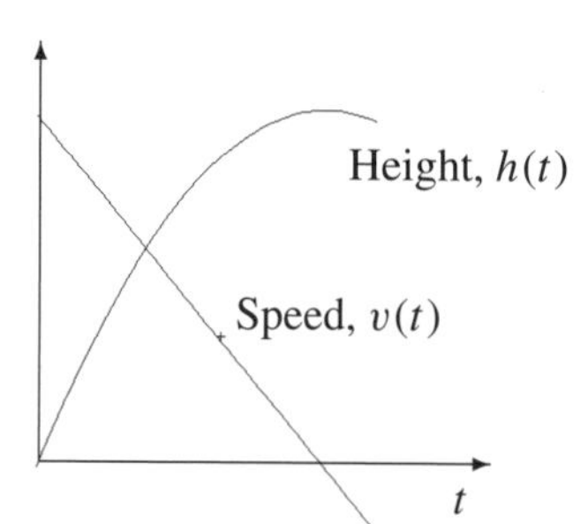

Figure 7.2 Illustrative behaviour of the vertical height h and speed v of a ball thrown in the air.

Many functions are difficult to differentiate *analytically* and require the application of numerical methods. Further, if a function is specified by a set of discrete values, such as observed measurements (see Table 7.1), then the option of using an analytical method is not available.

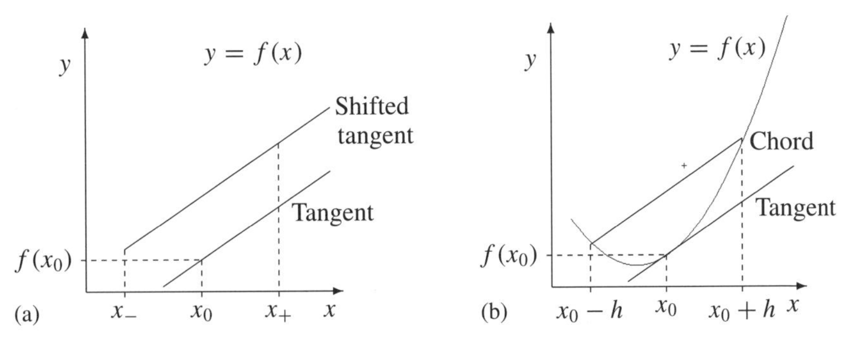

Figure 7.3 (a) Vertically shifted tangent. (b) Chordal approximation.

7.2 Motivation

The tangent line in Figure 7.1 is described by the equation

$$y_T(x) = f(x_0) + (x - x_0)f'(x_0). \tag{7.1}$$

Suppose that the line is shifted vertically by a distance Δy to give the straight line y_Δ defined by

$$\begin{aligned} y_\Delta &= y_T(x) + \Delta y \\ &= f(x_0) + \Delta y + (x - x_0)f'(x_0), \end{aligned} \tag{7.2}$$

(see Figure 7.3(a)). The *chord* y_Δ is parallel to the tangent y_T (the lines have the same slope) and cuts the graph of $y = f(x)$ at two points, say $x = x_-$ and $x = x_+$. The slope of the line y_Δ is

$$f'(x_0) = \frac{\mathrm{d}y_\Delta}{\mathrm{d}x} = y'_\Delta = \frac{f(x_+) - f(x_-)}{x_+ - x_-}.$$

This illustration supports the notion that the slope (first derivative) of a function can be represented by a combination of function values. Even if we 'tweak' the values of x_+ and x_- to be, for example, centred about $x = x_0$ (see Figure 7.3(b)), the slope of the chord y_C given by

$$y'_C = \frac{f(x_0 + \Delta x) - f(x_0 - \Delta x)}{2\Delta x} \tag{7.3}$$

is still likely to be 'close' to the slope of the tangent at $x = x_0$. The notion of 'closeness' is discussed with the subject of truncation error in Section 7.4.

7.3 Difference notation

Numerical methods for differentiation are based upon the use of two or more values of the function f tabulated on a set of nodes $\{x_i\}$ (see Figure 7.4). It is common to

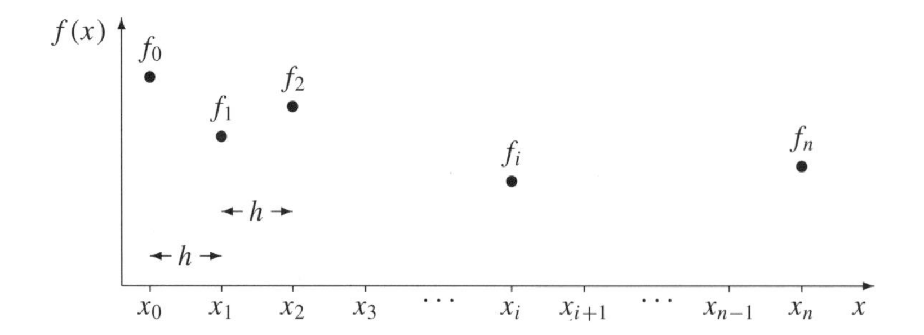

Figure 7.4 Equispaced nodes x_i and associated function values f_i.

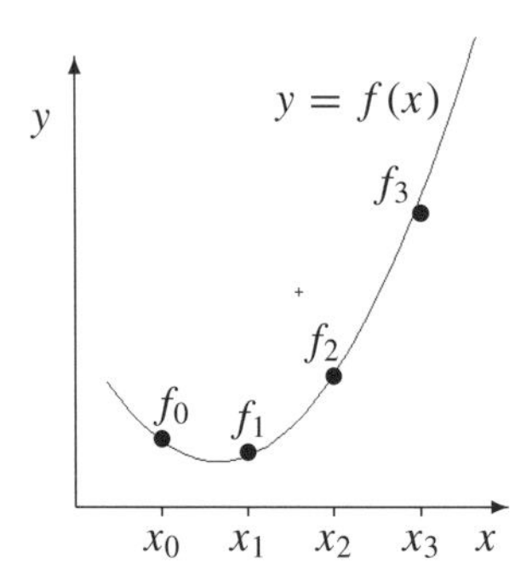

Figure 7.5 Function values f_i defined by the graph of $y = f(x)$.

use equispaced nodes $\ldots, x_{-1}, x_0, x_1, x_2, \ldots$, and given a *step size* h and an *origin* x_0, then $x_{-1} = x_0 - h$, $x_1 = x_0 + h$, $x_2 = x_1 + h = x_0 + 2h$, etc., and in general

$$x_i = x_0 + ih, \quad i = 0, \pm 1, \pm 2, \ldots \tag{7.4}$$

The f_i may be values of a function f taken from the graph of $y = f(x)$ (see Figure 7.5) or they may be measured experimental data[1].

7.3.1 Difference operators

Difference operators are used to manipulate a set of tabulated function values. A difference operator returns the difference of two (tabulated) function values. The three common difference operators[2] are defined in Table 7.2. A graphical representation is given in Figure 7.6.

There are several useful operator relations, such as $\Delta f_i = \nabla f_{i+1} = \delta f_{i+\frac{1}{2}}$ (see Exercise 7.1). Higher-order differences are defined recursively.

---

[1] The notation $f_i = f(x_i)$ means 'the value of the function f at $x = x_i$'.

[2] Further discussion of difference operators appears in Section 3.5.1.

Table 7.2 Common difference operators.

Operator name	Symbol	Definition
Forward difference	Δ	$\Delta f_i = f_{i+1} - f_i$
Backward difference	∇	$\nabla f_i = f_i - f_{i-1}$
Central difference	δ	$\delta f_i = f_{i+\frac{1}{2}} - f_{i-\frac{1}{2}}$

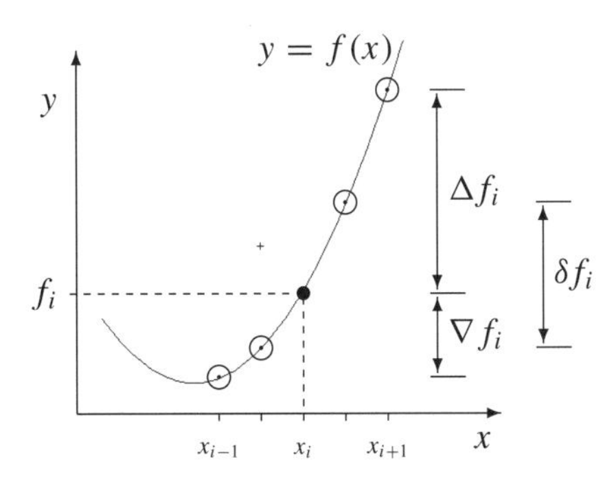

Figure 7.6 Graphical representation of (first-order) differences

Example 7.2

The second-order forward difference of f_i is

$$\begin{aligned}\Delta^2 f_i = \Delta(\Delta f_i) &= \Delta(f_{i+1} - f_i)\\ &= \Delta f_{i+1} - \Delta f_i\\ &= (f_{i+2} - f_{i+1}) - (f_{i+1} - f_i)\\ &= f_{i+2} - 2f_{i+1} - f_i.\end{aligned}$$

Similarly, for ∇ and δ, the second-order differences are

$$\begin{aligned}\nabla^2 f_i &= f_i - 2f_{i-1} + f_{i-2},\\ \delta^2 f_i &= f_{i+1} - 2f_i + f_{i-1}.\end{aligned}$$

The general recurrence formulae are

$$\Delta^k f_i = \Delta^{k-1}(\Delta f_i) = \Delta^{k-1} f_{i+1} - \Delta^{k-1} f_i, \tag{7.5}$$

$$\nabla^k f_i = \nabla^{k-1}(\nabla f_i) = \nabla^{k-1} f_i - \nabla^{k-1} f_{i-1}, \tag{7.6}$$

$$\delta^k f_i = \delta^{k-1}(\delta f_i) = \delta^{k-1} f_{i+\frac{1}{2}} - \delta^{k-1} f_{i-\frac{1}{2}}, \tag{7.7}$$

$k = 1, 2, \ldots$. Differences of order zero are defined as $\Delta^0 f_i = \nabla^0 f_i = \delta^0 f_i = f_i$.

7.4 Numerical differentiation

Given a table of values of x_i and associated function values f_i, application of the difference formulae of Section 7.3.1 can be used to generate estimates to the derivatives $f'(x), f''(x), \ldots$ at $x = x_0$, of varying degrees of accuracy.

Elementary calculus provides the standard result

$$f'(x_0) = \lim_{x_1 \to x_0} \frac{f(x_1) - f(x_0)}{x_1 - x_0} = \lim_{h \to 0} \frac{f_1 - f_0}{h} = \lim_{h \to 0} \frac{\Delta f_0}{h}. \quad \textbf{(7.8)}$$

This corresponds to passing a chord through the points (x_0, f_0) and (x_1, f_1), where $x_1 = x_0 + h$, and then refining the interval h (see Figure 7.7(a)).

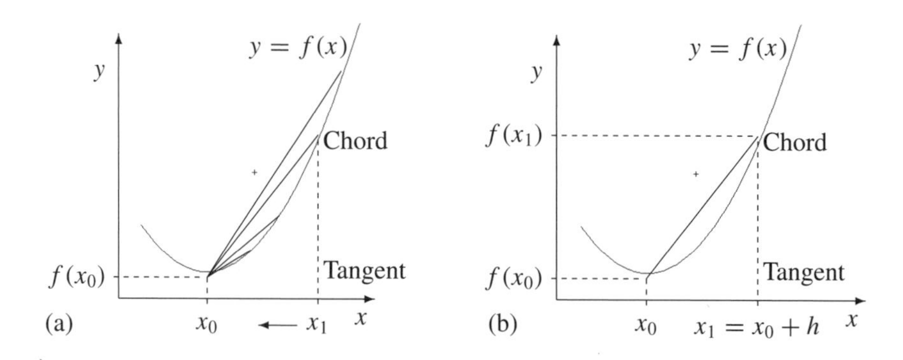

Figure 7.7 (a) The slope of a chord as the chord length approaches zero. (b) The first-derivative rule $(f_1 - f_0)/h$.

Example 7.3 Applying definition (7.8) to the function $f(x) = x^2$ gives

$$\begin{aligned}
\lim_{h \to 0} \frac{f(x+h) - f(x)}{h} &= \lim_{h \to 0} \frac{(x+h)^2 - x^2}{h} \\
&= \lim_{h \to 0} \frac{x^2 + 2xh + h^2 - x^2}{h} \\
&= \lim_{h \to 0} \frac{2xh + h^2}{h} \\
&= \lim_{h \to 0} 2x + h = \boxed{2x}
\end{aligned}$$

DERIVE Experiment 7.1

The limiting process illustrated in Example 7.3 may be implemented in *DERIVE* by way of the intrinsic function `lim`, specified by

```
lim(f,x,a)
  f   object function f(x)
  x   independent variable x
  a   limit point, x → a
 Author

              q(f,x):=(lim(f,x,x+h)-lim(f,x,x))/h
              h0(f,x):=lim(q(f,x),h,0)
```

q computes the quotient $(f(x+h) - f(x))/h$ and h0 takes the limit as $h \to 0$. To reproduce Example 7.3, Author and Simplify h0(x^2,x). Author h0(sinx,x) Simplify gives $\cos x$, etc.

On a computer the limiting process cannot be completed; the indeterminate quantity $\frac{0}{0}$ results. In practice a positive step size h is selected and $f'(x_0)$ is approximated by the formula (see Figure 7.7(b))

$$f'(x_0) \approx \frac{\Delta f_0}{h} = \frac{f_1 - f_0}{h}. \tag{7.9}$$

Example 7.4 The function $f(x) = e^x$ has first derivative $f'(x) = e^x$. At $x = 1$ the value of the derivative equals e (approximately 2.71828). Table 7.3 shows estimates to this value given by the differentiation rule (7.9) with successively refined values of the step size h.

Table 7.3 First-order first-derivative estimates to e^x at $x = 1$ using the formula $(f_1 - f_0)/h$.

Step h	Rule (7.9)	|Error|
0.1	2.858842	0.140560
0.05	2.787386	0.069104
0.025	2.752545	0.034263
0.0125	2.735342	0.017060

Evidently the error is a function of the step size h. Indeed, the error is monotonically decreasing with h, which indicates a *truncation error*.

DERIVE Experiment 7.2
The function

```
df11(f,x,x0,h):=(lim(f,x,x0+h)-lim(f,x,x0))/h
```

implements the first-derivative rule (7.9). Author `df11(expx,x,1,.1)` approX (working with Options Precision Digits set to 10) gives 2.858842, when rounded to 6 decimal places (see Table 7.3). To compute a sequence of derivative estimates using $m+1$ step sizes $h, h/2, \ldots, h/2^m$, Author

```
df11s(f,x,x0,h,m):=vector(df11(f,x,x0,h/2^i),i,0,m)
```

To reproduce Table 7.3, Author `df11s(expx,x,1,.1,3)` and approX the result.

To get a handle on the truncation error (see Section 2.5) we note that many differentiation formulae can be derived using Taylor's theorem. To obtain formula (7.9), $f(x_1)$ is expanded in a Taylor series about $x = x_0$

$$\begin{aligned} f(x_1) &= f(x_0+h) \\ &= f(x_0) + hf'(x_0) + \frac{h^2}{2} f''(\xi), \quad x_0 < \xi < x_1, \end{aligned}$$

where the remainder term in Taylor's theorem has been used (see Section 1.3). Rearranging the expansion gives the equation

$$\begin{aligned} f'(x_0) &= \frac{f(x_0+h) - f(x_0)}{h} - \frac{h}{2} f''(\xi) \\ &= \frac{f_1 - f_0}{h} - \frac{h}{2} f''(\xi), \quad x_0 < \xi < x_1, \end{aligned} \tag{7.10}$$

and using the formula $(f_1 - f_0)/h$ to estimate the derivative $f'(x_0)$ results in a truncation error $-\frac{1}{2}hf''(\xi)$. This result is supported by the behaviour of the error shown in Table 7.3. As h is halved so the error is approximately halved, that is 'Error' $\propto h$. Truncation error can be thought of as the 'closeness' of the numerical rule to the analytical derivative.

Rules of higher accuracy for estimating $f'(x_0)$ can be derived with truncation error terms by using Taylor's theorem.

Example 7.5 The difference of the expansions

$$\begin{aligned} f(x_0+h) &= f(x_0) + hf'(x_0) + \frac{h^2}{2} f''(x_0) + \frac{h^3}{6} f'''(\xi_+), \\ f(x_0-h) &= f(x_0) - hf'(x_0) + \frac{h^2}{2} f''(x_0) - \frac{h^3}{6} f'''(\xi_-), \end{aligned}$$

$x_0 < \xi_+ < x_1$ and $x_{-1} < \xi_- < x_0$, gives the equation[3]

$$f(x_0+h) - f(x_0-h) = 2hf'(x_0) + \frac{h^3}{6}\left[f'''(\xi_+) + f'''(\xi_-)\right].$$

Using the notation $f_i = f(x_0 + ih)$ and rearranging the formula gives the following differentiation rule and truncation error term

$$f'(x_0) = \frac{f_1 - f_{-1}}{2h} - \frac{h^2}{6}f'''(\xi), \quad x_{-1} < \xi < x_1. \tag{7.12}$$

This is a formal construction of the concept proposed by eqn (7.3).

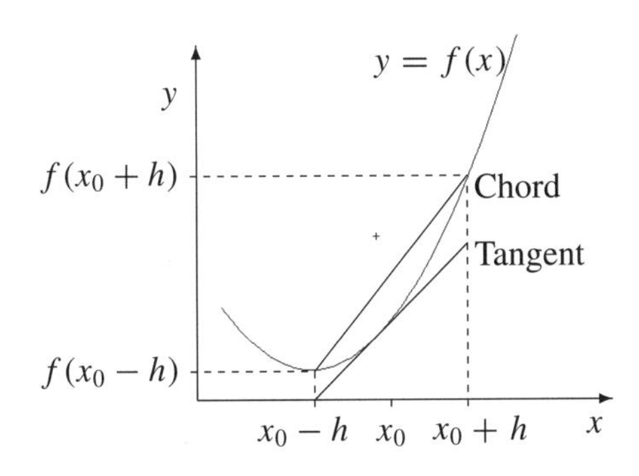

Figure 7.8 The first-derivative rule $(f_1 - f_{-1})/2h$.

Equation (7.12) states that the differentiation formula (see Figure 7.8)

$$f'(x_0) \approx \frac{f_1 - f_{-1}}{2h} \tag{7.13}$$

has a truncation error of $-\frac{h^2}{6}f'''(\xi)$.

Example 7.6

Repeating Example 7.5 with formula (7.13) yields the derivative approximations shown in Table 7.4. The error decreases monotonically with h, approximately quartering as h is halved. Thus, 'Error' $\propto h^2$ as predicted by the truncation error analysis

[3] The average z of n samples $z_1, z_2, \ldots, z_n$ taken from a certain population is

$$z = \frac{1}{n}[z_1 + \cdots + z_n] \Rightarrow nz = z_1 + \cdots + z_n. \tag{7.11}$$

For the differentiation rule the samples z_i are taken from the population of values of the third derivative of f, and $n = 2$. Hence, using eqn (7.11), $2f'''(\xi) = f'''(\xi_+) + f'''(\xi_-)$ where $f'''(\xi)$ represents an 'average' value of $f'''(x)$. More rigour is required to 'guarantee' the result and here we have simply given a flavour of the argument.

(7.12). Further, the magnitude of the error for each step size is smaller than the corresponding error for the rule (7.9) (see Table 7.3). That is, rule (7.13) is more accurate than rule (7.9). In describing the decreasing behaviour of the error we say that (7.9) is a first-order rule, 'Error' $\propto h$, and (7.13) is a second-order rule, 'Error' $\propto h^2$.

Table 7.4 Second-order first-derivative estimates to e^x at $x = 1$ using the formula $(f_1 - f_{-1})/2h$.

Step *h*	**Rule (7.13)**	**|Error|**
0.1	2.722815	0.004533
0.05	2.719415	0.001133
0.025	2.718565	0.000283
0.0125	2.718353	0.000071

The second-order rule (7.13) can be derived from the difference formulae discussed in Section 7.3.1. From elementary calculus

$$f'(x_0) = \lim_{h\to 0} \frac{f_1 - f_0}{h} = \lim_{h\to 0} \frac{\Delta f_0}{h}$$

and

$$f'(x_0) = \lim_{h\to 0} \frac{f_0 - f_{-1}}{h} = \lim_{h\to 0} \frac{\nabla f_0}{h}.$$

Adding these formulae leads to

$$2f'(x_0) \approx \frac{\Delta f_0}{h} + \frac{\nabla f_0}{h} \quad \Rightarrow \quad f'(x_0) \approx \frac{\Delta f_0 + \nabla f_0}{2h}.$$

The rule (7.13) is the average of two first-order rules whose leading errors cancel, giving rise to the improved accuracy of (7.13).

DERIVE Experiment 7.3

The function

```
df12(f,x,x0,h):=(lim(f,x,x0+h)-lim(f,x,x0-h))/(2h)
```

implements the first-derivative rule (7.13) and

```
df12s(f,x,x0,h,m):=vector(df12(f,x,x0,h/2^i),i,0,m)
```

computes a sequence of derivative estimates using $m + 1$ step sizes $h, h/2, \ldots, h/2^m$. Author `df12(expx,x,1,.1)` approX (working with Options Precision Digits set to 10) gives 2.722815 when rounded to six decimal places (see Table 7.4). To reproduce Table 7.4, Author `df12s(expx,x,1,.1,3)` and approX the result.

7.5 Higher-order derivatives

Numerical formulae for higher-order derivatives are not easily visualised and are most conveniently obtained by applying Taylor's theorem. The expansions

$$f(x_0 + h) = f(x_0) + hf'(x_0) + \frac{h^2}{2}f''(x_0) + \frac{h^3}{6}f'''(x_0) + \frac{h^4}{24}f''''(\xi_+)$$

and

$$f(x_0 - h) = f(x_0) - hf'(x_0) + \frac{h^2}{2}f''(x_0) - \frac{h^3}{6}f'''(x_0) + \frac{h^4}{24}f''''(\xi_-),$$

$x_{-1} < \xi_- < x_0 < \xi_+ < x_1$, can be added to give

$$f(x_0 + h) + f(x_0 - h) = 2f(x_0) + h^2 f''(x_0) + \frac{h^4}{24}[f''''(\xi_+) + f''''(\xi_-)].$$

Rearranging gives a second-order formula to estimate the second derivative (where $f_i = f(x_0 + ih)$)

$$\begin{aligned} f''(x_0) &= \frac{f_1 - 2f_0 + f_{-1}}{h^2} - \frac{h^2}{12}f''''(\xi) \\ &= \frac{\delta^2 f_0}{h^2} - \frac{h^2}{12}f''''(\xi), \quad x_{-1} < \xi < x_1. \end{aligned} \tag{7.14}$$

Example 7.7 For the function $f(x) = \mathrm{e}^x$, $f''(x) = \mathrm{e}^x$ which equals e at $x = 1$. Table 7.5 shows the result of applying rule (7.14) to estimate this derivative at $x = 1$.

Table 7.5 Second-order second-derivative estimates for e^x at $x = 1$ using the formula $\delta^2 f_0/h^2$.

Step h	Rule (7.14)	|Error|
0.1	2.720548	0.002266
0.05	2.718848	0.000566
0.025	2.718423	0.000141
0.0125	2.718317	0.000035

Again, the error monotonically *decreases* in a manner suggested by the truncation error $-\frac{h^2}{12}f''''(\xi)$.

DERIVE Experiment 7.4
The functions

```
df22(f,x,x0,h):=(lim(f,x,x0+h)-2lim(f,x,x0)+lim(f,x,x0-h))
   /h^2
df22s(f,x,x0,h,m):=vector(df22(f,x,x0,h/2^i),i,0,m)
```

implement the second-derivative rule (7.14) and compute a sequence of derivative estimates using $m+1$ step sizes $h, h/2, \ldots, h/2^m$. `Author df22(expx,x,1,.1)` `approX` (use `Options Precision Digits` set to 10) gives 2.720548 when rounded to six decimal places (see Table 7.5). To reproduce Table 7.5, `Author df22s(expx, x,1,.1,3)` and `approX` the result.

7.6 Total error for derivative rules

For the three differentiation rules, eqns (7.9), (7.13) and (7.14), the impression is that arbitrary precision can be obtained by simply refining the step size h. This hypothesis appears to be supported both by the numerical results presented in Tables 7.3, 7.4 and 7.5, and by the truncation error terms which are all proportional to positive powers of h (and thus decrease with h). It therefore appears, for example, that

$$\lim_{h\to 0}\left[f'(x_0) - \frac{f_1 - f_0}{h}\right] = 0,$$

The truncation error of eqn (7.10) is $-\frac{h}{2}f''(\xi)$ which certainly decreases as h gets smaller. Consequently, approximating $f'(x_0)$ by the first term in (7.10) we would expect the estimate to improve as $h \to 0$. However, this is not the case on a computer, as has already been suggested in Section 7.4.

Example 7.8 Table 7.3 is extended for estimates of $f'(x_0)$ using the first-order rule $(f_1 - f_0)/h$ with further reductions in step size h (see Table 7.6).

Below a certain value of h, the error *increases* due to *rounding errors* that arise because of finite machine precision. This is an example of a large relative error resulting in a loss of significant digits in the computed value, in this case $(f_1 - f_0)/h$. The rounding error in computing $f_1 - f_0$ is comparable to the size of $f_1 - f_0$, which is then divided by a small number h.

Table 7.6 First-order first-derivative estimates for e^x at $x = 1$ using the formula $(f_1 - f_0)/h$ and highly refined step sizes.

Step h	Rule (7.9)	|Error|
0.1	2.858842	0.140560
0.01	2.731914	0.013632
0.001	2.719826	0.001544
0.0001	2.719825	0.001544
0.00001	2.736505	0.018224
0.000001	2.569519	0.148763

Intuitively there exists an optimum value of h, say h_{opt}, that minimises the *total* effect of truncation and rounding errors. We analyse the rule

$$f'(x_0) \approx \frac{f_1 - f_0}{h}.$$

Truncation error: the maximum magnitude of the truncation error E_T is obtained from eqn (7.10):

$$E_T \le \max_{x_0 < x < x_1} \left| -\frac{h}{2} f''(x) \right| = \frac{h}{2} M_2$$

where $M_2 = \max |f''(x)|$, $x_0 < x < x_1$.

Rounding error: due to rounding on a finite-precision machine, the exact function values f_0 and f_1 are replaced by the values f_0^* and f_1^* where $|f_i - f_i^*| \le \frac{1}{2} \times 10^{-w}$ and w is the (decimal) *word length* of the machine (see Section 2.1.1). The maximum rounding error in the rule is

$$\begin{aligned} E_R &\le \max \left| \frac{f_1 - f_0}{h} - \frac{f_1^* - f_0^*}{h} \right| \\ &= \max \frac{1}{h} |(f_1 - f_1^*) - (f_0 - f_0^*)| \\ &\le \max\{|f_1 - f_1^*| + |f_0 - f_0^*|\} \\ &\le \frac{1}{h} \left[\frac{1}{2} \times 10^{-w} + \frac{1}{2} \times 10^{-w} \right] \\ &= \frac{10^{-w}}{h}. \end{aligned}$$

While the *truncation error decreases* with h, the *rounding error increases* (see Figure 7.9). The total error, $E(h)$, therefore has a *minimum* with respect to h. If

$$E(h) = E_T + E_R = \frac{h}{2} M_2 + \frac{10^{-w}}{h}$$

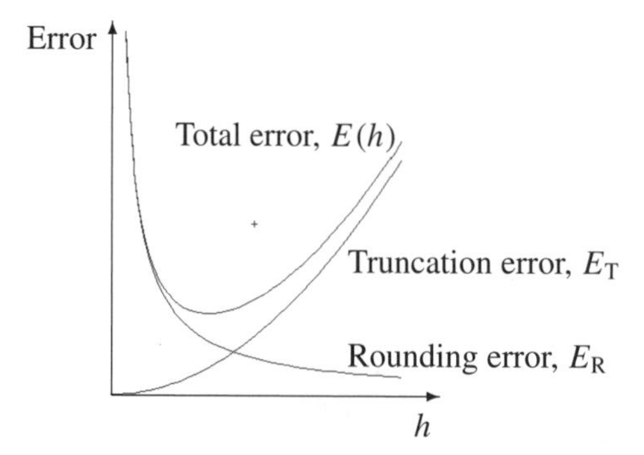

Figure 7.9 Truncation, rounding and combined errors for the first-order first-derivative rule $(f_1 - f_0)/h$.

then

$$\frac{\mathrm{d}E}{\mathrm{d}h} = \frac{M_2}{2} - \frac{10^{-w}}{h^2}.$$

A minimum of $E(h)$ satisfies the equation $\mathrm{d}E/\mathrm{d}h = 0$, that is

$$\frac{M_2}{2} - \frac{10^{-w}}{h^2} = 0,$$

from which

$$h^2 = \frac{2}{M_2} \times 10^{-w} \Rightarrow h = h_{\text{opt}} = \sqrt{\frac{2}{M_2} \times 10^{-w}}. \tag{7.15}$$

That this gives a *minimum* error is evident from the graph of $E(h)$ shown in Figure 7.9, and

$$E(h_{\text{opt}}) = \frac{M_2}{2}\sqrt{\frac{2}{M_2} \times 10^{-w}} + 10^{-w}\sqrt{\frac{M_2}{2} \times 10^{w}} = \sqrt{2M_2 \times 10^{-w}}. \tag{7.16}$$

Example 7.9 We relate eqns (7.15) and (7.16) to the behaviour seen in Table 7.6. A typical PC gives about seven decimal digits of accuracy, $w = 7$. For a small interval $[x_0, x_0 + h]$,

$$M_2 = \max |f''(x)| = \max_{x_0 < x < x_0 + h} |\mathrm{e}^x| \approx \mathrm{e}^{x_0}.$$

Since $x_0 = 1$, then

$$h_{\text{opt}} = \sqrt{\frac{2}{e} \times 10^{-7}} \simeq 0.000271, \quad E(h_{\text{opt}}) = \sqrt{2e \times 10^{-7}} \simeq 0.000737.$$

These values are in line with the numerical evidence presented in Table 7.6. No further accuracy can be extracted from the rule once h is reduced to a value of about 0.00027. If higher accuracy is required, a different rule must be chosen.

Note The minimum practical value of h for this example would be 0.0005.

Summary

The material in this chapter dealt with the estimation of derivatives of a function f defined by a set of function values, $\{f_i\}$ (the nature of differentiation was discussed in Section 7.1). A link between function values and differentiation was proposed in Section 7.2 and the difference operators required to develop the approximations were presented (see Section 7.3). Several first- and second-derivative rules were derived (see Sections 7.4 and 7.5), together with truncation error terms (by way of Taylor's theorem). The limitations of these rules were seen in Section 7.6 when rounding error was taken into account, and it was shown that an optimum step size exists which reduces the total error (truncation plus rounding) of each rule.

Exercises

7.1 Satisfy yourself that $\nabla f_{i+1} = \delta f_{i+\frac{1}{2}} = \Delta f_i$.

7.2 The telemetry data shown in the table below give the height h at various times t of a space rocket in upward vertical motion. Use equation (7.9) to complete the instantaneous vertical speeds v.

Time t (s)	0	4	8	12	16	20
Height h (km)	0	0.84	3.53	8.41	15.97	27.00
Speed v (km/s)						
Acceleration a (km/s^2)						

Apply the rule (7.9) to the speeds to obtain the instantaneous accelerations a. Approximately when does second-stage firing occur? Apply the rule (7.14) to the heights to directly obtain the accelerations. Comment on your answers, and prove it!

7.3 The following formula is proposed as an approximation to the first derivative of a tabulated function:

$$f_0' \approx \frac{-3f_0 + 4f_1 - f_2}{2h}.$$

Check the formula and obtain an expression for the truncation error by expansion (about x_0) in a Taylor series.

7.4 Show that the differentiation rule

$$f'(x_0) \approx a_0 f_0 + a_1 f_1 + a_2 f_2$$

is exact for all $f \in P_2$ if, and only if, it is exact for $f(x) = 1$, x and x^2, and find values of a_0, a_1 and a_2 so that the rule is exact for $f \in P_2$. [*Hint*: Let $x_0 = 0$, $x_1 = h$ and $x_2 = 2h$ without loss of generality.]

7.5 For the second-order rule (7.13) perform an error analysis to show that the optimum step size h_{opt} is given by

$$h = h_{\text{opt}} = \sqrt[3]{\frac{3}{2M_3}} \times 10^{-w},$$

where $M_3 = \max|f'''(x)|$.

7.6 Show that

$$f''(x_1) = \frac{f_2 - 2f_1 + f_0}{h^2} - \frac{h^2}{12} f''''(\xi), \quad \xi \in (x_0, x_2). \qquad \textbf{(7.17)}$$

If $|f''''(x)| \leq M_4$ and the f_i are in error by at most $\frac{1}{2} \times 10^{-w}$, show that a value of h close to

$$h = h_{\text{opt}} \sqrt[4]{\frac{24}{M_4}} \times 10^{-w} \qquad \textbf{(7.18)}$$

should minimise the combined truncation and rounding error.

7.7 For the data shown below $|f''''(x)| \leq 1$. Estimate $f''(0.5)$ using the rule (7.17) for all three possible choices of step size h. Given that $f(x) = \cos x$, investigate how the best value of h (that is, $h = 0.1$, 0.2 or 0.3) compares with the optimum value predicted using eqn (7.18).

x	0.2	0.3	0.4	0.5	0.6	0.7	0.8
$f(x)$	0.9801	0.9553	0.9211	0.8776	0.8253	0.7648	0.6967

8 Numerical integration

Have you ever thought about the effect of dropping an egg from the top of the Eiffel Tower onto the concrete base below? Read on!

Integration naturally follows *differentiation* in most introductory courses on calculus, often introduced as the *inverse* of differentiation. That is, a differentiation may be 'undone' by integrating, and vice versa. The topic of this chapter is concerned with obtaining *estimates* of the value of the *definite integral*

$$I = \int_a^b f(x)\,\mathrm{d}x, \tag{8.1}$$

where f is a function of one real variable x, and a and b define the *lower* and *upper limits* of the *domain of integration* on the x-axis. A formal evaluation of eqn (8.1) requires finding a *primitive* F which, when differentiated, gives the *original* function f, that is $\mathrm{d}F/\mathrm{d}x = f$. In this case

$$I = \int_a^b \frac{\mathrm{d}F}{\mathrm{d}x}\,\mathrm{d}x = F(b) - F(a). \tag{8.2}$$

Example 8.1 The primitive for $f(x) = x^2$ is $F(x) = \frac{x^3}{3}$. With $a = 1$ and $b = 2$

$$I = \int_1^2 f(x)\,\mathrm{d}x = F(2) - F(1) = \frac{2^3}{3} - \frac{1^3}{3} = \frac{7}{3}.$$

A major problem with integration is finding a primitive F, and for this reason the topic of *numerical integration* is required[1] in order to obtain *numerical estimates* to the value of I.

8.1 Preliminaries

A useful first step is to develop a visual appreciation of integration. It is often stated that the integral of a function f between limits $x = a$ and $x = b$ is equal to the area under the graph of $y = f(x)$ on the interval $a \leq x \leq b$.

Example 8.2 Consider the vertical motion of a mass falling under gravity g (in the absence of air resistance). The downward speed v of the mass is a function of time t, and for an initial speed u is given by $v(t) = u + gt$. The expression $u + gt$ is the integral of g with respect to t, where the *constant of integration* is determined from the initial speed, $v(0) = u$.

If $g = 9.8$ and $u = 0$, the speed at time $t = 10$ is $v(10) = 0+9.8\times 10 = 98$. Plotting acceleration vs. time (see Figure 8.1(a)) gives a horizontal line at a height $g = 9.8$. Dropping a perpendicular at $t = 10$ forms a rectangle of base 10 and height 9.8, with an area of 10×9.8, that is 98. The integral of the acceleration g, which gives the speed v, is equal to the area under the graph of $y = g$.

s, the distance travelled, is found by integrating the speed with respect to t, to give the formula $s(t) = ut + \frac{1}{2}gt^2$ (measured downwards from the mass's initial height h above the Earth's surface). At $t = 10$ the mass has travelled a distance $s(10) = \frac{1}{2} \times 9.8 \times 10^2 = 490$. A graph of speed vs. time (see Figure 8.1(b)) is a straight line, with slope 9.8, passing through the point (0, 0). Dropping a perpendicular at $t = 10$ forms a triangle with area equal to ('half-base times height') $\frac{1}{2} \times 10 \times 98$ (= 490). The integral of the speed (which gives the distance travelled) is the area under the graph of $y = v(t)$.

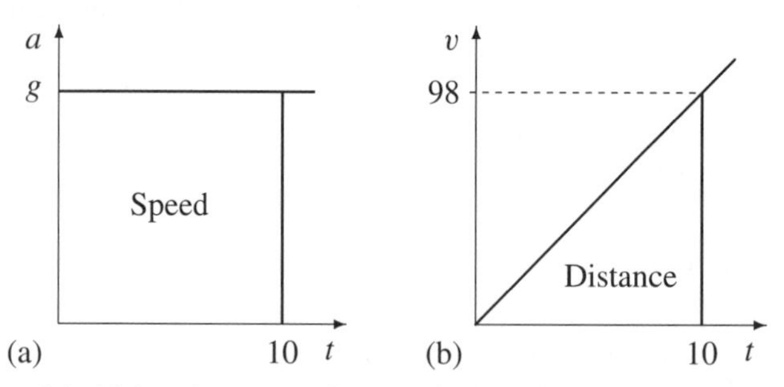

Figure 8.1 (a) Speed as area under a graph of acceleration vs. time. (b) Distance as area under a graph of speed vs. time.

[1] The term *quadrature* is often used in place of the phrase 'numerical integration'.

For the 'Eiffel Tower' egg, the initial height is 300 m. When $s = 300$, $t = \sqrt{300/4.9}$ and the speed of the egg, in m/s, is

$$v = 9.8\sqrt{\frac{300}{4.9}} \approx 76.7\,,$$

which is approximately 170 mph.

The value of the integral (8.1) equals the area under the graph of $y = f(x)$ (see Figure 8.2). *Numerical integration* can be thought of as the development of *numerical methods* for estimating the area of the shaded region.

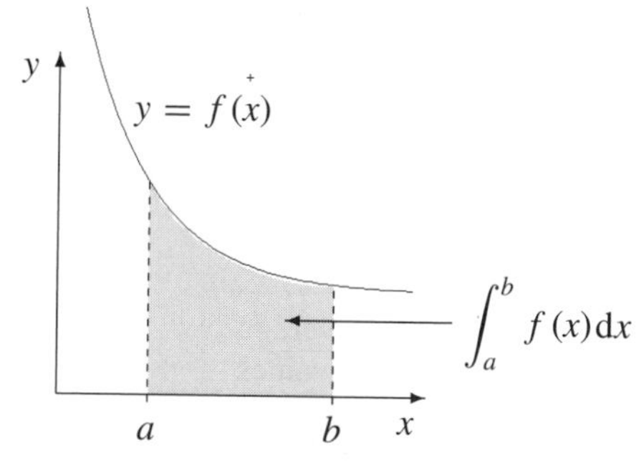

Figure 8.2 The definite integral as area under a curve.

8.1.1 Numerical integration

To estimate the area of a plane region you can draw the region on squared paper and count the number of complete squares enclosed by the region.

Example 8.3 The region in Figure 8.3 has a square mesh superimposed upon it. Seven complete squares are enclosed – the area of the region is approximately 7 squares. An improved estimate can be obtained by using smaller squares – *mesh refinement* – or by estimating the area of the partially enclosed squares around the perimeter of the region – *error estimation*.

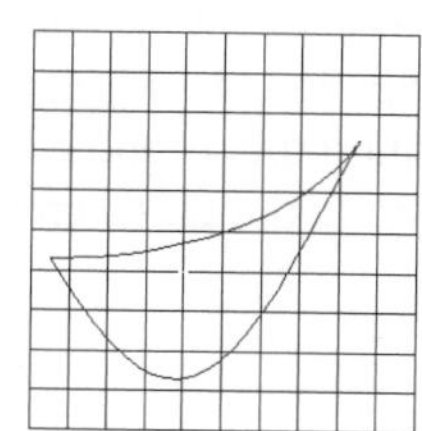

Figure 8.3 Estimating the area of a plane region.

'Square counting' forms the basis of the numerical methods developed here for estimating the integral (8.1). To assist implementation and analysis, a more formal framework is required.

Numerical integration uses a set of function values, $f_0, \ldots, f_n$, distributed in the integration domain $[a, b]$ at nodes $x_0, \ldots, x_n$. As we shall see, the outcome of 'square counting' is equivalent to

(a) replacing the *integrand* $f(x)$ in eqn (8.1) by a polynomial $p_n(x)$ passing through the $n + 1$ points (x_i, f_i), $i = 0, \ldots, n$, and then

(b) approximating the value of I by

$$R = \int_a^b p_n(x)\,\mathrm{d}x.$$

The result is a *numerical integration rule* R (or *quadrature rule*), together with a *truncation error* E,

$$I = \int_a^f f(x)\,\mathrm{d}x = \int_a^b p_n(x)\,\mathrm{d}x + E = \sum_{j=0}^{n} w_j f(x_j) + E = R + E. \qquad \textbf{(8.3)}$$

In other words

Exact value I	=	Weighted average of function values R	+	Truncation error E

For a *step size* $h = (b - a)/n$, the nodes are $x_j = x_0 + jh$, $j = 0, \ldots, n$ (where $x_0 = a$ and $x_n = b$). The *weights* $w_0, \ldots, w_n$ can be chosen so that the rule is *exact* if f is a polynomial of degree n. This aspect is discussed later in more detail and forms an important part of the analysis of *quadrature rules*. Rules based upon equispaced nodes are called *Newton–Côtes rules*.

ROGER CÔTES
(1682–1716)
English mathematician

LIFE

Côtes was a Fellow of Trinity College Cambridge in 1707. At the age of 26 he became the first Plumian Professor of Astronomy and Experimental Philosophy. From 1709 until 1713 much of Côtes's time was taken up editing the second edition of Newton's *Principia*. He conscientiously studied the work, gently but persistently arguing points with Newton.

WORK

Côtes only published one paper, *Logometria*, and was particularly pleased with his rectification of the logarithmic curve, as he makes clear in a letter to his friend William Jones in 1712. His work on logarithms led him to study the curve $r = a/q$ which he named the reciprocal spiral. Jones urged Côtes to publish his work in the *Phil. Trans. Roy. Soc.*, but Côtes resisted this, wishing to support Cambridge and publish with Cambridge University Press. His early death prevented publication.

Côtes discovered an important theorem on the *n*th roots of unity, anticipated the method of least squares and discovered a method of integrating rational fractions with binomial denominators. His substantial advances in the theory of logarithms, integral calculus and numerical methods (particularly interpolation and table construction) led Newton to say *'if he had lived we might have known something'*. Some of the work which Côtes hoped to publish with Cambridge University Press was published eventually by Thomas Simpson in *The Doctrine and Application of Fluxions* (1750).

8.2 Rectangle rules and Riemann sums

Consider a function f that is monotonically *increasing* on the interval $[a, b]$. For the region shown in Figure 8.4(b) two approximating rectangles are presented. The first, shown in Figure 8.4(a), uses the *left* function value, $f(a)$, and has an area equal to

$$R_L = [b - a]f(a). \tag{8.4}$$

R_L is the *left rectangle* estimate to I. The second rectangle, shown in Figure 8.4(c), uses the *right* function value, $f(b)$, and has an area equal to

$$R_R = [b - a]f(b). \tag{8.5}$$

R_R is the *right rectangle* estimate to I. Each rectangle estimates the shaded area I (the definite integral of f over the interval $[a, b]$). Geometrically R_L is an underestimate and R_R is an overestimate[2],

$$R_L \le I \le R_R. \tag{8.6}$$

R_L and R_R are *simple rectangle rules* and their accuracy is related to the interval size $b - a$ and to the behaviour of the function f over $[a, b]$.

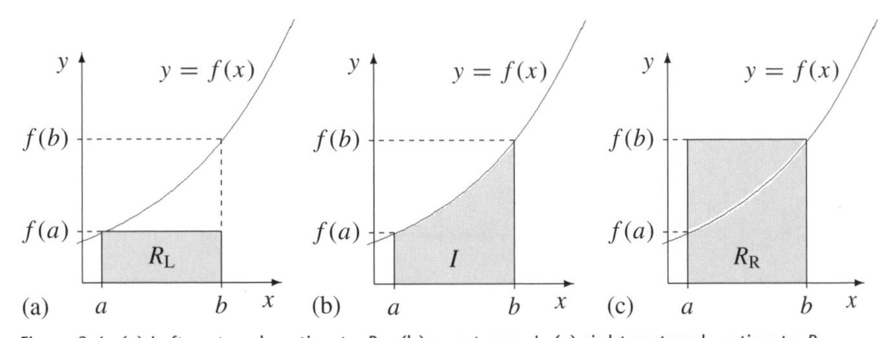

Figure 8.4 (a) Left rectangle estimate R_L, (b) exact area I, (c) right rectangle estimate R_R.

[2] If f is monotonically decreasing on $[a, b]$ then R_L provides an overestimate and R_R an underestimate, and $R_L \ge I \ge R_R$.

Example 8.4 illustrates the behaviour of R_L as b approaches a, for fixed a.

Example 8.4

For the function f defined by $f(x) = x^3$ on the interval $[1, b]$, Table 8.1 shows estimates provided by R_L for several values of b, with the observed difference between R_L and I (the primitive is $F(x) = x^4/4$). Clearly the area of the rectangle R_L is approaching (getting closer to) the area under $y = f(x)$, $1 \le x \le b$.

Table 8.1 Decreasing interval size for the simple rectangle rule R_L.

b	2	1.5	1.25	1.125	1.0625
I	3.75	1.0156	0.3604	0.1505	0.0686
R_L	1	0.5	0.25	0.125	0.0625
$\|I - R_L\|$	2.75	0.5156	0.1104	0.0255	0.0061

The accuracy of the rectangle rules improves as the interval size $b - a$ is reduced. However, the integration domain is not fixed, and each estimate in Table 8.1 approximates a different integral. It is essential that the domain $[a, b]$ remains fixed in any refinement process (see Section 8.2.2).

Example 8.5 illustrates the accuracy of the simple rectangle rule R_L as f approaches a constant form.

Example 8.5

The sequence of values in Table 8.2 shows the behaviour of R_L applied to x^3, x^2, x and 1. For each function the graph of $y = f(x)$ is monotonically increasing and passes through the point (1, 1). Consequently, the simple rectangle rule applied to each function over [1, 2] gives the same result, $R_L = 1$.

For the first three cases R_L underestimates the area below the curve $y = f(x)$, but as f becomes 'ever more constant' so the area under $y = f(x)$ approaches R_L. Once f is constant, the simple rectangle rule integrates f exactly.

Table 8.2 The simple R_L rule applied to the interval [1, 2] for $y = x^3$, $y = x^2$, $y = x$ and $y = 1$.

$f(x)$	x^3	x^2	x	1
I	3.75	2.1333	1.5	1
R_L	1	1	1	1
$\|I - R_L\|$	2.75	1.1333	0.5	0

The rule R_R exhibits a similar behaviour and it appears that the accuracy of the simple rectangle rules improves as f more closely resembles a constant function over the interval $[a, b]$. The questions to be answered are:

What is the error of R_L and R_R (see Section 8.2.1)?

How can a specified accuracy be achieved with R_L and R_R (see Section 8.2.2)?

8.2.1 Error analysis for rectangle rules

Taylor's theorem forms the basis of error analysis for numerical integration. For a function f and interval $[a, b]$ we consider the *truncation error* on using R_L or R_R to estimate I. All arithmetic is assumed to be exact and the error arises from representing the 'area under a curve' by a certain rectangle (see Tables 8.1 and 8.2).

For the left rectangle rule R_L, the truncation error E_L is (see Figure 8.5)

$$E_L = I - R_L = \int_a^b f(x)\,dx - (b-a)f(a). \tag{8.7}$$

If F is a primitive of f,

$$\begin{aligned} E_L &= F(b) - F(a) - (b-a)f(a) \\ &= F(a+h) - F(a) - hF'(a) \end{aligned}$$

where $h = b - a$. Applying Taylor's theorem to $F(a+h)$

$$\begin{aligned} E_L &= F(a) + hF'(a) + \frac{h^2}{2}F''(a) + \cdots - F(a) - hF'(a) \\ &= \frac{h^2}{2}F''(a) + \cdots \\ &= \frac{h^2}{2}F''(\xi), \quad a < \xi < b, \\ &= \frac{(b-a)^2}{2}f'(\xi), \quad a < \xi < b. \end{aligned} \tag{8.8}$$

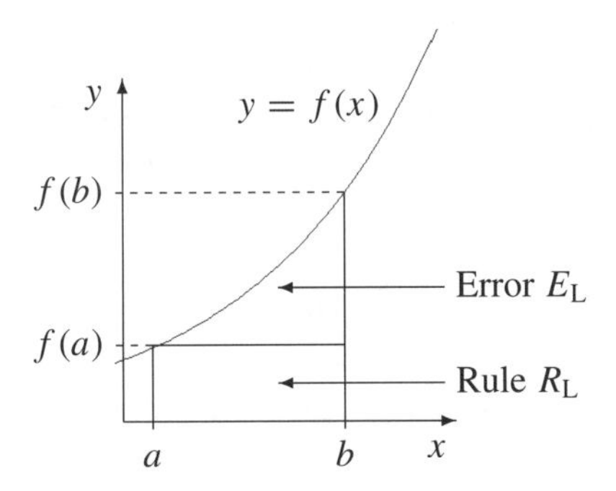

Figure 8.5 The rule R_L and its truncation error E_L.

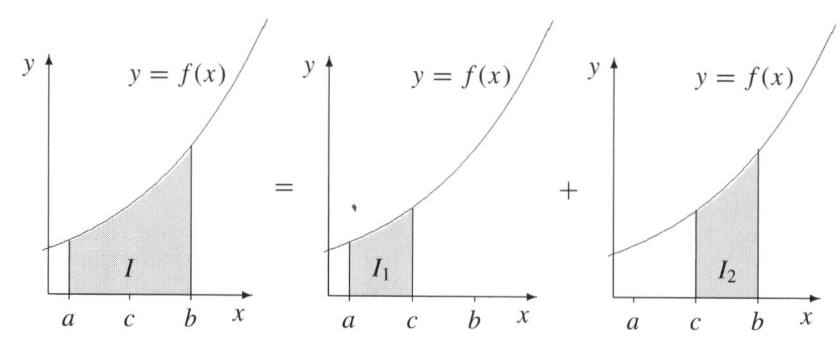

Figure 8.6 Decomposition of integration domain.

Equation (8.8) agrees with the intuition that the error reduces as the interval size h is reduced (see Table 8.1). If f is a *constant* function, the first derivative f' is identically zero (the error is zero) and R_L exactly integrates the function[3].

A further conclusion that might be drawn from eqn (8.8) is that halving the interval size h will quarter the error – that is, the rule is *second-order*. This observation appears to be supported by Table 8.1. However, reducing h requires more sub-intervals to span the interval $[a, b]$, which *must remain fixed*. This leads to *composite integration rules* and associated errors.

8.2.2 Composite rectangle rules

It has been observed that the smaller the interval $[a, b]$ the better the accuracy of R_L. The need is to 'refine the interval' while keeping the integration domain $[a, b]$ *fixed*.

If the interval $[a, b]$ is subdivided at an internal point $c \in (a, b)$ then

$$\int_a^b f(x)\,dx = \int_a^c f(x)\,dx + \int_c^b f(x)\,dx, \qquad \textbf{(8.9)}$$

that is $I = I_1 + I_2$ where I_1 and I_2 represent integrals over two spanning[4] sub-intervals of $[a, b]$ (see Figure 8.6).

Let $[a, b]$ be divided into N sub-intervals of equal size $h = (b - a)/N$. As an estimate to I the *left Riemann sum* is equal to the simple rectangle rule R_L applied to each sub-interval, described by a set of equispaced nodes $\{x_i\}$ where $x_i = a + ih$, $i = 0, \ldots, N$ (with $x_0 = a$ and $x_N = b$). Applied to the sub-interval $[x_{i-1}, x_i]$ the left rectangle estimate is hf_{i-1} (where $f_{i-1} = f(x_{i-1})$). Summing all N sub-estimates gives the left Riemann sum

$$R_{L_N} = hf_0 + hf_1 + \cdots + hf_{N-1} = h\sum_{i=0}^{N-1} f_i, \qquad \textbf{(8.10)}$$

also called the *composite left rectangle rule* (see Figure 8.7(a)).

---

[3] See Exercise 8.2 for the analysis relating to R_R.

[4] By 'spanning' we mean a pair of sub-intervals $[a, c]$ and $[c, b]$ such that $[a, c) \cap (c, b] = \emptyset$ and $[a, c] \cup [c, b] = [a, b]$.

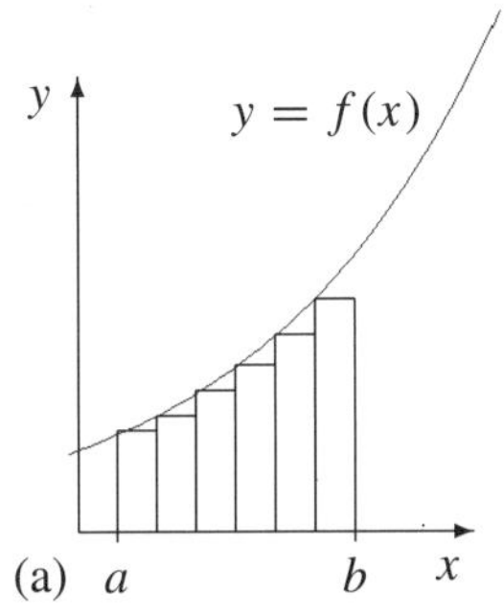

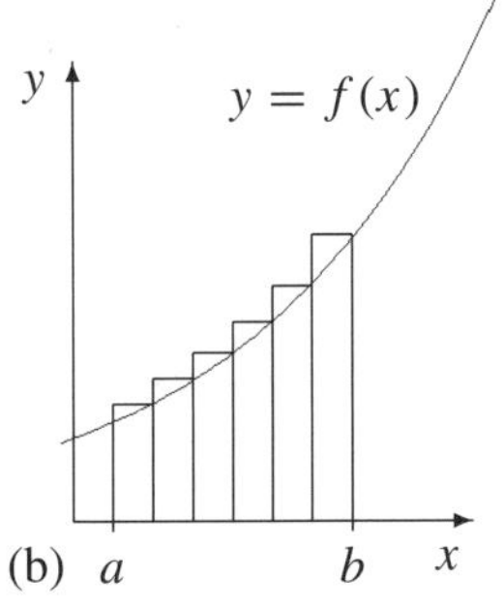

Figure 8.7 (a) Left Riemann sum, (b) right Riemann sum.

GEORG FRIEDRICH BERNHARD RIEMANN (1826–66)
German mathematician

LIFE

Riemann moved from Göttingen to Berlin in 1846 to study under Jacobi, Dirichlet and Eisenstein. In 1849 he returned to Göttingen and his PhD thesis, supervised by Gauss, was submitted in 1851. In his report on the thesis Gauss described Riemann as having a '*gloriously fertile originality*'. On Gauss's recommendation Riemann was appointed to a post in Göttingen. Gauss's chair at Göttingen was filled by Dirichlet (1855) and, after his death, by Riemann. Even at this time he was suffering from tuberculosis and he spent his last years in Italy in an attempt to improve his health.

WORK

Riemann's ideas concerning geometry of space had a profound effect on the development of modern theoretical physics and provided the concepts and methods used later in relativity theory. He was an original thinker, and a host of methods, theorems and concepts are named after him. Riemann's paper *Uber die Hypothesen welche der Geometrie zu Grunde liegen* (1854) became a classic of mathematics, and its results were incorporated into Einstein's relativistic theory of gravitation.

The Cauchy–Riemann equations (known before his time) and the concept of a Riemann surface appear in his doctoral thesis. He clarified the notion of integral by defining what we now call the Riemann integral. He is also famed for the still unsolved Riemann hypothesis.

It is instructive to see how the error in the composite rule (8.10) behaves as the sub-interval size h is reduced.

Example 8.6 The definite integral $\int_0^2 x^3\,\mathrm{d}x$ has the value 4. Table 8.3 shows several left Riemann sums using refined step sizes, h. Once h is sufficiently small, the error halves with each doubling of N (halving of h). This is 'at odds' with the error analysis (8.8) which implies that the error should quarter (see Section 8.2.3).

However, it is clear that mesh-halving (interval-doubling) does produce a convergent process (see Figure 8.8) whereby the area between the rectangular estimate and the graph of $y = f(x)$ steadily reduces.

Table 8.3 Convergence of R_{L_N}.

N	1	2	4	8	1	32	64	128
h	2	1	0.5	0.25	0.125	0.0625	0.0313	0.0156
R_{L_N}	0	1	2.25	3.0625	3.5156	3.7539	3.8760	3.9377
$\|I - R_{L_N}\|$	4	3	1.75	0.9375	0.4844	0.2461	0.1240	0.0623

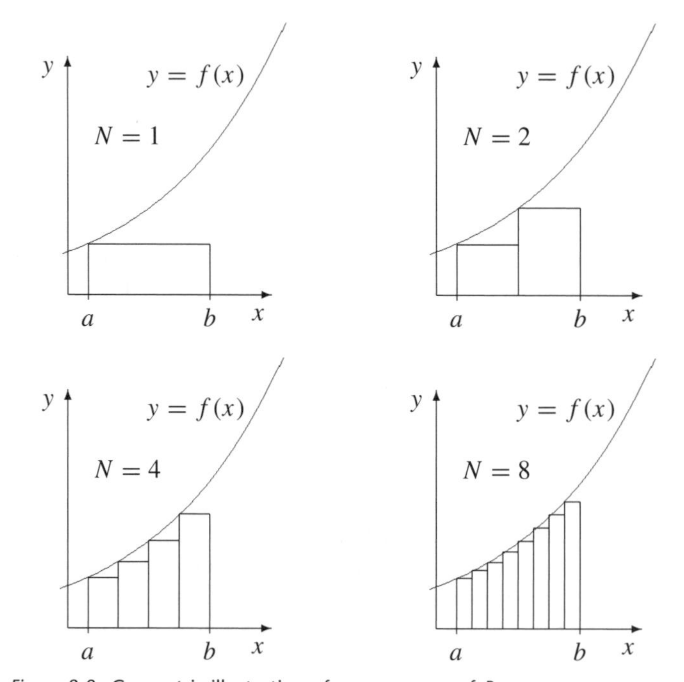

Figure 8.8 Geometric illustration of convergence of R_{L_N}.

DERIVE Experiment 8.1

To implement the left Riemann sum, Author the function

```
rln(f,a,b,h):=hsum(vector(f,x,a,b-h,h))
```

where f is the integrand $f(x)$, a and b are the lower/upper limits of integration, and h is the step size. f must be given in terms of the variable x. Author rln(x^3,0,2,.25) approX returns the value 3.0625 (see Table 8.3).

To generate a sequence of $m+1$ left Riemann sums with step sizes $h, h/2, \ldots, h/2^m$, Author

```
rls(f,a,b,h,m):=vector(rln(f,a,b,h/2^i),i,0,m)
```

m is the number of step size reductions. To reproduce the row R_{L_N} in Table 8.3, Author rls(x^3,0,2,2,7) and approX.

The corresponding functions for the right Riemann sum are

```
rrn(f,a,b,h):=hsum(vector(f,x,a+h,b,h))
rrs(f,a,b,h,m):=vector(rrn(f,a,b,h/2^i),i,0,m)
```

rln, rrn, rls and rrs can be used to experiment with various integrands, f, and intervals, $[a, b]$. To obtain the exact value of $\int_b^a f(x)\,dx$, use the intrinsic function int specified by

```
int(f,x,a,b)
f      integrand f(x)
x      independent variable
a, b   lower/upper limits of integration
```

and approX the result. Try a few integrands whose first derivative is undefined at one or more points in the domain $[a, b]$.

To visualise the left Riemann sum, Author the functions

```
zl(f,x,h):=[[x,0],[x,f],[x+h,f],[x+h,0]]
pl(f,a,b,h):=vector(zl(f,x,h),x,a,b-h,h)
```

To 'see' the R_{L_N} estimate to $\int_0^1 e^x dx$ using a step size of $h = 0.25$, Author expx and Plot Beside Plot Algebra then Author and approX the expression pl(expx,0,1,.25). Plot the result taking care to check that Options State is set to Connected and Small before pressing Plot for the second time[5].

The corresponding functions to visualise the right Riemann sum are

```
zr(f,x,h):=[[x-h,0],[x-h,f],[x,f],[x,0]]
pr(f,a,b,h):=vector(zr(f,x,h),x,a+h,b,h)
```

8.2.3 Composite error analysis

The error in the estimate R_{L_N} equals the sum of the N sub-interval errors,

$$E_{L_N} = \frac{h^2}{2} f'(\xi_1) + \frac{h^2}{2} f'(\xi_2) + \cdots + \frac{h^2}{2} f'(\xi_N),$$

[5] Setting Options Color Auto to No will stop the multicoloured plots!

where $x_{i-1} < \xi_i < x_i$. $\frac{h^2}{2}$ is a common factor and the above formula can be written

$$E_{\mathrm{L}_N} = \frac{h^2}{2}\sum_{i=1}^{N} f'(\xi_i), \quad x_{i-1} < \xi_i < x_i.$$

The sum represents N numbers drawn from the population of $f'(x)$, and can be written as $Nf'(\xi)$ where $f'(\xi)$ is the average of the N individual values of the first derivative. Thus

$$E_{\mathrm{L}_N} = \frac{h^2}{2}Nf'(\xi) = \frac{h}{2}(b-a)f'(\xi), \quad a < \xi < b. \tag{8.11}$$

For fixed $[a, b]$, as h is halved (the number of sub-intervals is doubled) so the truncation error is halved[6], that is $E_{\mathrm{L}_N} \propto h$ and the composite rule is *first order* (for sufficiently small h). This agrees with the values shown in Table 8.3. A similar result holds for the right Riemann sum.

In practice, the value of ξ is unknown and $f'(\xi)$ cannot be evaluated. For this reason E_{L_N} and E_{R_N} can only be used to determine *error bounds*,

$$|E_{\mathrm{L}_N}| = \left|\frac{h}{2}(b-a)f'(\xi)\right| \le \frac{h}{2}(b-a)\max_{a<x<b}|f'(x)|.$$

Example 8.7 The integral $\int_1^2 x^3\,\mathrm{d}x$ has the value 3.75. Table 8.4 shows a selection of estimates to I produced by the composite forms R_{L_N} and R_{R_N} with an indication of the truncation errors. In every case the observed error, $|I - R_{\mathrm{L}_N}|$ or $|I - R_{\mathrm{R}_N}|$, is less than the error bound.

Table 8.4 'Bracketing' convergence of R_{L_N} and R_{R_N}.

N	1	2	4	8	16	32
R_{L_N}	1	2.1875	2.9219	3.3242	3.5342	3.6414
$\lvert I - R_{\mathrm{L}_N}\rvert$	2.75	1.5625	0.8281	0.4258	0.2158	0.1086
$\lvert E_{\mathrm{L}_N}\rvert$	6	3	1.5	0.75	0.375	0.1875
R_{R_N}	8	5.6875	4.6719	4.1992	3.9717	3.8601
$\lvert I - R_{\mathrm{R}_N}\rvert$	4.25	1.9375	0.9219	0.4492	0.2217	0.1101
$\lvert E_{\mathrm{R}_N}\rvert$	6	3	1.5	0.75	0.375	0.1875

In Example 8.7 there is clear evidence of convergence of R_{L_N} (from below) and R_{R_N} (from above) towards I as N increases. If

$$I = \lim_{h\to 0} R_{\mathrm{L}_N} = \lim_{h\to 0} R_{\mathrm{R}_N}$$

[6] The sign of $f'(\xi)$ determines whether R_{L_N} is an over- or underestimate of I.

then f is said to be *Riemann integrable*. It is clear that N needs to be very large (h very small) in order that convergence is realised. Further, if f' is very large (the function f is 'steep') then the error term will not be very small. These difficulties are addressed in Section 8.3 onwards.

8.2.4 Rounding error

In Chapter 7 it was seen that numerical differentiation was highly sensitive to *rounding errors* in the data values f_i. The same is *not* true for numerical integration.

As an illustration, consider the left Riemann sum R_{L_N} using N sub-intervals of size h in which the data values f_i have an associated rounding error. In place of these exact values, the values f_i^* are used where $|f_i - f_i^*| < \varepsilon$. The left Riemann sum is then

$$R^*_{\mathrm{L}_N} = h \sum_{i=0}^{N-1} f_i^*. \tag{8.12}$$

The rounding error in the estimate $R^*_{\mathrm{L}_N}$ is

$$\begin{aligned} |\Delta R_{\mathrm{L}_N}| = \left|R_{\mathrm{L}_N} - R^*_{\mathrm{L}_N}\right| &= \left| h \sum_{i=0}^{N-1} f_i - h \sum_{i=0}^{N-1} f_i^* \right| \\ &\leq h \sum_{i=0}^{N-1} |f_i - f_i^*| \\ &\leq hN\varepsilon \\ &= (b-a)\varepsilon. \end{aligned} \tag{8.13}$$

The rounding error in $R^*_{\mathrm{L}_N}$ is of the same order as the rounding error in the data, and numerical integration is *not* sensitive to rounding errors in the data[7].

8.3 Trapezium rule

The remainder of this chapter focuses on refining the crude rectangle estimates to $\int_a^b f(x)\,\mathrm{d}x$.

The composite rectangle rules (Riemann sums) approximate the underlying function f by a piecewise constant function defined over a set of equispaced nodes $\{x_i\}$ on the interval $[a, b]$. On each sub-interval $[x_{i-1}, x_i]$ the approximant is based upon a single function value. No thought is given to continuity (apparent by the step-like appearance of R_{L_N} and R_{R_N} in Figure 8.7) – at each node x_i the two adjacent constant approximants have different values. Of course, the rules are exact for constant functions!

---

[7] This result holds for the general $(M+1)$-point quadrature rule $\sum w_i f_i$.

A simple improvement to R_L and R_R is to replace f by a straight line of arbitrary slope (a linear polynomial), which may be chosen to interpolate the function values at $x = a$ and $x = b$. In other words, the linear approximant matches the function f at the end points of the interval $[a, b]$ (see Figure 8.9). The area under the graph of $y = f(x)$ is now approximated by the area of a trapezium with base $b - a$ and sides of height $f(a)$ and $f(b)$,

$$T = \frac{(b-a)}{2}[f(a) + f(b)]. \tag{8.14}$$

T is the *simple trapezium rule* and uses two function values, whereas the simple rectangle rules R_L and R_R use just one. For the additional work is there any gain?

If f is a straight line then it is reasonable to expect T to integrate f exactly, that is T should be exact for constant functions, $f(x) \equiv 1$, and linear functions[8], $f(x) \equiv x$. This suggests an alternative method of derivation (see Section 8.3.4)

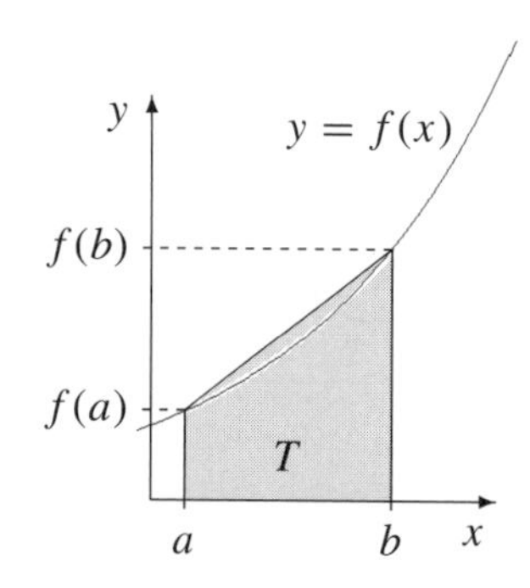

Figure 8.9 Simple trapezium approximation, T.

8.3.1 Error analysis

The analysis of errors is again based upon Taylor's theorem,

$$\begin{aligned} E_T = I - T &= \int_a^b f(x)\,dx - \frac{(b-a)}{2}[f(a) + f(b)] \\ &= F(b) - F(a) - \frac{(b-a)}{2}[F'(a) + F'(b)] \\ &= F(a+h) - F(a) - \frac{(b-a)}{2}[F'(a) + F'(a+h)], \end{aligned}$$

where F is a primitive ($F' = f$) and $h = b - a$. Then

$$\begin{aligned} E_T &= F(a) + hF'(a) + \frac{h^2}{2}F''(a) + \frac{h^3}{6}F'''(a) + \cdots - F(a) \\ &\quad -\frac{h}{2}F'(a) - \frac{h}{2}\left[F'(a) + hF''(a) + \frac{h^2}{2}F'''(a) + \cdots\right] \end{aligned}$$

---

[8] The monomials 1 and x are sufficient tests for 'exactness', since any constant a is simply $a \times 1$ and any linear term bx is simply $b \times x$.

$$= \left[\frac{1}{6} - \frac{1}{4}\right] h^3 F'''(a) + \cdots$$

$$= -\frac{h^3}{12} F'''(\xi), \quad a < \xi < b,$$

$$= -\frac{(b-a)^3}{12} f''(\xi), \quad a < \xi < b. \tag{8.15}$$

The error of the *simple* trapezium rule T decreases in a manner proportional to the cube of the interval size $b - a$ and is zero if f is a polynomial of degree at most 1 (since f'' is then identically zero). This is 'one better' than the rectangle rules, which integrate all constant polynomials exactly.

8.3.2 Composite trapezium rule

The improvement in accuracy of T over R_L and R_R is gained at little extra effort. To estimate I it is usual to apply a *composite* form of T, say T_N. Dividing $[a, b]$ into N sub-intervals of size $h = (b-a)/N$, defined by the nodes $x_i = x_0 + ih, i = 0, \ldots, N$ ($x_0 = a$, $x_N = b$), T_N is the sum of N applications of the simple trapezium rule on the sub-intervals $[x_0, x_1], \ldots, [x_{N-1}, x_N]$,

$$T_N = \frac{h}{2}[f(x_0) + f(x_1)] + \frac{h}{2}[f(x_1) + f(x_2)] + \cdots + \frac{h}{2}[f(x_{N-1}) + f(x_N)]$$

(see Figure 8.10). It is evident that each pair of adjacent sub-intervals share a value of $f(x)$, and so

$$T_N = \frac{h}{2}[f(x_0) + 2f(x_1) + \cdots + 2f(x_{N-1}) + f(x_N)]. \tag{8.16}$$

The composite trapezium rule (8.16) uses $N + 1$ function values which is just one more than the composite rectangle rules.

There is a graphical explanation of why T_N performs so well. For any function, no matter how 'twisty' its graph, any part of the curve can be well represented by a straight line if the sub-interval is 'sufficiently' small (see Figure 8.10).

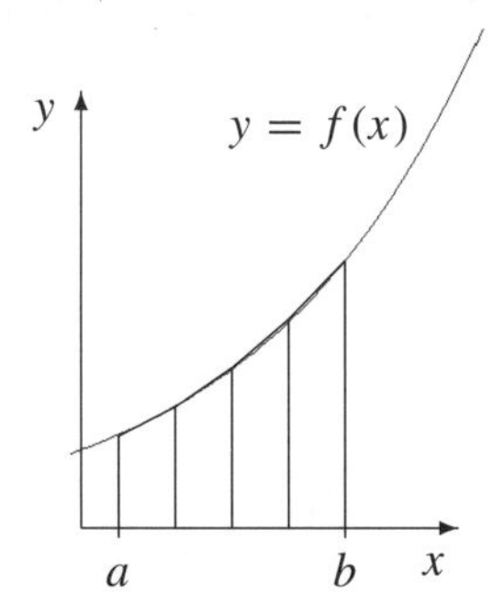

Figure 8.10 Composite trapezium rule, T_N.

Example 8.8 Consider again the integral $\int_0^2 x^3\,\mathrm{d}x$, with value 4. Table 8.5 shows a selection of T_N estimates with refined step size h.

Table 8.5 Convergence of T_N.

N	1	2	4	8	16	32	64	128
h	2	1	0.5	0.25	0.125	0.0625	0.0313	0.0156
T_N	8	5	4.25	4.0625	4.0156	4.0039	4.0010	4.0002
$\|I - T_N\|$	4	1	0.25	0.0625	0.0156	0.0039	0.00107	0.0002

The error quarters with each doubling of N (halving of h). This is 'at odds' with the error analysis (8.15) which implies that the error should reduce to one eighth.

DERIVE Experiment 8.2

Writing the composite trapezium rule as

$$T_n = \frac{h}{2}(f_0 + 2f_1 + 2f_2 + \cdots + 2f_{N-2} + 2f_{N-1} + f_N)$$

$$= h\left(\sum_{i=0}^{N} f_i - \frac{1}{2}(f_0 + f_N)\right)$$

suggests the following *DERIVE* function:

```
tn(f,a,b,h):=h(sum(vector(f,x,a,b,h))
   -sum(vector(f,x,a,b,b-a))/2)
```

Author `tn(x^3,0,2,.25)` approX gives 4.0625 (see Table 8.5). $m+1$ T_N-estimates with step sizes $h, h/2, \ldots, h/2^m$ are given by

```
ts(f,a,b,h,m):=vector(tn(f,a,b,h/2^i),i,0,m)
```

Table 8.5 may be reproduced by Author `ts(x^3,0,2,2,7)` approX.

The 'visuals' are defined by

```
z1(f,x,h):=[[x,0],[x,f],[0,0],[0,0]]
z2(f,x,h):=[[0,0],[0,0],[x,f],[x,0]]
pt(f,a,b,h):=vector(z1(f,x,h),x,a,b-h,h)
   +vector(z2(f,x,h),x,a+h,b,h)
```

To 'see' the T_N estimate to $\int_0^1 \mathrm{e}^x\,\mathrm{d}x$ using a step size of $h = 0.25$, Author `expx` and Plot Beside Plot Algebra then Author and approX the expression `pt(expx,k 0,1,.25)`. Plot Plot the result (see *DERIVE* Experiment 8.1 for further information).

8.3.3 Composite error analysis

The error of the estimate T_N equals the sum of N simple errors

$$E_{T_N} = -\frac{h^3}{12}f''(\xi_1) - \cdots - \frac{h^3}{12}f''(\xi_N).$$

$-\frac{h^3}{12}$ is a common factor and the above formula can be written

$$E_{T_N} = -\frac{h^3}{12}\sum_{i=1}^{N} f''(\xi_i), \quad x_{i-1} < \xi_i < x_i,$$

where $x_i = a + ih$. The term $\sum$ is the sum of N numbers drawn from the population of $f''(x)$ and can be written as $Nf''(\xi)$ where $f''(\xi)$ is the average of the N individual values of the second derivative. Thus

$$E_{T_N} = -\frac{h^3}{12}Nf''(\xi) = -\frac{h^2}{12}(b-a)f''(\xi), \quad a < \xi < b. \tag{8.17}$$

For fixed $[a, b]$, as h is halved (the number of sub-intervals is doubled) so the truncation error is quartered, that is $E_{T_N} \propto h^2$, and the composite trapezium rule is said to be *second order*. This is in agreement with the results displayed in Table 8.5.

In common with the rectangle rules, the analytic error formula (8.17) is impractical because ξ, and hence $f''(\xi)$, is unknown. Consequently, (8.17) can only be used to give an error bound,

$$\begin{aligned} |E_{T_N}| &\le \frac{h^2}{12}(b-a)\max_{a<x<b}|f''(x)| \\ &= \frac{h^2}{12}(b-a)M_2. \end{aligned} \tag{8.18}$$

From this formula we can conclude that T_N is exact for all linear polynomials, since the second derivative of a linear polynomial is identically zero – no error! Further, the rule is $O(h^2)$ since halving h quarters the error.

Example 8.9 Repeating the problem of evaluating the integral $I = \int_0^2 x^3\,\mathrm{d}x = 4$, Table 8.6 shows a selection of estimates to I produced by the composite form T_N with an indication of the truncation errors.

Table 8.6 Convergence of T_N.

N	1	2	4	8	16	32
T_N	4.5	3.9375	3.7969	3.7617	3.7529	3.7507
$\|I - T_N\|$	0.75	0.1875	0.0469	0.0117	0.0029	0.0007
$\|E_{T_N}\|$	1	0.25	0.0625	0.0156	0.0039	0.0010

Convergence is second order, and in each case the upper bound of $|E_{T_N}|$ given by eqn (8.18) is greater than the observed error $|I - T_N|$.

The bound (8.18) can be used to find the error in T_N given h or to find a step size h to ensure that the error in T_N is less than a specified value ε.

Example 8.10 It is required to evaluate the definite integral $\int_1^2 \frac{e^{-x}}{x}\,dx$. With $h = 0.005$ ($N = 200$), the composite trapezium rule gives the estimate $T_{200} = 0.17048475$. How many correct decimal places can be claimed?

M_2 in eqn (8.18) is required for $a = 1, b = 2$. We have

$$f(x) = \frac{e^{-x}}{x}, \quad f'(x) = -\frac{e^{-x}}{x}\left(1 + \frac{1}{x}\right), \quad f''(x) = \frac{e^{-x}}{x}\left(1 + \frac{2}{x} + \frac{2}{x^2}\right).$$

$f''(x)$ decreases monotonically with x, and $M_2 = |f''(1)| = \frac{5}{e}$. Hence

$$|E_{T_{200}}| \le \frac{(0.005)^2}{12} \times 1 \times \frac{5}{e} \simeq 3.83 \times 10^{-6}.$$

Five decimal places can be claimed.

To evaluate the integral to an accuracy of three decimal places, into how many sub-intervals should [1, 2] be subdivided? We require

$$|E_{T_N}| \le \frac{1}{2} \times 10^{-3} \Rightarrow \frac{h^2}{12} \times 1 \times \frac{5}{e} \le 5 \times 10^{-4}$$

from which

$$h^2 \simeq 12e \times 10^{-4} \Rightarrow h \simeq 0.057.$$

To achieve the specified accuracy (tolerance), a step size of $h = 0.05$ is appropriate. This translates to $N = 20$ sub-intervals – remember that N must be an *integer*.

In practice it may be the case that only the $N + 1$ function values $f_0, \ldots, f_N$ are available – in other words the second derivative $f''(x)$ needs to be estimated numerically. This can be done using the numerical differentiation techniques developed in Chapter 7.

8.3.4 Undetermined coefficients

We conclude this section with an alternative development of quadrature rules and their accuracy.

The rectangle rules are exact for $f \in P_0$ and the trapezium rule is exact for $f \in P_1$. This leads to the idea that integration rules may be constructed by forcing them to be exact for certain classes of polynomials.

Without loss of generality, consider the interval[9] $[0, h]$ which serves to simplify the algebra. The quadrature rules developed thus far estimate the value of $I = \int_0^h f(x)\,dx$ by means of a *weighted sum* of function values,

$$I = \int_0^h f(x)\,dx \approx R = \sum_{j=0}^{N} w_j f(x_j) \qquad \textbf{(8.19)}$$

where the *weights* w_j are to be determined and the nodes x_j are equispaced on $[0, h]$, including the end-points $x = x_0 = 0$ and $x = x_N = h$. For the simple rules discussed to date

		Nodes		Weights	
Rule	N	x_0	x_1	w_0	w_1
Left rectangle R_L	0	0		h	
Right rectangle R_R	0	h		h	
Trapezium T	1	0	h	$h/2$	$h/2$

The simplest integration rule of the form (8.19) on $[0, h]$ is

$$\int_0^h f(x)\,dx \approx w_0 f(x_0)$$

and should integrate *constant* functions exactly. The simplest constant function is $f(x) \equiv 1$ and the rule is defined by the weight w_0 such that

$$\int_0^h 1\,dx = w_0 \times 1 \quad \Rightarrow \quad h = w_0.$$

This is precisely the weight for R_L and R_R.

If the rule is to integrate constants, $f(x) \equiv 1$, *and* linear terms, $f(x) \equiv x$, two *undetermined coefficients* w_0 and w_1 are introduced,

$$\int_0^h f(x)\,dx \approx w_0 f(x_0) + w_1 f(x_1),$$

with $x_0 = 0$ and $x_1 = h$. The weights w_0 and w_1 are found so that the rule $w_0 f(0) + w_1 f(h)$ integrates 1 and x *exactly*.

[9] The interval $x \in [0, h]$ can be transformed to the interval $z \in [a, b]$ using the linear transformation $z = a + (b - a)x/h$.

Integrand $f(x)$	Integral $\int_0^h f(x)\,dx$		Rule $w_0 f(0) + w_1 f(h)$		
1	h	=	w_0	+	w_1
x	$h^2/2$	=			hw_1

The 'integrability' conditions result in two linear algebraic equations for w_0 and w_1. The solution is $w_0 = w_1 = h/2$, which are the weights for the trapezium rule T.

The rectangle and trapezium rules are examples of *closed Newton–Côtes* quadrature rules, closed in that the integrand f is sampled at the *end-points* of the integration domain.

8.4 Simpson's rule

The 'closed' rule discussed here takes the process of Sections 8.2 and 8.3 one step further. That is, is it possible to 'design' an integration rule that integrates the monomials 1, x and x^2 exactly ? To generate the simple rule consider the interval[10] $[-h, h]$ with three nodes at $x = -h$, $x = 0$ and $x = h$. The three weights w_0, w_1 and w_2 are required such that the rule $w_0 f(-h) + w_1 f(0) + w_2 f(h)$ integrates 1, x and x^2 exactly.

Integrand	Integral		Rule				
$f(x) \equiv 1$	$2h$	=	w_0	+	w_1	+	w_2
$f(x) \equiv x$	0	=	$-hw_0$	+		+	hw_2
$f(x) \equiv x^2$	$2h^3/3$	=	$h^2 w_0$	+		+	$h^2 w_2$

This represents three equations in three unknowns, w_0, w_1 and w_2, with solution $w_0 = w_2 = h/3$, $w_1 = 4h/3$. The resulting rule is

$$I \approx S = \frac{h}{3}[f(-h) + 4f(0) + f(h)], \tag{8.20}$$

called *Simpson's rule*. Applied to the general interval $[a, b]$ with three equispaced nodes x_0, x_1 and x_2, where $h = (b - a)/2$,

$$I \approx S = \frac{h}{3}[f(x_0) + 4f(x_1) + f(x_2)]. \tag{8.21}$$

[10] The interval $x \in [-h, h]$ is chosen to simplify the algebra and may be transformed to the interval $z \in [a, b]$ using the change of variable $z = [b + a + (b - a)x/h]/2$.

THOMAS SIMPSON
(1710–61)
English mathematician

LIFE

His first job was as a weaver. At this time he taught mathematics privately and from 1737 he began to write texts on mathematics. Simpson was the most distinguished of a group of itinerant lecturers who taught in the London coffee houses. In 1754 he became editor of the *Ladies Diary*. The following description of Simpson by Charles Hutton (made 35 years after Simpson's death) is interesting: *'It has been said that Mr Simpson frequented low company, with whom he used to guzzle porter and gin: but it must be observed that the misconduct of his family put it out of his power to keep the company of gentlemen, as well as to procure better liquor'.* It would be fair to note that others described Simpson's conduct as 'irreproachable'.

WORK

Simpson is best remembered for his work on interpolation and numerical methods of integration. He also worked on probability theory and published *The Nature and Laws of Chance* (1740). Much of Simpson's work in this area was based on earlier work of De Moivre. Simpson also worked on the problem of the minimum length of path that must be used to join three points. He worked on the theory of errors and aimed to prove that the arithmetic mean was better than a single observation. Simpson published the two-volume work *The Doctrine and Application of Fluxions* in 1750. It contains work of Côtes.

It is not difficult to show that Simpson's rule exactly integrates all polynomials of degree at most 2. Let p_2 be a quadratic polynomial

$$p_2(x) = A + Bx + Cx^2$$

where A, B and C are constants. In other words, a quadratic polynomial can be written as a linear combination of the constant, linear and quadratic monomials 1, x and x^2. Further,

$$\int_a^b p_2(x)\,\mathrm{d}x = \int_a^b (A + Bx + Cx^2)\,\mathrm{d}x = A\int_a^b \mathrm{d}x + B\int_a^b x\,\mathrm{d}x + C\int_a^b x^2\,\mathrm{d}x.$$

Since 1, x and x^2 can be integrated exactly by Simpson's rule then $p_2(x)$ is integrated exactly. This analysis can also be applied to the rectangle and trapezium rules.

8.4.1 Error analysis

The error analysis again uses Taylor's theorem. Using eqn (8.20), the truncation error is

$$\begin{aligned} E_S &= F(h) - F(-h) - \frac{h}{3}[F'(-h) + 4F'(0) + F'(h)] \\ &= -\frac{h^5}{90}F^{(5)}(0) + \cdots \\ &= -\frac{h^5}{90}F^{(5)}(\xi), \quad -h < \xi < h, \\ &= -\frac{h^5}{90}f^{(4)}(\xi), \quad -h < \xi < h. \end{aligned} \tag{8.22}$$

Simpson's rule is exact for cubics (in addition to quadratics) since the fourth derivative of a cubic is identically zero.

8.4.2 Composite rule and error analysis

Simpson's composite rule requires that the integration domain $[a, b]$ is divided into an *even* number of sub-intervals, say $2N$, since each application of the simple rule requires two sub-intervals. Defining the sub-intervals by the node set $x_0, x_1, x_2, \ldots, x_{2N-2}, x_{2N-1}, x_{2N}$, the composite rule is

$$\begin{aligned} S_{2N} &= \frac{h}{3}[f(x_0) + 4f(x_1) + 2f(x_2) + 4f(x_3) + \cdots \\ &\quad + 2f(x_{2N-2}) + 4f(x_{2N-1}) + f(x_{2N})]. \end{aligned} \tag{8.23}$$

The composite error is, again, the sum of N 'simple' errors,

$$|E_{S_{2N}}| = N\frac{h^5}{90}\left|f^{(4)}(\xi)\right| = \frac{h^4}{180}(b-a)\left|f^{(4)}(\xi)\right| \tag{8.24}$$

where $2Nh = b - a$. A practical error bound is given by

$$|E_{S_{2N}}| \le \frac{h^4}{180}(b-a)M_4 \tag{8.25}$$

where $M_4 = \max|f^{(4)}(x)|$, $a < x < b$. If f is given discretely then $f^{(4)}(x)$ cannot be evaluated and for M_4 we take

$$M_4 = \max\left|\frac{\Delta^4 f}{h^4}\right| \tag{8.26}$$

where Δ^4 denotes *fourth differences*. This approximation to M_4 should also be used if $f^{(4)}(x)$ is difficult/tedious to evaluate.

Example 8.11 For the integral $I = \int_0^1 e^{\sin x}\, dx$ the estimate $S_{10} = 1.63187$ is obtained. How many correct decimal places can be claimed ? Differentiating $f(x) = e^{\sin x}$ four times gives

$$f^{(4)}(x) = e^{\sin x}\{3(\sin x - \cos^2 x)^2 - 2(\cos^2 x + 1)^2 - 2\}.$$

$f^{(4)}(x)$ has a maximum magnitude of about 6.4 on the interval [0, 1]. Here the fourth derivative, and hence M_4, are not easy to evaluate despite the 'simple' appearance of $f(x)$. Using fourth-order differences (entries from column Δf onwards are in terms of the last digit in the $f(x)$ column), inspecting the final column, $\max|\Delta^4 f| = 60 \times 10^{-5} = 6 \times 10^{-4}$. Hence, from (8.26), $M_4 \approx 6 \times 10^{-4}/(0.1)^4 = 6$ (reasonable agreement with the analytical value). The error bound (8.25) is, approximately, 3.3×10^{-6} so that all five decimal places in S_{10} are correct (the exact value of I is 1.6318696084...).

A comparison of the error bound and the observed error is given overleaf. The error reduces by a factor of approximately 1/16 each time the step size is halved – very rapid convergence – due to the factor h^4 in the error formula (8.24).

x	$f(x)$	Δf	$\Delta^2 f$	$\Delta^3 f$	$\Delta^4 f$
0.0	1.00000				
		10499			
0.1	1.10499		980		
		11479		−54	
0.2	1.21978		926		−48
		12405		−102	
0.3	1.34383		824		−48
		13229		−150	
0.4	1.47612		674		−60
		13903		−210	
0.5	1.61515		464		−53
		14367		−263	
0.6	1.75882		201		−52
		14568		−315	
0.7	1.90450		−114		−52
		14451		−367	
0.8	2.04901		−481		−21
		13973		−388	
0.9	2.18874		−869		
		13104			
1.0	2.31978				

h	$2N$	$\lvert I - S_{2N}\rvert$	$\max\lvert E_{S_{2N}}\rvert$
0.1	10	2.3×10^{-6}	3.6×10^{-6}
0.05	20	1.4×10^{-7}	2.2×10^{-7}
0.025	40	9.0×10^{-9}	1.4×10^{-8}
0.01	100	2.0×10^{-10}	3.6×10^{-10}
0.005	200	0.5×10^{-11}	2.2×10^{-11}

DERIVE Experiment 8.3

S_{2N} may be implemented by `Author`

```
s2n(f,a,b,h):=h/3(sum(vector(f,x,a,b,b-a))
   +4sum(vector(f,x,a+h,b-h,2h))
   +2sum(vector(f,x,a+2h,b-2h,2h)))
```

`Author s2n(x^3,0,2,.25) approX` gives 4 – the exact value, why? `Author s2n(x^4,0,2,.25) approX` gives 6.40104 (close to the exact value 6). $m+1$ estimates with step sizes $h, h/2, \ldots, h/2^m$ are given by

```
ss(f,a,b,h,m):=vector(s2n(f,a,b,h/2^i),i,0,m)
```

'Visuals' are hardly worth the effort since the piecewise quadratic interpolation provided by S_{2N} is virtually indistinguishable from the integrand (to screen graphics resolution).

8.5 Mid-point rule

To illustrate the benefit of carefully choosing the sampling points on the interval $[a, b]$, consider again the simple left and right rectangle rules R_{L} and R_{R} – one rule underestimates I and the other overestimates I. Suppose that an intermediate value of x is used in the interval (a, b), say $x = c$ (see Figure 8.11). It is clear that the area under f is overestimated on the sub-interval $[a, c]$ and is underestimated on the sub-interval $[c, b]$. The over- and underestimates will cancel each other (to some extent).

A natural 'guess' for c is to use the mid-point of the interval $[a, b]$. Without loss of generality let the interval $[a, b]$ be $[0, h]$. The area of the rectangle based upon the mid-point $x = h/2$ gives the *mid-point rule*

$$I \approx M = hf\left(\frac{h}{2}\right). \tag{8.27}$$

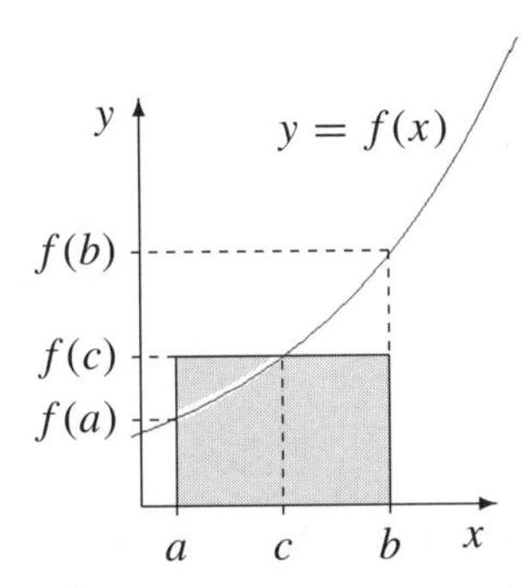

Figure 8.11 Simple mid-point rule, M.

The mid-point rule is an *open Newton–Côtes* rule in the sense that it only samples the function f at *internal* nodes in the interval $[a, b]$.

8.5.1 Error analysis

To appreciate the significance of eqn (8.27) we again resort to an error analysis using primitives and Taylor's theorem. The truncation error is

$$
\begin{aligned}
E_{\mathrm{M}} &= F(h) - F(0) - hF'(h/2) \\
&= F(0) + hF'(0) + \frac{h^2}{2}F''(0) + \frac{h^3}{6}F'''(0) + \cdots - F(0) \\
&\quad - h\left[F'(0) + \frac{h}{2}F''(0) + \frac{h^2}{8}F'''(0) + \cdots\right] \\
&= \left[\frac{1}{6} - \frac{1}{8}\right] h^3 F'''(0) + \cdots \\
&= \frac{h^3}{24} f''(\xi), \quad 0 < \xi < h. \qquad \textbf{(8.28)}
\end{aligned}
$$

By simply moving the node for the simple rectangle-type rules to the middle of the integration interval $[0, h]$, the order of the truncation error is improved from $O(h^2)$ to $O(h^3)$, which is the same as the simple trapezium rule, but now using a *single* function value.

8.5.2 Composite rule and error

Dividing the interval $[a, b]$ into N sub-intervals of size $h = (b - a)/N$ and applying the simple mid-point rule to each sub-interval gives the *composite mid-point rule*

$$
M_N = h \sum_{j=1}^{N} f\left(x_j - \frac{h}{2}\right), \qquad \textbf{(8.29)}
$$

with error

$$E_{M_N} = \frac{h^3}{24} \sum_{j=1}^{N} f''(\xi_j) = \frac{h^3}{24} N f''(\xi) = \frac{h^2}{24}(b-a) f''(\xi), \tag{8.30}$$

where $a < \xi < b$. The order of the composite error is identical to the trapezium result. The difference is the sign. If $f''(x)$ is one-signed on the interval $[a, b]$ then T_N and M_N provide a 'bracketing' interval for I.

Example 8.12 Revisiting Example 8.8, $\int_0^2 x^3\,dx$, Table 8.7 shows a selection of M_N estimates with refined step size h. The trapezium estimates from Table 8.5 are included for comparison.

The error *quarters* with each doubling of N (halving of h), which is consistent with eqn (8.30) and compares with the behaviour of T_N. max $|E_{M_N}|$ as defined by h^2 does bound the observed errors. M_N underestimates and T_N overestimates the true value. Since $f(x) = x^3$, then $f''(x) = 6x$ is one-signed on $0 \leq x \leq 2$, so T_N and M_N bracket the exact value. Also $|I - M_N| \propto \frac{1}{2}|I - T_N|$ which is consistent with eqns (8.17) and (8.30).

Table 8.7 Convergence of M_N.

N	1	2	4	8	16	32	64	128
h	2	1	0.5	0.25	0.125	0.0625	0.0313	0.0156
T_N	8	5	4.25	4.0625	4.0156	4.0039	4.0010	4.0002
$\|I - T_N\|$	4	1	0.25	0.0625	0.0156	0.0039	0.00107	0.0002
M_N	2	3.5	3.875	3.9688	3.9922	3.9980	3.9995	3.9999
$\|I - M_N\|$	2	0.5	0.125	0.0312	0.0078	0.0020	0.0005	0.0001
$\|E_{M_N}\|$	4	1	0.25	0.0625	0.0156	0.0039	0.0010	0.0002

DERIVE Experiment 8.4

To implement the composite mid-point rule, M_N, Author

```
mn(f,a,b,h):=hsum(vector(f,x,a+h/2,b-h/2,h))
```

where f is the integrand $f(x)$, a and b are the lower/upper limits of integration, and h is the step size. f must be given in terms of the variable x. Author `mn(x^3,0,2,.25)` approX returns the value 3.9688 (see Table 8.7).

$m + 1$ values of M_N, with step sizes $h, h/2, \ldots, h/2^m$, are defined by

```
ms(f,a,b,h,m):=vector(mn(f,a,b,h/2^i),i,0,m)
```

m is the number of step size reductions. To reproduce the row M_N in Table 8.7, Author ms(x^3,0,2,2,7) and approX. Try a few integrands whose first derivative is undefined at one or more points in the domain $[a, b]$.

To visualise M_N, Author

```
zm(f,x,h):=[[x-h/2,0],[x-h/2,f],[x+h/2,f],[x+h/2,0]]
pm(f,a,b,h):=vector(zm(f,x,h),x,a+h/2,b-h/2,h)
```

To 'see' the M_N estimate to $\int_0^1 e^x \, dx$ using a step size of $h = 0.25$, Author expx and Plot Beside Plot Algebra then Author and approX the expression pm(expx,0,1,.25). Plot Plot the result (see *DERIVE* Experiment 8.1 for further information).

8.6 Romberg integration

The preceding sections should have convinced you that substantial improvements in numerical integration accuracy can be obtained by either (a) refining the mesh size h or, more effectively, (b) by increasing the order of the method (forcing higher-order monomials to be integrated exactly: rectangle, trapezium, mid-point, Simpson's). At some point, however, the computational effort/complexity becomes restrictive. An additional technique is required to further enhance accuracy.

The discussion here is based upon the trapezium rule. Let the integration domain $[a, b]$ be divided by three equispaced nodes $x_0 = a$, $x_1 = (a + b)/2$ and $x_2 = b$ at intervals of size h. Two successive trapezium estimates using one and two subintervals respectively (see Figure 8.12) are

$$T_1 = \frac{2h}{2}\left[f(x_0) + f(x_2)\right] \quad \text{and} \quad T_2 = \frac{h}{2}\left[f(x_0) + 2f(x_1) + f(x_2)\right].$$

On including the truncation error for these estimates we can write

$$I = T_1 - \frac{(2h)^2}{12} f''(x_0) - G(2h)^4 - \cdots$$

$$I = T_2 - \frac{h^2}{12} f''(x_0) - Gh^4 - \cdots$$

where G is independent of the step size h. A combination of the estimates T_1 and T_2 can be used to eliminate the leading term in the error expansion. Four times the second estimate minus the first estimate gives

$$I = \frac{1}{3}[4T_2 - T_1] + 4Gh^4 + O(h^6). \tag{8.31}$$

Taken as an estimate to I, the value $(4T_2 - T_1)/3$ has a leading error of $O(h^4)$ (in place of $O(h^2)$ associated with the basic estimates T_1 and T_2). Expanding this

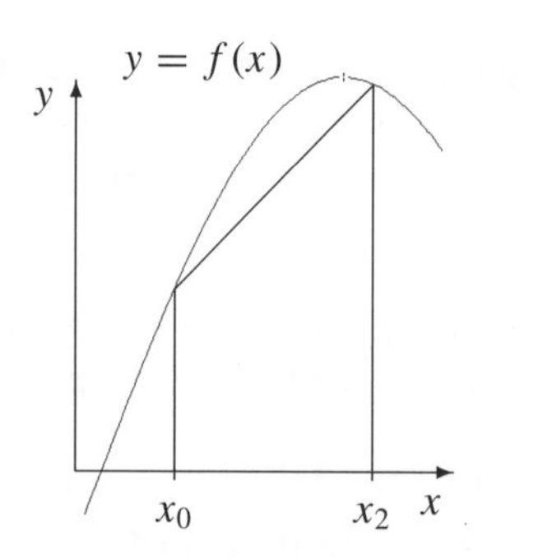

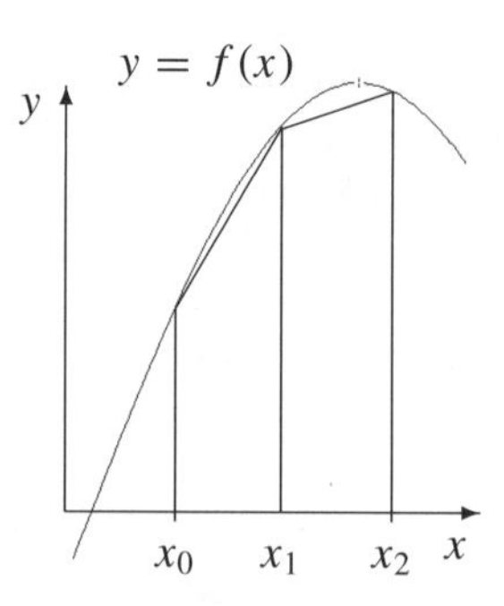

Figure 8.12 Successive trapezium estimates T_1 and T_2.

estimate

$$I \approx \frac{1}{3}[4T_2 - T_1] = \frac{1}{3}\left[4\left\{\frac{h}{2}(f_0 + 2f_1 + f_2)\right\} - \frac{2h}{2}(f_0 + f_2)\right]$$
$$= \frac{h}{3}[f_0 + 4f_1 + f_2]$$

shows it to be the Simpson estimate S_2 using two sub-intervals of size $h = (b-a)/2$. This process can be carried out for any two trapezium estimates T_N and T_{2N} to give the more accurate Simpson's estimate S_{2N}.

Trapezoidal	**Simpson**	
T_1		
	S_2	
T_2		
	S_4	In general $S_{2N} = \frac{1}{3}(4T_{2N} - T_N)$
T_4		
	S_8	
T_8		

Moving down columns of the table gives increased accuracy by *mesh refinement*. Moving across columns gives improved accuracy by *increased order* (reduced truncation error).

There is no reason why this process cannot be continued. Combinations of successive Simpson estimates can be constructed to eliminate the leading $O(h^4)$ term in the truncation error expansion. Using eqn (8.31), two successive Simpson estimates with truncation error terms can be written as

$$I = S_2 + 4Gh^4 + O(h^6)$$
$$I = S_4 + 4G\left(\frac{h}{2}\right)^4 + O(h^6)$$

Sixteen times S_4 minus S_2 leads to the estimate

$$I \approx \frac{1}{15}[16S_4 - S_2], \tag{8.32}$$

known as *Bode's rule*. The truncation error in eqn (8.32) is $O(h^6)$.

Trapezoidal	Simpson	Bode	
T_1			
	S_2		In general $S_{2N} = \frac{1}{3}(4T_{2N} - T_N)$
T_2		B_4	
	S_4		
T_4		B_8	In general $B_{4N} = \frac{1}{15}(16S_{4N} - S_{2N})$
	S_8		
T_8			

This process is called *extrapolation* and if the integrand f is sufficiently differentiable then it can be shown that *repeated extrapolation* gains two orders of accuracy at each step. This process of *extrapolation to the limit*, or *Richardson extrapolation*, applied to the trapezium rule is known as *Romberg integration*.

LEWIS FRY RICHARDSON
(1881–1953)
English scientist

LIFE

Richardson attended Newcastle Preparatory School, where his favourite subject was the study of Euclid. He went to school in York (1894–98) and then spent two years in Newcastle at the Durham College of Science where he studied mathematics, physics, chemistry, botany and zoology. He completed his education at King's College Cambridge, graduating with a First Class degree in Natural Science (1903). Richardson held many posts. He worked in the National Physical Laboratory (1903–04,1907–09) and the Meteorological Office (1913–16), and he held university posts at University College Aberystwyth (1905–06) and Manchester College of Technology (1912–13). He was a chemist with National Peat Industries (1906–07) and in charge of the physical and chemical laboratory of the Sunbeam Lamp Company (1909–12). Richardson was head of the Physics Department at Westminster Training College (1920–29) and Principal of Paisley College of Technology (1929–40).

WORK

Richardson was the first to apply mathematics (the method of finite differences) to predicting the weather in *Weather Prediction by Numerical Process* (1922). Another application of mathematics by Richardson was in his study of the causes of war in *Generalized Foreign Politics* (1939), *Arms and Insecurity* (1949) and *Statistics of Deadly Quarrels* (1950). He made contributions to the calculus and to the theory of diffusion (eddy diffusion in the atmosphere). The 'Richardson number' is a fundamental quantity involving gradients of temperature and wind velocity.

Starting with trapezium estimates $T_N, T_{2N}, \ldots$ the results are laid out in a tabular form with the final result appearing in the right-most column.

Example 8.13 Estimate the value of $\int_0^1 e^{\sin x}\,dx$ using Romberg integration.

N	Trapezium $k=1$	Simpson $k=2$	Bode $k=3$	$k=4$
1	1.65988841			
		1.63006033		
2	1.63751735		1.63189067	
		1.63177627		1.63186955
4	1.63321154		1.63186988	
		1.63186403		
8	1.63220091			

The trapezium estimates are set out in the first column and each improved estimate is obtained from the two adjacent values to the left. The final value is accurate to 6×10^{-8}.

Observations *When applicable, Romberg integration is highly efficient. For Example 8.13 just 9 function values are required in column 1, plus a small amount of arithmetic to compute the Romberg table. To achieve a similar accuracy using T_N requires 1332 function values, and S_{2N} requires 29 function values. It is worth noting that since the estimate T_{2N} uses the $N+1$ function values required by T_N, so the first column of a Romberg table can be generated very efficiently*[11].

DERIVE Experiment 8.5

A Romberg table requires (a) a function to compute m composite trapezium estimates $T_1, \ldots, T_{2^{m-1}}$ (column 1) and (b) an iterative process to compute successive Romberg columns from

$$r_{i,k} = \frac{4^{k-1} r_{i,k-1} - r_{i-1,k-1}}{4^{k-1} - 1}, \quad i = k, \ldots, m, \quad k = 2, \ldots, m, \tag{8.33}$$

[11] See, for example, [7], Section 4.6.

where $r_{i,1} = T_{2^{i-1}}$. First, Author

```
tn(f,a,b,h):=h(sum(vector(f,x,a,b,h))
   -sum(vector(f,x,a,b,b-a))/2)
r1(f,a,b,m):=vector(tn(f,a,b,(b-a)/2^i),i,0,m-1)
```

`tn` computes the composite trapezium estimate with a step size h and `r1` defines the first column of the Romberg table. Now Author

```
e1(x,i):=element(x,i)
q(x,i,j):=(4^(j-1)e1(x,i)-e1(r,i-1))/(4^(j-1)-1)
rj(r,j,m):=append(vector(0,i,1,j-1),vector(q(r,i,j),i,j,m))
```

`q` implements the right-hand side of eqn (8.33) and `rj` computes column j of the Romberg table from the previous column, denoted by `r`. The complete table is obtained with

```
rom(f,a,b,m):=
    iterates([j+1,rj(x,j+1,m)],[j,x],[1,r1(f,a,b,m)],m-1)'
```

To produce the table in Example 8.13, Author `rom(expsinx,0,1,4)` and `approX`.

Convergence of Romberg integration is observed (or not if the 'differentiability' conditions are not met) by looking down columns.

Example 8.14 Consider the integral $\int_0^1 x^{\frac{1}{4}}\,\mathrm{d}x$ which has the exact value 0.8. Using a set of trapezium estimates, the Romberg table for this problem is

$k = 1$	$k = 2$	$k = 3$	$k = 4$	$k = 5$
0.500000				
0.670448	0.727264			
0.744652	0.769387	0.772195		
0.776508	0.787126	0.788309	0.788565	
0.790067	0.794587	0.795085	0.795192	0.795218

The final result is quite inaccurate. The reason is that no derivative of the integrand $x^{\frac{1}{4}}$ exists at $x = 0$, which lies in the integration domain. Consequently, terms in the error expansion for T_N do not exist.

This table can be obtained using the function `rom` (see *DERIVE* Experiment 8.5). Author `rom(x^(1/4),0,1,5)` and `approX`.

Many more sophisticated (but generally more complicated) numerical integration rules exist, the most famous being the *Gaussian* rules[12].

Summary

The theme of this chapter concerned the development of simple numerical methods for estimating the value of the integral $\int_a^b f(x)\,\mathrm{d}x$. In Section 8.1 a graphical interpretation of integration was discussed and the general form (8.3) for the numerical methods was introduced. Sections 8.2–8.5 dealt with specific common methods: rectangle rules, trapezium rule, Simpson's rule and mid-point rule. In each case, the simple form was introduced together with a simple error analysis. The methods were generalised to a composite form. The ideas of improved accuracy through mesh refinement (composite rules) or increased order (truncation error) were discussed, as were the effects of rounding errors (see Section 8.2.4) and the use of undetermined coefficients as a method of developing quadrature rules (see Section 8.3.4). Section 8.6 discussed an alternative method of high-accuracy quadrature, called Romberg integration, based upon the idea of extrapolation.

Exercises

8.1 What are the values of the weight w_0 and abscissa x_0 for the right rectangle rule R_{R} on the interval $a \leq x \leq b$?

8.2 Repeat the error analysis that led to eqn (8.8) for the *right* rectangle rule R_{R}. Hence deduce the conditions on f that ensure I is bounded by the estimates R_{L} and R_{R}.

8.3 How many sub-intervals (function values) are required by the composite right-rectangle rule to compute the integral of Example 8.13 to an accuracy of 6×10^{-8}?

8.4 For the tabulated function

x	0.1	0.2	0.3	0.4	0.5
$f(x)$	0.9950	0.9801	0.9553	0.9211	0.8776

apply the composite trapezium rule to estimate

$$\int_{0.1}^{0.5} f(x)\,\mathrm{d}x.$$

Using a difference table, estimate $|E_{T_4}|$ (work to four decimal places).

[12] See, for example, [18], Section 7.3. For a comprehensive treatment of numerical integration see [8].

8.5 There are many special functions in mathematics that are described in terms of an integral, two of which are the *error function* erf, and the *Debye function*, defined by[13]

$$\operatorname{erf}(x) = \frac{2}{\sqrt{\pi}} \int_0^x e^{-u^2}\, du,$$

$$\mathcal{D}(x, n) = \int_0^x \frac{u^n}{e^u - 1}\, du.$$

Use the composite trapezium rule to compute the value of each function, accurate to two decimal places, for $x = \frac{1}{2}$ and $n = 2$ (in the case of $\mathcal{D}$).

8.6 Why might the mid-point rule be preferred to the trapezium rule for estimating the value of

$$\int_0^1 \frac{\sin x}{x}\, dx.$$

The exact value is 0.946082 to six decimal places.

8.7 Use Simpson's rule with $h = 1$ and $h = 0.5$ to estimate

$$\int_0^2 e^{-x}\, dx.$$

In each case obtain an error bound. Apply Richarson's extrapolation to determine a better estimate. Compare the observed error of all three estimates, working to four decimal places.

8.8 Evaluate

$$\int_0^{0.5} \sqrt{1 - x^2}\, dx$$

(a) by direct integration,
(b) using Simpson's rule with six sub-intervals (work to 4 d.p.), and
(c) by expanding $\sqrt{1 - x^2}$ in ascending powers of x (to 4 d.p.).

8.9 Apply the composite trapezium rule, with $n = 5, 10, 20, 40, 80$ and 160 intervals, to estimate the value of the integral

$$\int_0^{\frac{\pi}{2}} \sin(29x)\, dx.$$

Is convergence as expected? The exact value is $\frac{1}{29}$ (equals 0.889396 to 6 d.p.).

---

[13] Copious tables of values of these and other functions exist. See, for example, [1].

8.10 Use Romberg integration to estimate

$$\int_0^2 x^2 e^{-x^2}\, dx$$

as accurately as possible, working to four decimal places. You will need to start the process with the estimates T_1, T_2, T_4 and T_8.

8.11 What is the first extrapolation in the Romberg process were it based upon the mid-point rule? Deduce the form of subsequent extrapolations and apply this modified process to the problem of Exercise 8.10, based upon the estimates M_1, M_2, M_4 and M_8.

9 Systems of linear algebraic equations

The topic of this chapter concerns the systematic solution of a *system of linear algebraic equations*

$$
\begin{array}{ccccccccc}
a_{11}x_1 & + \cdots + & a_{1j}x_j & + \cdots + & a_{1n}x_n & = & b_1 \\
\vdots & & \vdots & & \vdots & & \vdots \\
a_{i1}x_1 & + \cdots + & a_{ij}x_j & + \cdots + & a_{in}x_n & = & b_i \\
\vdots & & \vdots & & \vdots & & \vdots \\
a_{n1}x_1 & + \cdots + & a_{nj}x_j & + \cdots + & a_{nn}x_n & = & b_n
\end{array}
\tag{9.1}
$$

(9.1) denotes *n equations in n unknowns*, $x_1, \ldots, x_n$. The aim is to find the set(s) of values of $x_1, \ldots, x_n$ that simultaneously satisfy all n equations.

9.1 Preliminaries and matrix notation

To represent and manipulate matrices[1] in a compact and efficient manner, a notation is required.

A *matrix* is a rectangular array of objects, typically numbers, collected together under a single name. An individual object is called an *element*. We denote by A the square matrix having *n rows* and *n columns*

$$
A = \begin{bmatrix}
a_{11} & \cdots & a_{1j} & \cdots & a_{1n} \\
\vdots & & \vdots & & \vdots \\
a_{i1} & \cdots & a_{ij} & \cdots & a_{in} \\
\vdots & & \vdots & & \vdots \\
a_{n1} & \cdots & a_{nj} & \cdots & a_{nn}
\end{bmatrix}. \tag{9.2}
$$

[1] It is assumed that the reader is familiar with the basic concepts of vectors, matrices and associated operations such as addition, multiplication and inversion. For those requiring further information, see [3, 5].

The element a_{ij} is referenced by two *indices*, i and j, that specify the row and column in which the element is located.

A *column vector* $\boldsymbol{x}$ is a matrix having one column with n elements $x_1, \ldots, x_n$,

$$\boldsymbol{x} = \begin{bmatrix} x_1 \\ \vdots \\ x_i \\ \vdots \\ x_n \end{bmatrix}. \tag{9.3}$$

The element x_i is identified by the index i which specifies its location within the vector. Defining the column vector $\boldsymbol{b}$ of n elements $b_1, \ldots, b_n$, as

$$\boldsymbol{b} = \begin{bmatrix} b_1 \\ \vdots \\ b_i \\ \vdots \\ b_n \end{bmatrix}, \tag{9.4}$$

the system (9.1) can be compactly written as

$$A\boldsymbol{x} = \boldsymbol{b}. \tag{9.5}$$

Two classes of method for solving linear systems of the form (9.5) are considered here[2] – *direct methods* and *iterative methods*.

9.2 Direct methods: Gauss elimination

A *direct method* finds a solution vector $\boldsymbol{x}$ of the system $A\boldsymbol{x} = \boldsymbol{b}$ in a finite number of operations by transforming the system into an equivalent system $A^+\boldsymbol{x} = \boldsymbol{b}^+$ that is 'easier' to solve and that has the *same* solution, $\boldsymbol{x}$.

We start by looking at simple linear systems, and then build in layers of complexity until we reach the general system $A\boldsymbol{x} = \boldsymbol{b}$. This should assist in the understanding of the subsequent solution techniques.

9.2.1 Diagonal systems

The *single* linear equation

$$a_{11}x_1 = b_1 \tag{9.6}$$

has solution $x_1 = b_1/a_{11}$. The matrix form of the equation is $\boldsymbol{Ax} = \boldsymbol{b}$ where $\boldsymbol{A} = [a_{11}]$ is a 1×1 matrix, $\boldsymbol{x} = [x_1]^{\mathrm{T}}$ and $\boldsymbol{b} = [b_1]^{\mathrm{T}}$.

[2] See [10] for a comprehensive text on matrix computations.

Example 9.1 The equation $5x_1 = 2$ has solution $x_1 = 2/5 = 0.4$.

The formal solution of eqn (9.6) is $x_1 = \boldsymbol{A}^{-1}b_1$ where the *matrix inverse* $\boldsymbol{A}^{-1}$ is simply $1/a_{11}$. This is easily verified, since $\boldsymbol{A}\boldsymbol{A}^{-1} = a_{11}/a_{11} = 1$ (the 1×1 identity matrix).

The *two* linear equations $a_{11}x_1 = b_1$, $a_{22}x_2 = b_2$ have the solution $x_1 = b_1/a_{11}$, $x_2 = b_2/a_{22}$. The matrix form of this system is

$$\begin{bmatrix} a_{11} & 0 \\ 0 & a_{22} \end{bmatrix} \begin{bmatrix} x_1 \\ x_2 \end{bmatrix} = \begin{bmatrix} b_1 \\ b_2 \end{bmatrix}, \tag{9.7}$$

that is $\boldsymbol{A}\boldsymbol{x} = \boldsymbol{b}$, where $\boldsymbol{A}$ is a 2×2 *diagonal matrix.*

Example 9.2 The equations $5x_1 = 2$ and $4x_2 = 7$ can be written in matrix form as

$$\begin{bmatrix} 5 & 0 \\ 0 & 4 \end{bmatrix} \begin{bmatrix} x_1 \\ x_2 \end{bmatrix} = \begin{bmatrix} 2 \\ 7 \end{bmatrix},$$

and have solution $x_1 = 2/5 = 0.4$, $x_2 = 7/4 = 1.75$.

Once again it should be noted that the solution vector $\boldsymbol{x}$ equals the matrix–vector product $\boldsymbol{A}^{-1}\boldsymbol{b}$.

Example 9.3 For Example 9.2 the matrix inverse is

$$\boldsymbol{A}^{-1} = \begin{bmatrix} \frac{1}{5} & 0 \\ 0 & \frac{1}{4} \end{bmatrix}, \quad \text{since} \quad \boldsymbol{A}\boldsymbol{A}^{-1} = \begin{bmatrix} 5 & 0 \\ 0 & 4 \end{bmatrix} \begin{bmatrix} \frac{1}{5} & 0 \\ 0 & \frac{1}{4} \end{bmatrix} = \begin{bmatrix} 1 & 0 \\ 0 & 1 \end{bmatrix},$$

and

$$\boldsymbol{x} = \boldsymbol{A}^{-1}\boldsymbol{b} = \begin{bmatrix} \frac{1}{5} & 0 \\ 0 & \frac{1}{4} \end{bmatrix} \begin{bmatrix} 2 \\ 7 \end{bmatrix} = \begin{bmatrix} \frac{2}{5} \\ \frac{7}{4} \end{bmatrix}.$$

The general form of a diagonal linear system is

$$\begin{bmatrix} a_{11} & & & & \\ & \ddots & & & \\ & & a_{ii} & & \\ & & & \ddots & \\ & & & & a_{nn} \end{bmatrix} \begin{bmatrix} x_1 \\ \vdots \\ x_i \\ \vdots \\ x_n \end{bmatrix} = \begin{bmatrix} b_1 \\ \vdots \\ b_i \\ \vdots \\ b_n \end{bmatrix}, \tag{9.8}$$

with *immediate* solution $x_i = b_i/a_{ii}$, $i = 1, \ldots, n$.

Observation *The simplicity of the solution for* diagonal systems *is due to the fact that the equations are* uncoupled *(no equation depends upon the solution of any other equation in the system). Each equation in the system may be solved in isolation.*

9.2.2 Triangular systems and back substitution

A simple *coupled* system of linear equations is defined by an *upper* (or *lower*) *triangular* coefficient matrix $\boldsymbol{A}$.

Example 9.4

The linear equations

$$\begin{aligned} 3x + 2y &= 11 \\ 3y &= 12 \end{aligned}$$

can be written in matrix form as

$$\begin{bmatrix} 3 & 2 \\ 0 & 3 \end{bmatrix} \begin{bmatrix} x \\ y \end{bmatrix} = \begin{bmatrix} 11 \\ 12 \end{bmatrix}.$$

The coefficient matrix is upper triangular – all elements *below* the diagonal are zero.

The attractiveness of the upper-triangular system is that it can be solved easily. The last line represents one equation in one unknown with solution $y = 12/3 = 4$. The value of y can be substituted back into the first equation to give $3x + 8 = 11$, that is $3x = 3$. This now represents one equation in one unknown with solution $x = 1$.

Formally, the solution of an upper-triangular system is again denoted by $\boldsymbol{x} = \boldsymbol{A}^{-1}\boldsymbol{b}$. For this example

$$\boldsymbol{A}^{-1} = \frac{1}{9} \begin{bmatrix} 3 & -2 \\ 0 & 3 \end{bmatrix}$$

and

$$\boldsymbol{x} = \boldsymbol{A}^{-1}\boldsymbol{b} = \frac{1}{9} \begin{bmatrix} 3 & -2 \\ 0 & 3 \end{bmatrix} \begin{bmatrix} 11 \\ 12 \end{bmatrix} = \begin{bmatrix} 1 \\ 4 \end{bmatrix}.$$

The general form of an $n \times n$ *upper-triangular system* is

$$\begin{bmatrix} a_{11} & \cdots & a_{1i} & \cdots & a_{1n} \\ 0 & \ddots & \vdots & & \vdots \\ & & a_{ii} & \cdots & a_{in} \\ & & & \ddots & \vdots \\ 0 & & & 0 & a_{nn} \end{bmatrix} \begin{bmatrix} x_1 \\ \vdots \\ x_i \\ \vdots \\ x_n \end{bmatrix} = \begin{bmatrix} b_1 \\ \vdots \\ b_i \\ \vdots \\ b_n \end{bmatrix}. \tag{9.9}$$

Example 9.5 A 4×4 upper-triangular system can be written as

$$\begin{aligned} a_{11}x_1 + a_{12}x_2 + a_{13}x_3 + a_{14}x_4 &= b_1 \\ a_{22}x_2 + a_{23}x_3 + a_{24}x_4 &= b_2 \\ a_{33}x_3 + a_{34}x_4 &= b_3 \\ a_{44}x_4 &= b_4 \end{aligned}$$

The fourth equation has solution $x_4 = b_4/a_{44}$.

x_4 is now known in the third equation, which can be written as $a_{33}x_3 = b_3 - a_{34}x_4$ with solution $x_3 = (b_3 - a_{34}x_4)/a_{33}$.

x_3 and x_4 are now known in the second equation, which can be written as $a_{22}x_2 = b_2 - a_{23}x_3 - a_{24}x_4$ with solution $x_2 = (b_2 - a_{23}x_3 - a_{24}x_4)/a_{22}$.

Finally, x_2, x_3, and x_4 are now known in the first equation, which can be written as $a_{11}x_1 = b_1 - a_{12}x_2 - a_{13}x_3 - a_{14}x_4$ with solution $x_1 = (b_1 - a_{12}x_2 - a_{13}x_3 - a_{14}x_4)/a_{11}$.

The solution process outlined in Example 9.5 is called *back substitution*. In general

$$\begin{aligned} x_n &= \frac{b_n}{a_{nn}}, \\ x_i &= \frac{b_i - \sum_{j=i+1}^{n} a_{ij}x_j}{a_{ii}}, \qquad i = n-1, \ldots, 1. \end{aligned} \tag{9.10}$$

Algorithm 9.1 Back substitution

Function: Solution of upper-triangular system

Line	Description	
1	**Problem**	`input` a_{ij}, b_i, n
2	Last element	$x_n = b_n/a_{nn}$
3	Loop back	`for` $i = n-1$ `to` 1 `step` -1
4		$s = b_i$
5		`for` $j = i+1$ `to` n
6		$s = s - a_{ij}x_j$
7		`end`
8	Element i	$x_i = s/a_{ii}$
9		`end`
10	**Solution**	`output` $x_1, \ldots, x_n$

Line 2 computes element x_n from equation n. Loop 3 traverses equations $n-1$ to 1 in reverse order. For equation i, line 4 initialises element x_i to b_i (the right-hand side), loop 5 subtracts the known values of elements x_{i+1} to x_n from the right-hand side, and line 8 divides by the diagonal element a_{ii} to give x_i.

The general $n \times n$ *lower-triangular system* is

$$\begin{bmatrix} a_{11} & 0 & & 0 \\ \vdots & \ddots & & \\ a_{i1} & \cdots & a_{ii} & \\ \vdots & & \vdots & \ddots & 0 \\ a_{n1} & \cdots & a_{ni} & \cdots & a_{nn} \end{bmatrix} \begin{bmatrix} x_1 \\ \vdots \\ x_i \\ \vdots \\ x_n \end{bmatrix} = \begin{bmatrix} b_1 \\ \vdots \\ b_i \\ \vdots \\ b_n \end{bmatrix}, \tag{9.11}$$

and *forward substitution* can be used to solve it, starting at the first equation and 'rippling' the solution through to equation n.

9.2.3 Square systems and forward elimination

Section 9.2.2 showed that if a system has a triangular structure then a systematic approach exists for solving the system – back (or forward) substitution. Here a systematic approach for transforming a *general* system to triangular form is developed, called *forward elimination*. Coupled with back substitution, this provides a general direct system solver.

First, two 'acceptable' operations are established. 'Acceptable' is taken to mean that the solution of the system remains unaltered by the operation.

I An equation can be multiplied by a non-zero constant.

Example 9.6

The equation $5x_1 = 2$ has the solution $x_1 = 2/5 = 0.4$. Multiplying the equation by, say, 3 gives $15x_1 = 6$. The solution of the modified equation is $x_1 = 6/15 = 0.4$.

The two equations $x_1 - x_2 = 1$ and $2x_1 + x_2 = 5$ have solution $x_1 = 2$, $x_2 = 1$. Multiplying the first equation by, say, 4 results in $4x_1 - 4x_2 = 4$ which is still satisfied by $x_1 = 2$, $x_2 = 1$.

II Non-zero multiples of two equations can be added.

Example 9.7

Adding the two equations in Example 9.6 gives the equation $3x_1 = 6$, which is still satisfied by the solution $x_1 = 2$.

Armed with I and II, a systematic *elimination method* can be developed to transform the matrix

$$\mathbf{A} = \begin{bmatrix} a_{11} & \cdots & a_{1j} & \cdots & a_{1n} \\ \vdots & & \vdots & & \vdots \\ a_{i1} & \cdots & a_{ij} & \cdots & a_{in} \\ \vdots & & \vdots & & \vdots \\ a_{n1} & \cdots & a_{nj} & \cdots & a_{nn} \end{bmatrix}$$

to the upper-triangular form

$$A^+ = \begin{bmatrix} a_{11}^+ & \cdots & a_{1i}^+ & \cdots & a_{1n}^+ \\ 0 & \ddots & \vdots & & \vdots \\ & & a_{ii}^+ & \cdots & a_{in}^+ \\ & & & \ddots & \vdots \\ 0 & & & 0 & a_{nn}^+ \end{bmatrix}.$$

To simplify the manipulation, the *augmented matrix* is introduced. For the $n \times n$ system $\boldsymbol{Ax} = \boldsymbol{b}$, the $n \times (n+1)$ *augmented matrix* $[\boldsymbol{A} : \boldsymbol{b}]$ is obtained by adding a column to $\boldsymbol{A}$ consisting of the vector $\boldsymbol{b}$. The reason for introducing this matrix is that manipulations using I and II can be performed on $[\boldsymbol{A} : \boldsymbol{b}]$ – the vector $\boldsymbol{x}$ can temporarily be 'discarded'.

Example 9.8

The three linear systems

$$\begin{aligned} x - 2y + 3z &= 2 \\ 2x + y - 5z &= -1, \\ 3x - y + 2z &= 0 \end{aligned} \qquad \begin{aligned} x_1 - 2x_2 + 3x_3 &= 2 \\ 2x_1 + x_2 - 5x_3 &= -1, \\ 3x_1 - x_2 + 2x_3 &= 0 \end{aligned} \qquad \begin{aligned} u - 2v + 3w &= 2 \\ 2u + v - 5w &= -1 \\ 3u - v + 2w &= 0 \end{aligned}$$

have the *same* solution vector $[-\frac{7}{20} \; -\frac{31}{20} \; -\frac{1}{4}]^{\mathrm{T}}$. In other words, the solution vectors

$$\begin{bmatrix} x \\ y \\ z \end{bmatrix}, \quad \begin{bmatrix} x_1 \\ x_2 \\ x_3 \end{bmatrix}, \quad \text{and} \quad \begin{bmatrix} u \\ v \\ w \end{bmatrix}$$

are *transparent* to the solution process.

To develop the general elimination process it is useful to first work through a specific illustration.

Example 9.9

The solution of the 4×4 linear system

$$\begin{aligned} w + x + y + z &= 3 \\ 2w - x - y + 2z &= 12 \\ w + 3x - 2y - z &= -9 \\ -w - x + y + 4z &= 17 \end{aligned}$$

is $[w \; x \; y \; z]^{\mathrm{T}} = [1 \; -2 \; 0 \; 4]^{\mathrm{T}}$. The augmented matrix is

$$\begin{bmatrix} 1 & 1 & 1 & 1 & : & 3 \\ \mathbf{2} & -1 & -1 & 2 & : & 12 \\ \mathbf{1} & 3 & -2 & -1 & : & -9 \\ \mathbf{-1} & -1 & 1 & 4 & : & 17 \end{bmatrix} \begin{matrix} E_1 \\ E_2 \\ E_3 \\ E_4 \end{matrix}$$

Elimination proceeds column-by-column. The first stage is to 'zero' the *first* column below the diagonal element a_{11}. Multiples of the first row (equation) E_1 are subtracted from subsequent rows, E_2, E_3 and E_4. To eliminate the 2 in row E_2, 2 times row E_1 is subtracted,

$$\begin{bmatrix} 1 & 1 & 1 & 1 & : & 3 \\ 0 & -3 & -3 & 0 & : & 6 \\ 1 & 3 & -2 & -1 & : & -9 \\ -1 & -1 & 1 & 4 & : & 17 \end{bmatrix} \begin{matrix} E_1 \\ E_2^{(1)} = E_2 - 2E_1 \\ E_3 \\ E_4 \end{matrix}$$

(make sure that you can identify all changes). To eliminate the 1 in row E_3, 1 times row E_1 is subtracted,

$$\begin{bmatrix} 1 & 1 & 1 & 1 & : & 3 \\ 0 & -3 & -3 & 0 & : & 6 \\ 0 & 2 & -3 & -2 & : & -12 \\ -1 & -1 & 1 & 4 & : & 17 \end{bmatrix} \begin{matrix} E_1 \\ E_2^{(1)} = E_2 - 2E_1 \\ E_3^{(1)} = E_3 - E_1 \\ E_4 \end{matrix}$$

And finally, to eliminate the -1 in row E_4, -1 times E_1 is subtracted from E_4 to give

$$\begin{bmatrix} 1 & 1 & 1 & 1 & : & 3 \\ 0 & -3 & -3 & 0 & : & 6 \\ 0 & \mathbf{2} & -3 & -2 & : & -12 \\ 0 & \mathbf{0} & 2 & 5 & : & 20 \end{bmatrix} \begin{matrix} E_1 \\ E_2^{(1)} = E_2 - 2E_1 \\ E_3^{(1)} = E_3 - E_1 \\ E_4^{(1)} = E_4 + E_1 \end{matrix}$$

E_1 is the *pivotal row* and the diagonal element a_{11} is the *pivot*. The multiples of E_1 subtracted are computed by dividing the element to be eliminated, a_{i1}, by the pivot, that is $m_{i1} = a_{i1}/a_{11}$ where m_{i1} is the *multiplier*. The superscript $^{(1)}$ denotes equations (rows) that have been modified *once*.

All the operations performed satisfy conditions I and II and the system represented by the modified augmented matrix, namely

$$\begin{aligned} w + x + y + z &= 3 \\ -3x - 3y \quad &= 6 \\ 2x - 3y - 2z &= -12 \\ 2y + 5z &= 20 \end{aligned}$$

has the same solution as the original linear system (you can verify this by direct substitution of the solution vector).

The elimination process moves to the *second* column to zero elements below the diagonal element a_{22} (the new pivot), achieved by subtracting multiples of row E_2 (the new pivotal row) from rows E_3 and E_4. The multipliers are $m_{32} = a_{32}/a_{22} = -2/3$ and $m_{42} = a_{42}/a_{22} = 0$ (the last element in column 2 is already zero and no further action on it is required). The updated augmented matrix is

$$\begin{bmatrix} 1 & 1 & 1 & 1 & : & 3 \\ 0 & -3 & -3 & 0 & : & 6 \\ 0 & 0 & -5 & -2 & : & -8 \\ 0 & 0 & \mathbf{2} & 5 & : & 20 \end{bmatrix} \begin{matrix} E_1 \\ E_2^{(1)} = E_2 - 2E_1 \\ E_3^{(2)} = E_3^{(1)} + \frac{2}{3}E_2^{(1)} \\ E_4^{(2)} = E_4^{(1)} \end{matrix}$$

The superscript $^{(2)}$ denotes rows that have been modified twice. The final stage of the elimination process moves to column 3 and 'zeros' the single element below the pivot a_{33}. The multiplier is $m_{43} = a_{43}/a_{33} = -5/2$ and subtracting m_{43} times row E_3 (the pivotal row) from row E_4 results in the final *upper-triangular* augmented matrix

$$\begin{bmatrix} 1 & 1 & 1 & 1 & : & 3 \\ 0 & -3 & -3 & 0 & : & 6 \\ 0 & 0 & -5 & -2 & : & -8 \\ 0 & 0 & 0 & \frac{21}{5} & : & \frac{84}{5} \end{bmatrix} \begin{matrix} E_1 \\ E_2^{(1)} = E_2 - 2E_1 \\ E_3^{(2)} = E_3^{(1)} + \frac{2}{3}E_2^{(1)} \\ E_4^{(3)} = E_4^{(2)} + \frac{2}{5}E_3^{(2)} \end{matrix}$$

The solution is now obtained with back substitution. From equation $E_4^{(3)}$, $21x_4/5 = 84/5$, that is $x_4 = 4$. Equation $E_3^{(2)}$ now reads $-5x_3 - 2 \times 4 = -8$, that is $x_3 = 0$. Equation $E_2^{(1)}$ becomes $-3x_2 - 3 \times 0 + 0 \times 4 = 6$, that is $x_2 = -2$. Finally, the first equation, E_1, becomes $x_1 + 1 \times -2 + 1 \times 0 + 1 \times 4 = 3$, that is $x_1 = 1$.

In the general elimination procedure, for columns k from 1 to $n-1$, the elements below the pivot a_{kk} are set to zero. For each row i from $k+1$ to n the multiplier $m_{ik} = a_{ik}/a_{kk}$ is calculated and m_{ik} times the pivotal row E_k is subtracted from row E_i (through columns j from $k+1$ to $n+1$). The final column is the vector $\boldsymbol{b}$.

Algorithm 9.2 Forward elimination

Function: Reduction to upper-triangular system

Line	Description	
1	**Problem**	input a_{ij}, b_i, n
2	Column	for $k = 1$ to $n - 1$
3	Equation	for $i = k + 1$ to n
4	Multiplier	$m_{ik} = a_{ik}/a_{kk}$
5	Update E_i	for $j = k + 1$ to n
6		$a_{ij} = a_{ij} - m_{ik}a_{kj}$
7		end
8		$b_i = b_i - m_{ik}b_k$
9		end
10		end
11	**Solution**	output Upper-triangular matrix

Loop 2 traverses columns 1 to $n-1$. For column k, loop 3 eliminates elements below the pivot a_{kk} (line 4 computes the multiplier m_{ik} for equation E_i and loop 5 to line 8 update the equation by subtracting m_{ik} times E_k).

DERIVE Experiment 9.1

To simulate forward elimination requires some effort (using the functions `force0` and `pivot` from the utility file `vector.mth`).

`force0(a,i,j,p)`

`a`	matrix $\boldsymbol{A}$
`i, j`	row and column indices
`p`	pivotal row index

`pivot(a,i,j)`

`a`	matrix $\boldsymbol{A}$
`i, j`	row and column indices

`force0` zeros element a_{ij} of $\boldsymbol{A}$ using the pivotal row p and `pivot` zeros column j of $\boldsymbol{A}$ below row i. To access these functions, `Transfer Load Utility vector`. To zero the first column of the matrix

$$\boldsymbol{A} = \begin{bmatrix} 1 & 2 & 3 \\ 2 & 3 & 2 \\ 1 & 3 & -1 \end{bmatrix}$$

`Author a:=[[1,2,3],[2,3,2],[1,3,-1]]` and `Author` and `Simplify pivot(a,1,1)`. To visualise the process, define the functions

```
dim(a):=dimension(a)
a0(a,i,j,p):=if(i=j,a,
    if(i=j+1,force0(a,i,j,p),a0(force0(a,i,j,p),i-1,j,p)))
col0(a,k):=vector(a0(a,i,k,k),i,k,dim(a)
```

`a0` zeros elements a_{p+1j} to a_{ij} and `col0` displays the intermediate steps of the `pivot` function. To 'see' the previous `pivot` application, `Author` and `Simplify col0(a,1)`. Further, `col0(pivot(a,1,1),2)` displays the elimination of column 2.

To visualise the column-by-column process, define

```
c0(a,j,k):=if(j=k,pivot(a,j,k),c0(pivot(a,j,j),j+1,k))
felim(a):=append([a],vector(c0(a,1,k),k,1,dim(a)-1))
```

`c0` eliminates columns 1 to k and `felim` displays the elimination process. The elimination process for the matrix $\boldsymbol{A}$ is then obtained with `Author felim(a) Simplify`

Apply these functions to the augmented matrix of Example 9.9.

9.2.3.1 Determinacy

Forward elimination can tell us something about the form of the solution to $\boldsymbol{Ax} = \boldsymbol{b}$, in addition to finding the solution vector $\boldsymbol{x}$. If the final row of the augmented matrix (after forward elimination) has the form

$$0\,0\,\cdots\,0\,a^*_{nn}\;:\;b^*_n \tag{9.12}$$

($a^*_{nn}x_n = b^*_n$), the original system has a *unique* solution $\boldsymbol{x}$, which may be obtained with back substitution.

Example 9.10 The 4×4 system of Example 9.9 has a unique solution – the last row of the augmented matrix is 0 0 0 21/5:84/5, that is $21x_4 = 84$.

If the final row of the augmented matrix has the form

$$0\,0\,\cdots\,0\,0\;:\;b^*_n \tag{9.13}$$

where $b^*_n \neq 0$, the corresponding equation $0x_1+0x_2+\cdots+0x_n = b^*_n$ has *no* solution.

Example 9.11 The 3×3 system

$$\begin{aligned} x+2y+3z &= 1\\ 2x\;+y-2z &= 1\\ 3x\qquad -7z &= 2 \end{aligned}$$

produces the elimination process

$$\begin{bmatrix} 1 & 1 & 3 & : & 1\\ 2 & 1 & -2 & : & 1\\ 3 & 0 & -7 & : & 2 \end{bmatrix} \rightarrow \begin{bmatrix} 1 & 1 & 3 & : & 1\\ 0 & -3 & -8 & : & -1\\ 0 & -6 & -16 & : & -1 \end{bmatrix}$$

$$\rightarrow \begin{bmatrix} 1 & 1 & 3 & : & 1\\ 0 & -3 & -8 & : & -1\\ 0 & 0 & 0 & : & 1 \end{bmatrix}.$$

The last equation has no solution.

If the final row of the augmented matrix has the form

$$0\,0\,\cdots\,0\,0\;:\;0 \tag{9.14}$$

the corresponding equation $0x_1 + 0x_2 + \cdots + 0x_n = 0$ has an *infinity* of solutions.

Example 9.12 The 3×3 system

$$\begin{aligned} x + 2y + 3z &= 1 \\ 2x \;\; + y - 2z &= 1 \\ 3x \qquad - 7z &= 1 \end{aligned}$$

produces the elimination process

$$\begin{bmatrix} 1 & 1 & 3 & : & 1 \\ 2 & 1 & -2 & : & 1 \\ 3 & 0 & -7 & : & 1 \end{bmatrix} \rightarrow \begin{bmatrix} 1 & 1 & 3 & : & 1 \\ 0 & -3 & -8 & : & -1 \\ 0 & -6 & -16 & : & -2 \end{bmatrix}$$

$$\rightarrow \begin{bmatrix} 1 & 1 & 3 & : & 1 \\ 0 & -3 & -8 & : & -1 \\ 0 & 0 & 0 & : & 0 \end{bmatrix}.$$

The system represents *two* equations in *three* unknowns. One unknown can be assigned an *arbitrary* value. If $z = a$ then the second equation is $-3y - 8a = -1$ with solution $y = (1 - 8a)/3$. The first equation becomes $x + 2(1 - 8a)/3 + 3a = 1$ with solution $x = (7a - 1)/3$. The solution vector is $\boldsymbol{x} = [(7a - 1)/3 \; (1 - 8a)/3 \; a]^{\mathrm{T}}$.

9.2.4 Gauss elimination: implementation

The combination of forward elimination and back substitution forms what is known as *Gauss elimination*.

CARL FRIEDRICH GAUSS
(1777–1855)
German mathematician

LIFE

At seven, Gauss started elementary school and his potential was noticed almost immediately. His teacher, Büttner, was amazed when Gauss summed the integers from 1 to 100 instantly by spotting that the sum was 50 pairs of numbers, each pair summing to 101. In 1788 Gauss began his education at the Gymnasium where he learnt High German and Latin. After receiving a stipend from the Duke of Brunswick-Wolfenbüttel, Gauss entered Brunswick Collegium Carolinum (1792). At the academy Gauss independently discovered Bode's law, the binomial theorem, the arithmetic-geometric mean, the law of quadratic reciprocity and the prime number theorem.

In 1795 Gauss left Brunswick to study at Göttingen University. His only known friend amongst the students was Bolyai. They met in 1799 and corresponded for many years. Gauss left Göttingen (1798) without a diploma but had already made one of his most important discoveries – the construction of a regular 17-gon by ruler and compass.

This was the most major advance in the field since the time of Greek mathematics and was published as Section VII of his famous work *Disquisitiones Arithmeticae*.

Gauss returned to Brunswick where he received a degree in 1799. After the Duke of Brunswick had agreed to continue Gauss's stipend, he requested that Gauss submit a doctoral dissertation to the University of Helmstedt. He already knew Pfaff, who was chosen to be his advisor. Gauss's dissertation was a discussion of the fundamental theorem of algebra. With his stipend to support him, Gauss did not need to find a job, so devoted himself to research. He published the book *Disquisitiones Arithmeticae* in the summer of 1801. There were seven sections, all but the last section being devoted to number theory.

In June 1801, the astronomer Zach published the orbital positions of Ceres, a 'small planet' discovered by Piazzi on 1 January 1801. Unfortunately, Piazzi had only been able to observe nine degrees of its orbit before it disappeared behind the Sun. Zach published several predictions of its position, including one by Gauss which differed greatly from the others. When Ceres was rediscovered by Zach on 7 December 1801 it was almost exactly where Gauss had predicted. Although he did not disclose his methods at the time, Gauss had used his least-squares approximation method.

In June 1802 Gauss visited Olbers who had discovered Pallas in March of that year and Gauss investigated its orbit. Gauss began corresponding with Bessel and Germain.

Gauss married Johanna Ostoff in 1805. Despite having a happy personal life for the first time, his benefactor, the Duke of Brunswick, was killed fighting for the Prussian army. In 1807 Gauss left Brunswick to become director of the Göttingen observatory.

In 1808 his father died, and a year later Gauss's wife Johanna died after giving birth to their second son (who died soon after). Gauss remarried the next year, to Minna, the best friend of Johanna. Gauss's work never seemed to suffer from his personal tragedy.

The period 1817–32 was particularly distressing. He took in his sick mother, who stayed until her death (1839), while he was arguing with his wife and her family about whether they should go to Berlin. He had been offered a position at Berlin University and Minna and her family were keen to move there. Gauss never liked change and decided to stay in Göttingen. In 1831 Gauss's second wife died after a long illness.

Gauss spent the years from 1845 to 1851 updating the Göttingen University widow's fund. This work gave him practical experience in financial matters, and he went on to make his fortune through shrewd investments in bonds issued by private companies.

Two of Gauss's last doctoral students were Moritz Cantor and Dedekind. Gauss presented his golden jubilee lecture in 1849, fifty years after his diploma had been granted by Hemstedt University. It was appropriately a variation on his dissertation of 1799. From the mathematical community only Jacobi and Dirichlet were present, but Gauss received many messages and honours.

From 1850 onwards Gauss's work was again nearly all of a practical nature, although he did approve Riemann's doctoral thesis and heard his probationary lecture. His last known scientific exchange was with Gerling. He discussed a modified Foucalt pendulum in 1854. He was also able to attend the opening of the new railway link between Hanover and Göttingen, but this proved to be his last outing. His health deteriorated slowly, and Gauss died in his sleep early in the morning of 23 February 1855.

WORK

Gauss worked in a wide variety of fields in both mathematics and physics, including number theory, analysis, differential geometry, geodesy, magnetism, astronomy and optics. His work has had an immense influence in many areas.

He published his second book, *Theoria motus corporum coelestium in sectionibus conicis Solem ambientium* in 1809, a major two-volume treatise on the motion of celestial bodies. In the first volume he discussed differential equations, conic sections and elliptic orbits, while in the second volume, the main part of the work, he showed how to estimate and then to refine the estimation of a planet's orbit.

Much of Gauss's time was spent on a new observatory, completed in 1816, but he still found the time to work on other subjects. His publications during this time include *Disquisitiones generales circa seriem infinitam*, a rigorous treatment of series and an introduction of the hypergeometric function, *Methodus nova integralium valores per approximationem inveniendi*, a practical essay on approximate integration, *Bestimmung der Genauigkeit der Beobachtungen*, a discussion of statistical estimators, and *Theoria attractionis corporum sphaeroidicorum ellipticorum homogeneorum methodus nova tractata*. The latter work was inspired by geodesic problems and was principally concerned with potential theory.

Gauss had been asked (1818) to carry out a geodesic survey of the state of Hanover to link up with the existing Danish grid. He took personal charge of the survey, making measurements during the day and reducing them at night, using his extraordinary mental capacity for calculations. Because of the survey, Gauss invented the heliotrope, which worked by reflecting the Sun's rays using a design of mirrors and a small telescope. Gauss often wondered if he would have been better advised to have pursued some other occupation, but he published over 70 papers between 1820 and 1830.

In 1822 Gauss won the Copenhagen University Prize with *Theoria attractionis...* together with the idea of mapping one surface onto another so that the two *'are similar in their smallest parts'*. This paper (1825) led to the later publication of *Untersuchungen über Gegenstände der Höheren Geodšie* (1843 and 1846). *Theoria combinationis observationum erroribus minimis obnoxiae* (1823) and its supplement (1828) were devoted to mathematical statistics, in particular to the least-squares method.

From the early 1800s Gauss had an interest in the question of the possible existence of a non-Euclidean geometry. He discussed this topic with Bolyai and in correspondence with Gerling and Schumacher. In a book review (1816) he discussed proofs which deduced the axiom of parallels from the other Euclidean axioms, suggesting that he believed in the existence of non-Euclidean geometry, although he was rather vague. Gauss confided in Schumacher, telling him that he believed his reputation would suffer if he admitted in public that he believed in the existence of such a geometry.

Gauss had a major interest in differential geometry and published many papers. *Disquisitiones generales circa superficies curva* (1828) was his major work in this field, arisng from his geodesic interests, but containing such geometrical ideas as Gaussian curvature. The paper also includes Gauss's famous theorem *'If an area in E^3 can be developed (i.e. mapped isometrically) into another area of E^3, the values of the Gaussian curvatures are identical in corresponding points'.*

In 1831, Wilhelm Weber arrived in Göttingen as physics professor. Gauss had known Weber since 1828 and supported his appointment. Gauss had worked on physics before 1831, publishing *Uber ein neues allgemeines Grundgesetz der Mechanik,* which contained the principle of least constraint, and *Principia generalia theoriae figurae fluidorum in statu aequilibrii,* which discussed forces of attraction. These papers were based on Gauss's potential theory, which proved of great importance in his work on physics.

In 1832, Gauss and Weber began investigating the theory of terrestrial magnetism after von Humboldt attempted to obtain Gauss's assistance in making a grid of magnetic observation points around the Earth. By 1840 he had written three important papers on the subject: *Intensitas vis magneticae terrestris ad mensuram absolutam revocata* (1832), *Allgemeine Theorie des Erdmagnetismus* (1839) and *Allgemeine Lehrsätze in Beziehung auf die im verkehrten Verhältnisse des Quadrats der Entfernung wirkenden Anziehungs und Abstossungskräfte* (1840). These papers dealt with the current theories on terrestrial magnetism, including Poisson's ideas, absolute measure for magnetic force and an empirical definition of terrestrial magnetism. Dirichlet's principle was mentioned without proof. Gauss used the Laplace equation to aid him with his calculations, and ended up specifying a location for the magnetic South Pole.

Humboldt had devised a calendar for observations of magnetic declination. Once Gauss's new magnetic observatory (completed in 1833 – free of all magnetic metals) had been built, he proceeded to alter many of Humboldt's procedures, not pleasing Humboldt greatly. Gauss's changes obtained more accurate results with less effort.

Gauss and Weber achieved much in their six years together. They discovered Kirchhoff's laws and built a primitive telegraph device which could send messages over a distance of 5000 ft. However, this was just an enjoyable pastime for Gauss. He was more interested in the task of establishing a world-wide net of magnetic observation points. This occupation produced many concrete results. The *Magnetischer Verein* and its journal were founded, and the atlas of geomagnetism was published, while Gauss and Weber's own journal in which their results were published ran from 1836 to 1841. In 1837, Weber was forced to leave Göttingen when he became involved in a political dispute and, from this time, Gauss's activity gradually decreased.

There are a few details to stress regarding implementation. If at any stage of the forward elimination procedure the pivot a_{kk} is zero, the multipliers $m_{ik} = a_{ik}/a_{kk}$ are not defined and the procedure will fail.

A simple strategy to overcome this difficulty is, at each stage of the elimination process, to search column k, from row $k+1$, until a non-zero element is found, say a_{pk} in row E_p. Rows k and p are then swapped *before* the elimination of column k (below the diagonal).

Example 9.13 The 3 × 3 system

$$\begin{aligned} 2y + 3z &= 13 \\ x + y \ + z &= 6 \\ 2x \qquad + z &= 5 \end{aligned}$$

is reduced with the elimination process

$$\begin{bmatrix} 0 & 2 & 3 & : & 13 \\ 1 & 1 & 1 & : & 6 \\ 2 & 0 & 1 & : & 5 \end{bmatrix} \rightarrow \begin{bmatrix} 1 & 1 & 1 & : & 6 \\ 0 & 2 & 3 & : & 13 \\ 2 & 0 & 1 & : & 5 \end{bmatrix}$$

$$\rightarrow \begin{bmatrix} 1 & 1 & 1 & : & 6 \\ 0 & 2 & 3 & : & 13 \\ 0 & -2 & -1 & : & -7 \end{bmatrix}$$

$$\rightarrow \begin{bmatrix} 1 & 1 & 1 & : & 6 \\ 0 & 2 & 3 & : & 13 \\ 0 & 0 & 2 & : & 6 \end{bmatrix}$$

The first step swaps rows 1 and 2 so that the pivot a_{11} is non-zero.

If no $p \in \{k, \ldots, n)\}$ exists such that $a_{pk} \neq 0$ then the method fails and the system $\boldsymbol{Ax} = \boldsymbol{b}$ has no solution (see Section 9.2.3.1 on determinacy).

9.2.4.1 Induced errors and partial pivoting

There are still several aspects of Gauss elimination that must be understood. Whilst the method clearly fails if at any stage during forward elimination a non-zero pivot cannot be found, the algorithm has a more subtle failing relating to its implementation on a finite-precision machine. Under certain conditions the process can be sensitive to rounding errors, termed *induced errors*.

Example 9.14 Solve the following 3 × 3 system with Gauss elimination using 4-digit floating-point arithmetic.

$$0.8526x + 0.3713y + 0.1947z = 0.9834$$

$$0.4763x + 0.2082y + 0.8149z = 0.8516$$

$$0.3123x + 0.7369y + 0.5132z = 0.7982$$

Backward substitution gives $z = 0.4277$, $y = 0.3750$ and $x = 0.8924$. The solution satisfies E_1 and E_2 but gives a large error when substituted into E_3 and so, at this stage, $[x, y, z]^{\mathrm{T}}$ *cannot* be accepted as a satisfactory solution.

Operation	Coefficients x	y	z	RHS
E_1	0.8526	0.3713	0.1947	0.9834
E_2	0.4763	0.2082	0.8149	0.8516
E_3	0.3123	0.7369	0.5132	0.7982
E_1	0.8526	0.3713	0.1947	0.9834
$E_2^{(1)} = E_2 - 0.5586E_1$	0	0.0008000	0.7061	0.3023
$E_3^{(1)} = E_3 - 0.3663E_1$	0	0.6009	0.4419	0.4380
E_1	0.8526	0.3713	0.1947	0.9834
$E_2^{(1)}$	0	0.0008000	0.7061	0.3023
$E_3^{(2)} = E_3^{(1)} - 751.1E_2^{(1)}$	0	0	−530.0	−226.7

The major error occurred when y was computed from $E_2^{(1)}$ as

$$\frac{0.0003000}{0.0008000}$$

which results in a small pivot dividing into an equally small numerator, both of which resulted from subtracting two almost equal numbers. This is a classical situation where significant figures are lost in computed results and should be avoided.

The source of the problem in Example 9.14 lies in the small value of the pivot $a_{22} = 0.0008$. The multiplier m_{32} is very large and a very large multiple of $E_2^{(1)}$ is added to row 3, thereby swamping its contribution (see Example 2.20, where it is shown that adding numbers of widely differing magnitude can lead to loss of significance). This is only a problem when the arithmetic is not exact.

To navigate around this weakness we may apply a modified version of the 'column search and row swapping' procedure adopted for the case of a zero pivot (see Section 9.2.4). At each stage of the forward elimination process, row interchanges are made to *maximise* the magnitude of the pivot a_{kk}. This requires searching column k, through rows k to n, to find the element a_{pk} of *maximum magnitude*. Rows k and p are then swapped before elimination takes place. This procedure is called *partial pivoting*.

Example 9.15 Re-solve Example 9.14 using Gauss elimination with *partial pivoting* and 4-digit floating-point arithmetic. The element of maximum modulus in column 1 lies in row 1 and no swap is made for the first elimination phase. For the second phase, the largest value in column 2 appears in row 3, and so these rows are swapped before the elimination. The process is summarised below. Backward substitution gives $z = 0.4276$, $y = 0.4144$ and $x = 0.8752$ which satisfies all three original equations within round-off.

Operation	Coefficients			RHS
	x	y	z	
E_1	0.8526	0.3713	0.1947	0.9834
E_2	0.4763	0.2082	0.8149	0.8516
E_3	0.3123	0.7369	0.5132	0.7982
E_1	0.8526	0.3713	0.1947	0.9834
$E_2^{(1)} = E_2 - 0.5586E_1$	0	0.0008000	0.7061	0.3023
$E_3^{(1)} = E_3 - 0.3663E_1$	0	0.6009	0.4419	0.4380
E_1	0.8526	0.3713	0.1947	0.9834
$E_2^{(1)} = E_3^{(1)}$	0	0.6009	0.4419	0.4380
$E_3^{(1)} = E_2^{(1)}$	0	0.0008000	0.7061	0.3023
E_1	0.8526	0.3713	0.1947	0.9834
$E_2^{(1)}$	0	0.6009	0.4419	0.4380
$E_3^{(2)} = E_3^{(1)} - 0.001331E_2^{(1)}$	0	0	0.7055	0.3017

The forward elimination algorithm with partial pivoting is given below.

Algorithm 9.3 Forward elimination with partial pivoting

Function: Stable reduction to upper-triangular form

Line	Description	
1	**Problem**	input a_{ij}, b_i, n
2	Column	for $k = 1$ to $n - 1$
3	Maximise pivot	$p = k$
4		for $i = k + 1$ to n
5		if $\|a_{ik}\| > \|a_{pk}\|$ then $p = i$
6		end
7	Swap rows	if $p > k$ then
8		for $j = k$ to n
9		$t = a_{kj}, a_{kj} = a_{pj}, a_{pj} = t$
10		end
11		$t = b_k, b_k = b_p, b_p = t$
12		end
13	Equation	for $i = k + 1$ to n
14	Multiplier	$m_{ik} = a_{ik}/a_{kk}$
15	Update E_i	for $j = k + 1$ to n
16		$a_{ij} = a_{ij} - m_{ik}a_{kj}$

```
17                          end
18                          b_i = b_i − m_ik b_k
19                          end
20                      end
21      Solution        output Upper-triangular matrix
```

Lines 3–12 search for the maximum pivot in column k and move the 'containing' row to the pivotal row. Lines 13–19 then execute the elimination of column k, as in Algorithm 9.2 (lines 3–9).

9.2.4.2 Inherent errors and ill-conditioning

Sometimes the solution of a system is sensitive to small changes in the coefficients of the system. This is termed *inherent instability* and is hard to remove by simply choosing/modifying the solution method.

Example 9.16 Solve the 3×3 system of Example 9.15 with the coefficient of x in the first equation now taken as 0.8612 (a change of approximately 1%). Partial pivoting is again used to *avoid* small pivotal values and *induced instability*. Backward substitution gives $z = 0.4334$, $y = 0.4147$ and $x = 0.8651$ which satisfies all three original equations within round-off *and* represents a shift of 1.2%, 0.07% and 1.4% in the values of the solution variables x, y and z. This system is *well-conditioned* – small changes in $a_{i,j}$ imply small (insensitive) changes in the solution.

	Coefficients			
Operation	x	y	z	**RHS**
E_1	<u>0.8612</u>	0.3713	0.1947	0.9834
E_2	0.4763	0.2082	0.8149	0.8516
E_3	0.3123	0.7369	0.5132	0.7982
E_1	0.8612	0.3713	0.1947	0.9834
$E_2^{(1)} = E_2 - 0.5531E_1$	0	<u>0.002800</u>	0.7072	0.3077
$E_3^{(1)} = E_3 - 0.3626E_1$	0	0.6023	0.4426	0.4416
E_1	0.8612	0.3713	0.1947	0.9834
$E_2^{(1)} = E_3^{(1)}$	0	<u>0.6023</u>	0.4426	0.4416
$E_3^{(1)} = E_2^{(1)}$	0	0.002800	0.7072	0.3077
E_1	0.8612	0.3713	0.1947	0.9834
$E_2^{(1)}$	0	0.6023	0.4426	0.4416
$E_3^{(2)} = E_3^{(1)} - 0.004649E_2^{(1)}$	0	0	<u>0.7051</u>	0.3056

Example 9.17 The system

$$1.01x + y = 1$$
$$x + y = 2$$

has the (exact) solution $x = -100$, $y = 102$. The perturbed system

$$1.02x + y = 1$$
$$x + y = 2$$

has the (exact) solution $x = -50$, $y = 52$. In other words, a 1% change in $a_{1,1}$ has the effect of changing the solution by some 50%. Such a system is said to be *ill-conditioned* and needs to be pre-conditioned before solving[3].

9.2.4.3 Operation count

The final discussion concerns the performance of the elimination algorithm. This needs to be addressed in order to ensure (a) that a computer program to simulate the algorithm will complete its task in a 'reasonable' time and (b) that rounding errors do not build up and swamp the solution.

The quality or performance of a numerical algorithm is judged in terms of storage, effect of rounding error and the operation count. The *operation count* refers to the number of (arithmetic) floating-point operations required by the algorithm to complete its task. For Gauss elimination this quantity can be developed by fairly simple means.

For the analysis that follows, it is assumed that the four common arithmetic operations of addition, subtraction, multiplication and division consume comparable CPU times, say τ_f.

(1) *Forward elimination:* Referring to Algorithm 9.3, the innermost loop 14 requires 2 flops (1$-$ and 1$\times$) per element a_{ij}, repeated $n-k$ times. The total operation count for loop 14 is $2(n-k)$ flops. Either side of the loop the multiplier m_{ik} and b_i update are computed with 3 flops. The total cost so far is $2(n-k)+3$ flops. This workload is contained within loop 12 which traverses $n-k$ equations. The total cost of loop 12 is then $(n-k)[2(n-k)+3]$ flops, and is executed for columns k from 1 to $n-1$, giving a total cost (in flops) of

$$\mathrm{FE_{flop}} = \sum_{k=1}^{n-1}(n-k)[2(n-k)+3].$$

[3] A rough guide to ill-conditioning is given by the determinant, $|\boldsymbol{A}|$, of the coefficient matrix $\boldsymbol{A}$. Small $|\boldsymbol{A}|$ implies an ill-conditioned system. For the 2×2 system here, $|\boldsymbol{A}| = 0.01$. For the 3×3 system of Example 9.16, $|\boldsymbol{A}| \simeq -0.36$ (which is quite reasonable in determinant terms).

With $s = n - k$,

$$\mathrm{FE}_{\mathrm{flop}} = \sum_{s=1}^{n-1}(2s^2 + 3s) = 2\frac{(n-1)n(2n-1)}{6} + 3\frac{(n-1)n}{2}$$

$$= \frac{4n^3 + 3n^2 - 7n}{6}. \qquad \mathbf{(9.15)}$$

(2) *Back substitution*: Loop 14 is executed $n - i$ times at a cost of 2 flops per pass. The total workload for this loop is $2(n - i)$ flops. Either side of the loop, s is initialised and x_i is computed using 1 flop, giving a subtotal of $2(n-i)+1$ flops. This workload is contained within loop 12 which is executed $n - 1$ times at a total cost of

$$\sum_{i=1}^{n-1}[2(n-i)+1]$$

flops. To this is added the single flop required to compute x_n giving a total operation count for back substitution of

$$\mathrm{BS}_{\mathrm{flop}} = 1 + \sum_{i=1}^{n-1}[2(n-i)+1] = n^2. \qquad \mathbf{(9.16)}$$

Adding expressions (9.15) and (9.16) gives the total operation count for Gauss elimination (*without* pivoting) as

$$\mathrm{GE}_{\mathrm{flop}}\frac{4n^3 + 9n^2 - 7n}{6}. \qquad \mathbf{(9.17)}$$

For large n the effort increases approximately like $2n^3/3$ (in other words, it is the forward elimination that dominates the computer time). However, this seemingly enormous effort can be put in context by comparison with *Cramer's rule* (using adj $\boldsymbol{A}$) which requires about $(1 + \mathrm{e})(n + 1)!$ flops.

GABRIEL CRAMER
(1704–52)
Swiss mathematician

LIFE

Gabriel Cramer was elected to the Royal Society of London in 1749.

WORK

Gabriel Cramer worked on analysis and determinants. He became professor of mathematics at Geneva and wrote on work related to physics, also on geometry and the history of mathematics. Cramer is best known for his work on determinants (1750) but also made contributions to the study of algebraic curves (1750).

Example 9.18 If $n = 60$ equations are to be solved on a machine working at 10^6 flops/s (quite reasonable), *Gauss elimination* requires approximately $2 \times 60^3/3 \times 10^{-6}$ s, that is 0.144 s of CPU time. *Cramer's rule* requires about $(1+e)61! \times 10^{-6}$ s, that is 1.89×10^{78} s, or 5.98×10^{70} years. The age of the Solar System is only 4×10^9 years (approx.) – enough time to solve 23 equations by Cramer's rule!

Notes In cases where the coefficient matrix A has a special structure, such as *tridiagonal*, that structure may be utilised to modify the 'direct' solution method and give a more efficient (fewer flops) algorithm. Two common cases are *symmetric systems* (using Choleski decomposition[4]) and *tridiagonal systems* (using Thomas's algorithm[5]).

9.3 Iterative solution methods

Iterative methods work on the original system and, starting with an initial guess $\boldsymbol{x}^{(0)}$ to the solution vector $\boldsymbol{x}$, generate a *vector sequence* $\{\boldsymbol{x}^{(k)}\}$ where $\boldsymbol{x}^{(k)}$ estimates $\boldsymbol{x}$.

From Section 6.3.2 you will recall that iterative solutions of $f(x) = 0$ are based upon the equivalent form $x = g(x)$, motivating the recurrence relation $x_{n+1} = g(x_n)$. Iterative solvers for linear systems are based upon a modified version of the original system $\boldsymbol{Ax} = \boldsymbol{b}$, obtained by a *matrix decomposition* of $\boldsymbol{A}$. The steps are

(1) develop a recurrence relation, based upon the system $\boldsymbol{Ax} = \boldsymbol{b}$, to *generate* a vector sequence $\{\boldsymbol{x}^{(k)}\}$ and

(2) determine the conditions under which the vector sequence converges,

$$\lim_{k\to\infty} \boldsymbol{x}^{(k)} = \boldsymbol{x}.$$

9.3.1 Iterative methods: general

Broadly speaking the term *matrix decomposition* implies that a specified matrix can be written as the sum of two matrices of the same size. The coefficient matrix $\boldsymbol{A}$ is decomposed into two matrices $\boldsymbol{E}$ and $\boldsymbol{F}$ such that

$$\boldsymbol{A} = \boldsymbol{E} + \boldsymbol{F}. \quad \textbf{(9.18)}$$

The linear system (9.5) can then be written in the equivalent form

$$(\boldsymbol{E} + \boldsymbol{F})\boldsymbol{x} = \boldsymbol{b},$$

---

[4] See [18], Section 9.8.

[5] See [18], Section 9.9.

which can be arranged to

$$\boldsymbol{E}\boldsymbol{x} = -\boldsymbol{F}\boldsymbol{x} + \boldsymbol{b}. \tag{9.19}$$

In other words

$$\boldsymbol{x} = -\boldsymbol{E}^{-1}\boldsymbol{F}\boldsymbol{x} + \boldsymbol{E}^{-1}\boldsymbol{b} \tag{9.20}$$

(analogous to the form $x = g(x)$ for the equation $f(x) = 0$). Given a vector $\boldsymbol{x}^{(0)}$, a vector sequence $\{\boldsymbol{x}^{(k)}\}$ can be generated by the recurrence relation

$$\boldsymbol{x}^{(k+1)} = -\boldsymbol{E}^{-1}\boldsymbol{F}\boldsymbol{x}^{(k)} + \boldsymbol{E}^{-1}\boldsymbol{b}, \quad k = 0, 1, 2, \ldots \tag{9.21}$$

Different decompositions, $\boldsymbol{E}$ and $\boldsymbol{F}$, of the matrix $\boldsymbol{A}$ will yield different recurrence relations.

Example 9.19 Generate a vector sequence from the 2×2 system

$$\begin{aligned} 5x \quad - y &= 3 \\ -x + 10y &= 19 \end{aligned} \tag{9.22}$$

The coefficient matrix and vector are

$$\boldsymbol{A} = \begin{bmatrix} 5 & -1 \\ -1 & 10 \end{bmatrix}, \quad \boldsymbol{b} = \begin{bmatrix} 3 \\ 19 \end{bmatrix}.$$

A valid decomposition of the form $\boldsymbol{A} = \boldsymbol{E} + \boldsymbol{F}$ is given by

$$\boldsymbol{E} = \begin{bmatrix} 5 & 0 \\ 0 & 10 \end{bmatrix} \quad \text{and} \quad \boldsymbol{F} = \begin{bmatrix} 0 & -1 \\ -1 & 0 \end{bmatrix}.$$

The derived components of eqn (9.20) are

$$\boldsymbol{E}^{-1} = \begin{bmatrix} \frac{1}{5} & 0 \\ 0 & \frac{1}{10} \end{bmatrix}, \quad \boldsymbol{E}^{-1}\boldsymbol{F} = \begin{bmatrix} 0 & -\frac{1}{5} \\ -\frac{1}{10} & 0 \end{bmatrix}, \quad \boldsymbol{E}^{-1}\boldsymbol{b} = \begin{bmatrix} \frac{3}{5} \\ \frac{19}{10} \end{bmatrix},$$

and the recurrence relation (9.21) becomes

$$\boldsymbol{x}^{(k+1)} = -\begin{bmatrix} 0 & -\frac{1}{5} \\ -\frac{1}{10} & 0 \end{bmatrix} \boldsymbol{x}^{(k)} + \begin{bmatrix} \frac{3}{5} \\ \frac{19}{10} \end{bmatrix}, \quad k = 0, 1, 2, \ldots$$

With $\boldsymbol{x}^{(0)} = \boldsymbol{0}$ the first two iterations are

$$\boldsymbol{x}^{(1)} = \begin{bmatrix} \frac{3}{5} \\ \frac{19}{10} \end{bmatrix} = \begin{bmatrix} 0.6 \\ 1.9 \end{bmatrix},$$

$$\boldsymbol{x}^{(2)} = -\begin{bmatrix} 0 & -\frac{1}{5} \\ -\frac{1}{10} & 0 \end{bmatrix} \begin{bmatrix} \frac{3}{5} \\ \frac{19}{10} \end{bmatrix} + \begin{bmatrix} \frac{3}{5} \\ \frac{19}{10} \end{bmatrix} = \begin{bmatrix} 0.98 \\ 1.96 \end{bmatrix}.$$

The next few iterates are shown in Table 9.1. It is clear that the vector sequence $\{\boldsymbol{x}^{(k)}\}$ *is* convergent, with limit $[1, 2]$ (to machine precision) after 10 iterations. Substituting the limit into the original system shows that [1, 2] is indeed the solution of the system – vector sequences can be used to solve linear systems.

Table 9.1 Vector sequence based upon eqn (9.22)

k	Iterate $x^{(k)}$	
3	0.992	1.998
4	0.9996	1.9992
5	0.99984	1.99996
6	0.999992	1.999984
7	0.999997	1.999999
8	1	2

The key to this process is ensuring that the vector sequence generated by (9.21) does indeed converge to the solution vector $\boldsymbol{x}$.

Example 9.20 If the equations in the 2×2 system of Example 9.19 are written down in reverse order,

$$\begin{aligned} -x + 10y &= 19 \\ 5x \quad - y &= 3 \end{aligned} \tag{9.23}$$

the solution vector is still $\boldsymbol{x} = [1, 2]^{\mathrm{T}}$ (check this by substitution). To generate a vector sequence from this ordering we have

$$\boldsymbol{A} = \begin{bmatrix} -1 & 10 \\ 5 & -1 \end{bmatrix}, \quad \boldsymbol{b} = \begin{bmatrix} 19 \\ 3 \end{bmatrix},$$

and a valid decomposition of the form $\boldsymbol{A} = \boldsymbol{E} + \boldsymbol{F}$ is given by

$$\boldsymbol{E} = \begin{bmatrix} -1 & 0 \\ 0 & -1 \end{bmatrix} \quad \text{and} \quad \boldsymbol{F} = \begin{bmatrix} 0 & 10 \\ 5 & 0 \end{bmatrix}.$$

The derived components of eqn (9.20) are

$$\boldsymbol{E}^{-1} = \begin{bmatrix} -1 & 0 \\ 0 & -1 \end{bmatrix}, \quad \boldsymbol{E}^{-1}\boldsymbol{F} = \begin{bmatrix} 0 & -10 \\ -5 & 0 \end{bmatrix}, \quad \boldsymbol{E}^{-1}\boldsymbol{b} = \begin{bmatrix} -19 \\ -3 \end{bmatrix},$$

and the recurrence relation (9.21) is

$$\boldsymbol{x}^{(k+1)} = -\begin{bmatrix} 0 & -10 \\ -5 & 0 \end{bmatrix} \boldsymbol{x}^{(k)} + \begin{bmatrix} -19 \\ -3 \end{bmatrix}, \quad k = 0, 1, 2, \ldots$$

With $\boldsymbol{x}^{(0)} = \boldsymbol{0}$ a selection of iterations is shown in Table 9.2. It is clear that this vector sequence is *not* convergent, yet the reordered system has the *same* solution as that of Example 9.19.

Table 9.2 Vector sequence based upon eqn (9.23).

k	Iterate $x^{(k)}$	
0	0	0
2	11	98
4	−539	−4802
6	20961	240198

Examples 9.19 and 9.20 imply that the choice of $\boldsymbol{E}$ and $\boldsymbol{F}$ is important, and we require an idea of how to choose $\boldsymbol{E}$ and $\boldsymbol{F}$ in order to ensure a *convergent sequence* $\{\boldsymbol{x}^{(k)}\}$.

9.3.1.1 Error analysis

The usual definition of a convergent iterative method is one in which successive errors are reduced. If the error after k iterations is $\boldsymbol{e}^{(k)} = \boldsymbol{x} - \boldsymbol{x}^{(k)}$, subtracting the recurrence relation (9.21) from the system (9.20) gives

$$\boldsymbol{x} - \boldsymbol{x}^{(k+1)} = -\boldsymbol{E}^{-1}\boldsymbol{F}\boldsymbol{x} + \boldsymbol{E}^{-1}\boldsymbol{F}\boldsymbol{x}^{(k)} = -\boldsymbol{E}^{-1}\boldsymbol{F}(\boldsymbol{x} - \boldsymbol{x}^{(k)})$$

(the terms $\boldsymbol{E}^{-1}\boldsymbol{b}$ cancel), that is

$$\boldsymbol{e}^{(k+1)} = -\boldsymbol{E}^{-1}\boldsymbol{F}\boldsymbol{e}^{(k)}. \qquad \textbf{(9.24)}$$

Equation (9.24) is a recurrence relation for the error vector (compare this with the relation $e_{k+1} = g'(\xi)e_k$ for the scalar scheme $x_{k+1} = g(x_k)$). For convergence $\boldsymbol{e}^{(k)}$ should approach $\boldsymbol{0}$ as k increases, that is

$$\lim_{k\to\infty} \boldsymbol{e}^{(k)} = \boldsymbol{0}. \qquad \textbf{(9.25)}$$

If the 'size' of $\boldsymbol{e}^{(k+1)}$ is less than the 'size' of $\boldsymbol{e}^{(k)}$ then condition (9.25) will be satisfied.

The 'size' of a vector or matrix can be measured by using a *norm*[6], that is a scalar value derived from a vector/matrix that is non-negative and equals zero if and only if all elements of the vector/matrix are zero.

Definition The *maximum norm* (∞-norm) is defined as follows:

(1) If $\boldsymbol{x}$ is a vector with n elements then

$$\|\boldsymbol{x}\|_\infty = \max_{1\le i\le n} |x_i|. \qquad \textbf{(9.26)}$$

(2) If A is an $n \times n$ matrix then

$$\|A\|_\infty = \max_{1\le i\le n} \sum_{j=1}^{n} |a_{ij}| \qquad \text{(maximum row sum)}. \qquad \textbf{(9.27)}$$

---

[6] For a thorough discussion on norms see, for example, [18], Chapter 10.

The ∞-norm equals the element of largest magnitude (for vectors) or the maximum absolute row sum (for matrices).

Example 9.21 For the vectors/matrices

$$[1\ 2\ 5],\quad [1\ 2\ -5],\quad \begin{bmatrix} 1 \\ 2 \\ -5 \end{bmatrix},\quad \begin{bmatrix} 1 & 2 \\ 3 & 4 \end{bmatrix},\quad \begin{bmatrix} 1 & 2 \\ -3 & 4 \end{bmatrix},$$

the ∞-norms are 5, 5, 5, 7 and 7, respectively.

DERIVE Experiment 9.2
DERIVE does not have intrinsic norm functions, but they are simple to define: `inormv` (vector) and `inormm` (matrix). Author

```
dim(x):=dimension(x)
e1(x,i):=element(x,i)
e2(x,i,j):=element(x,i,j)
absv(x):=vector(abs(e1(x,i)),i,1,dim(x))
inormv(x):=max(absv(x))
rsum(x):=vector(sum(abs(e2(x,i,j)),j,1,dim(x)),i,1,dim(x))
inormm(x):=max(rsum(x))
```

`dim` gives the size of a vector/matrix, `e1` and `e2` extract a specified element from a vector/matrix, `absv` removes minus signs from vector elements and `rsum` computes the absolute row sums of a matrix.

Apply these functions to the vectors and matrices of Example 9.21.

Using eqns (9.26) and (9.27),

$$\begin{aligned} \|\boldsymbol{Ax}\|_\infty = \max_{1\le i\le n}\left|\sum_{j=1}^{n} a_{ij}x_j\right| &\le \max_{1\le i\le n}\sum_{j=1}^{n}|a_{ij}|\,|x_j| \\ &\le \max_{1\le i\le n}\sum_{j=1}^{n}|a_{ij}| \times \max_{1\le i\le n}|x_i| \\ &= \|\boldsymbol{A}\|_\infty\,\|\boldsymbol{x}\|_\infty. \end{aligned} \tag{9.28}$$

Example 9.22 Given

$$\boldsymbol{A} = \begin{bmatrix} 5 & -1 \\ 2 & 3 \end{bmatrix},\quad \boldsymbol{x} = \begin{bmatrix} 2 \\ -3 \end{bmatrix},\quad \boldsymbol{Ax} = \begin{bmatrix} 13 \\ -5 \end{bmatrix},$$

then $\|\boldsymbol{x}\|_\infty = 3$, $\|\boldsymbol{A}\|_\infty = 6$ and $\|\boldsymbol{Ax}\|_\infty = 13 \le \|\boldsymbol{A}\|_\infty\|\boldsymbol{x}\|_\infty = 18$.

On applying (9.28) to the recurrence relation (9.24),

$$\|e^{(k+1)}\|_\infty = \| -E^{-1}Fe^{(k)}\|_\infty \le \|E^{-1}F\|_\infty \, \|e^{(k)}\|_\infty.$$

$\|e^{(k+1)}\|_\infty$ will be less than $\|e^{(k)}\|_\infty$ if[7]

$$\|E^{-1}F\|_\infty < 1. \tag{9.29}$$

This result is valid for *any* norm.

Example 9.23 For Example 9.19

$$E^{-1}F = \begin{bmatrix} 0 & -\frac{1}{5} \\ -\frac{1}{10} & 0 \end{bmatrix}.$$

$\|E^{-1}F\|_\infty = \max\{\frac{1}{5}, \frac{1}{10}\} = \frac{1}{5}$, which is less than 1, and the sequence generated by the corresponding recurrence relation *converges*.

For Example 9.20,

$$E^{-1}F = \begin{bmatrix} 0 & -10 \\ -5 & 0 \end{bmatrix}$$

and $\|E^{-1}F\|_\infty = \max\{10, 5\} = 5$, which is greater than 1. The sequence generated by the recurrence relation *does not* converge.

In practice, the iterative process is terminated once the measure $\|e^{(k)}\|$ is sufficiently small. Iterative methods have the advantage of being easy to program and they naturally reduce the *residual* $r^{(k)} = b - Ax^{(k)}$ to zero.

9.3.2 Iterative methods: specific cases

Here several of the more common iterative solvers are developed (based upon specific decomposition matrices E and F).

9.3.2.1 Jacobi iteration

KARL GUSTAV JACOB JACOBI (1804–51)
German mathematician

LIFE

Jacobi's father was a banker and his family were prosperous. He received a good education at the University of Berlin. He obtained his PhD in 1825 and taught

[7] A stronger result concerning convergence is obtained using the eigenvalues of the matrix $E^{-1}F$ – see [18], p. 287.

mathematics at the University of Königsberg from 1826 until his death, being appointed to a chair in 1832. Jacobi's reputation as an excellent teacher attracted many students. He introduced the seminar method to teach students the latest advances in mathematics.

WORK

Jacobi founded the theory of elliptic functions based on four theta functions. His *Fundamenta nova theoria functionum ellipticarum* (1829) and its supplements made basic contributions to the theory of elliptic functions.

Jacobi proved that if a single-valued function of one variable is doubly periodic then the ratio of the periods is imaginary (1834). This result prompted much further work in this area, in particular by Liouville and Cauchy.

Jacobi carried out important research in partial differential equations of the first order and applied them to the differential equations of dynamics. He worked on determinants and studied the functional determinant now called the Jacobian (it appears first in a 1815 paper of Cauchy) and wrote a long memoir *De determinantibus functionalibus* (1841) devoted to this determinant. He proves, among many things, that if a set of n functions in n variables are functionally related then the Jacobian is identically zero, while if the functions are independent the Jacobian cannot be identically zero.

The simplest iterative scheme is developed by isolating the diagonal terms in the original system (9.1) to give

$$
\begin{aligned}
a_{11}x_1 &= b_1 - \square - a_{12}x_2 - \quad - a_{1n}x_n \\
a_{22}x_2 &= b_2 - a_{21}x_1 - \square - \quad - a_{2n}x_n \\
\vdots \quad & \quad \vdots \quad \vdots \qquad\qquad \vdots \\
a_{nn}x_n &= b_n - a_{n1}x_1 - a_{n2}x_2 - \cdots \quad \square
\end{aligned}
\tag{9.30}
$$

Given the estimates $x_1^{(k)}, \ldots, x_n^{(k)}$ to the solution vector $x_1, \ldots, x_n$, an 'improved' set of estimates $x_1^{(k+1)}, \ldots, x_n^{(k+1)}$ is obtained from (9.30),

$$
\begin{aligned}
a_{11}x_1^{(k+1)} &= b_1 - \square - a_{12}x_2^{(k)} - \quad - a_{1n}x_n^{(k)} \\
a_{22}x_2^{(k+1)} &= b_2 - a_{21}x_1^{(k)} - \square - \quad - a_{2n}x_n^{(k)} \\
\vdots \quad & \quad \vdots \quad \vdots \qquad\qquad \vdots \\
a_{nn}x_n^{(k+1)} &= b_n - a_{n1}x_1^{(k)} - a_{n2}x_2^{(k)} - \cdots \quad \square
\end{aligned}
\tag{9.31}
$$

If the iterative process (9.31) converges, it does so for *any initial guess* $\boldsymbol{x}^{(0)}$. Typically, $\boldsymbol{x}^{(0)} = \boldsymbol{0}$. Further, for the method to work, *all* diagonal elements of the coefficient matrix $\boldsymbol{A}$ must be non-zero, that is $a_{ii} \neq 0, i = 1, \ldots, n$.

A more formal approach is to consider the triple decomposition of $\boldsymbol{A}$:

$$\boldsymbol{A} = \boldsymbol{D} + \boldsymbol{L} + \boldsymbol{U} \tag{9.32}$$

where

$$\boldsymbol{A} = \begin{bmatrix} a_{11} & a_{12} & \cdots & a_{1n} \\ a_{21} & a_{22} & \cdots & a_{2n} \\ \vdots & \vdots & & \vdots \\ a_{n1} & a_{n2} & \cdots & a_{nn} \end{bmatrix}, \quad \boldsymbol{D} = \begin{bmatrix} a_{11} & 0 & \cdots & 0 \\ 0 & a_{22} & \ddots & \vdots \\ \vdots & \ddots & & 0 \\ 0 & \cdots & 0 & a_{nn} \end{bmatrix},$$

$$\boldsymbol{L} = \begin{bmatrix} 0 & \cdots & \cdots & 0 \\ a_{21} & \ddots & & \vdots \\ \vdots & \ddots & \ddots & \vdots \\ a_{n1} & \cdots & a_{nn-1} & 0 \end{bmatrix}, \quad \boldsymbol{U} = \begin{bmatrix} 0 & a_{12} & \cdots & a_{1n} \\ \vdots & \ddots & \ddots & \vdots \\ \vdots & & \ddots & a_{n-1n} \\ 0 & \cdots & \cdots & 0 \end{bmatrix}.$$

$\boldsymbol{D}$ is a *diagonal matrix*, $\boldsymbol{L}$ is *strictly lower triangular* and $\boldsymbol{U}$ is *strictly upper triangular*. On comparing eqns (9.32) and (9.18), the simplest iterative scheme is generated by taking $\boldsymbol{E} = \boldsymbol{D}$ and $\boldsymbol{F} = \boldsymbol{L} + \boldsymbol{U}$, whereby the recurrence relation (9.21) becomes

$$\boldsymbol{D}\boldsymbol{x}^{(k+1)} = -(\boldsymbol{L} + \boldsymbol{U})\boldsymbol{x}^{(k)} + \boldsymbol{b}. \tag{9.33}$$

At each step the right-hand side is known and eqn (9.33) represents the diagonal system of equations (9.31) with immediate solution

$$\boldsymbol{x}^{(k+1)} = -\boldsymbol{D}^{-1}(\boldsymbol{L} + \boldsymbol{U})\boldsymbol{x}^{(k)} + \boldsymbol{D}^{-1}\boldsymbol{b}. \tag{9.34}$$

Equation (9.34) is the *Jacobi iteration method* and, on expanding the matrix–vector products, can be written in the element-by-element form

$$x_i^{(k+1)} = \frac{1}{a_{ii}} \left[b_i - \sum_{j=1}^{i-1} a_{ij} x_j^{(k)} - \sum_{j=i+1}^{n} a_{ij} x_j^{(k)} \right]. \tag{9.35}$$

The first term in [ ] represents the vector $\boldsymbol{b}$, the second term is the contribution from $\boldsymbol{Lx}$ and the third term is the contribution from $\boldsymbol{Ux}$. Each estimate $x_i^{(k+1)}$ is determined solely in terms of the previous iterate $\boldsymbol{x}^{(k)}$.

Example 9.24 For the 2×2 system of Example 9.19,

$$\boldsymbol{A} = \begin{bmatrix} 5 & -1 \\ -1 & 10 \end{bmatrix}, \quad \boldsymbol{b} = \begin{bmatrix} 3 \\ 19 \end{bmatrix}.$$

The triple decomposition $\boldsymbol{A} = \boldsymbol{D} + \boldsymbol{L} + \boldsymbol{U}$ is defined by

$$\boldsymbol{D} = \begin{bmatrix} 5 & 0 \\ 0 & 10 \end{bmatrix}, \quad \boldsymbol{L} = \begin{bmatrix} 0 & 0 \\ -1 & 0 \end{bmatrix}, \quad \boldsymbol{U} = \begin{bmatrix} 0 & -1 \\ 0 & 0 \end{bmatrix},$$

($\boldsymbol{E} = \boldsymbol{D}$ and $\boldsymbol{F} = \boldsymbol{L} + \boldsymbol{U}$) and the recurrence relation (9.34) is

$$
\begin{aligned}
\boldsymbol{x}^{(k+1)} &= -\boldsymbol{D}^{-1}(\boldsymbol{L}+\boldsymbol{U})\boldsymbol{x}^{(k)} + \boldsymbol{D}^{-1}\boldsymbol{b} \\
&= -\begin{bmatrix} \frac{1}{5} & 0 \\ 0 & \frac{1}{10} \end{bmatrix} \begin{bmatrix} 0 & -1 \\ -1 & 0 \end{bmatrix} \boldsymbol{x}^{(k)} + \begin{bmatrix} \frac{3}{5} \\ \frac{19}{10} \end{bmatrix} \\
&= \begin{bmatrix} 0 & \frac{1}{5} \\ \frac{1}{10} & 0 \end{bmatrix} \boldsymbol{x}^{(k)} + \begin{bmatrix} \frac{3}{5} \\ \frac{19}{10} \end{bmatrix}, \qquad k = 0, 1, 2, \ldots
\end{aligned}
$$

This is the same update formula as Example 9.19. For convenience, selected iterates are again presented here (Table 9.3), starting with the initial guess $\boldsymbol{x}^{(0)} = \boldsymbol{0}$. It is clear that the vector sequence $\{\boldsymbol{x}^{(k)}\}$ is convergent, with limiting value $[1, 2]$ (to 6 decimal places) after 8 iterations, which is the exact solution of the system.

That the method is convergent can be validated by inspecting the norm of $\boldsymbol{E}^{-1}\boldsymbol{F}$, that is the norm of $\boldsymbol{D}^{-1}(\boldsymbol{L}+\boldsymbol{U})$. By inspection, the maximum row sum equals 1/5. This value is less than 1 and the convergence criterion (9.29) is satisfied.

Table 9.3 Jacobi vector sequence.

k	Iterate $x^{(k)}$	
0	0	0
1	0.6	1.9
2	0.98	1.96
3	0.992	1.998
4	0.9996	1.9992
5	0.99984	1.99996
6	0.999992	1.999984
7	0.999997	1.999999
8	1	2

DERIVE Experiment 9.3

The intrinsic function `iterates` can be used to implement Jacobi iteration. First, some auxiliary functions to compute $\boldsymbol{D}$, $\boldsymbol{L}$ and $\boldsymbol{U}$ from the matrix $\boldsymbol{A}$ – Author

```
dim(a):=dimension(a)
e2(a,i,j):=element(a,i,j)
d(a):=vector(vector(if(k=j,e2(a,k,j),0),j,1,dim(a)),k,1,dim(a))
l(a):=vector(vector(if(k>j,e2(a,k,j),0),j,1,dim(a)),k,1,dim(a))
u(a):=vector(vector(if(k<j,e2(a,k,j),0),j,1,dim(a)),k,1,dim(a))
```

The iteration matrix $-\boldsymbol{D}^{-1}(\boldsymbol{L}+\boldsymbol{U})$ and vector $\boldsymbol{D}^{-1}\boldsymbol{b}$ are given by

```
j(a):=-d(a)^-1.(l(a)+u(a))
c(a,b):=d(a)^-1.b
```

and n steps of the iteration process are implemented with

```
jac(j,c,x,n):=iterates(j.x+c,x,vector(0,k,1,dim(x))',n)
```

where $\boldsymbol{x}^{(0)} = \boldsymbol{0}$. x represents the solution vector $\boldsymbol{x}$.

To reproduce Example 9.24, Author

```
a:=[[5,-1],[-1,10]]
b:=[3,19]'
jac(j(a),c(a,b),[x,y],8)
```

and approX. If $\boldsymbol{A}$ is 3×3, replace `[x,y]` by `[x,y,z]`. If $\boldsymbol{A}$ is 4×4, replace `[x,y]` by `[w,x,y,z]`, etc.

9.3.2.2 Gauss–Seidel iteration

Here we show how to improve the basic Jacobi iteration method (see Section 9.3.2.1) by reducing the number of iterations required to attain convergence (without increasing the operation count per step).

PHILIPP LUDWIG VON SEIDEL
(1821–96)
German mathematician

LIFE
Seidel entered the University of Berlin in 1840 and studied under Dirichlet and Encke. He moved to Königsberg, where he studied under Bessel, Jacobi and Neumann. Seidel obtained his doctorate from Munich (1846) and went on to become professor.

WORK
Seidel wrote on dioptics and mathematical analysis. His work on lenses identified mathematically five coefficients describing the aberration of a lens, now called 'Seidel sums'. He introduced the concept of non-uniform convergence and applied probability to astronomy.

In eqns (9.31) the improved estimate $x_1^{(k+1)}$ is evaluated in the first equation. It makes sense to then replace the value $x_1^{(k)}$ in equations 2 to n by this improved estimate for the first element. Similarly, once $x_2^{(k+1)}$ is evaluated from equation 2 it can be used to replace the value $x_2^{(k)}$ in equations 3 to n.

Continuing in this way, the system (9.31) is modified to the form

$$\begin{aligned}
a_{11}x_1^{(k+1)} &= b_1 - \square - a_{12}x_2^{(k)} - \cdots - a_{1n}x_n^{(k)}\\
a_{22}x_2^{(k+1)} &= b_2 - a_{21}x_1^{(k+1)} - \square - \cdots - a_{2n}x_n^{(k)}\\
&\vdots\\
a_{nn}x_n^{(k+1)} &= b_n - a_{n1}x_1^{(k+1)} - a_{n2}x_2^{(k+1)} - \cdots \quad \square
\end{aligned} \tag{9.36}$$

In other words, in each equation the *most recent* estimates to $x_1, \ldots, x_{i-1}$ are used to obtain the improved estimate $x_i^{(k+1)}$.

In terms of the matrices $\boldsymbol{D}$, $\boldsymbol{L}$ and $\boldsymbol{U}$, system (9.36) can be written as

$$\boldsymbol{D}\boldsymbol{x}^{(k+1)} = -\boldsymbol{L}\boldsymbol{x}^{(k+1)} - \boldsymbol{U}\boldsymbol{x}^{(k)} + \boldsymbol{b},$$

that is

$$(\boldsymbol{D}+\boldsymbol{L})\boldsymbol{x}^{(k+1)} = -\boldsymbol{U}\boldsymbol{x}^{(k)} + \boldsymbol{b}, \tag{9.37}$$

with formal solution

$$\boldsymbol{x}^{(k+1)} = -(\boldsymbol{D}+\boldsymbol{L})^{-1}\boldsymbol{U}\boldsymbol{x}^{(k)} + (\boldsymbol{D}+\boldsymbol{L})^{-1}\boldsymbol{b}, \quad k = 0, 1, 2, \ldots \tag{9.38}$$

At each iteration the right-hand side is known and eqn (9.37) represents a *lower-triangular* system of equations which can be solved by *forward substitution.* The update formula is known as the *Gauss–Seidel iteration method.*

On expanding the matrix–vector products, eqn (9.37) can be written in the element-by-element form

$$x_i^{(k+1)} = \frac{1}{a_{ii}}\left[b_i - \sum_{j=1}^{i-1} a_{ij}x_j^{(k+1)} - \sum_{j=i+1}^{n} a_{ij}x_j^{(k)}\right]. \tag{9.39}$$

The first term in [ ] represents the vector $\boldsymbol{b}$, the second term is the contribution from $\boldsymbol{L}\boldsymbol{x}$ and the third term is the contribution from $\boldsymbol{U}\boldsymbol{x}$. Gauss-Seidel iteration is a *successive correction* method, whereas Jacobi iteration is a *simultaneous correction* method.

Example 9.25 Re-solving the 2×2 system of Example 9.24, the coefficients are

$$\boldsymbol{A} = \begin{bmatrix} 5 & -1 \\ -1 & 10 \end{bmatrix}, \quad \boldsymbol{b} = \begin{bmatrix} 3 \\ 19 \end{bmatrix}.$$

The triple decomposition $\boldsymbol{A} = \boldsymbol{D} + \boldsymbol{L} + \boldsymbol{U}$ is given by

$$\boldsymbol{D} = \begin{bmatrix} 5 & 0 \\ 0 & 10 \end{bmatrix}, \quad \boldsymbol{L} = \begin{bmatrix} 0 & 0 \\ -1 & 0 \end{bmatrix}, \quad \boldsymbol{U} = \begin{bmatrix} 0 & -1 \\ 0 & 0 \end{bmatrix},$$

($\boldsymbol{E} = \boldsymbol{D} + \boldsymbol{L}$ and $\boldsymbol{F} = \boldsymbol{U}$). The recurrence relation (9.38) is

$$\begin{aligned} \boldsymbol{x}^{(k+1)} &= -(\boldsymbol{D}+\boldsymbol{L})^{-1}\boldsymbol{U}\boldsymbol{x}^{(k)} + (\boldsymbol{D}+\boldsymbol{L})^{-1}\boldsymbol{b} \\ &= \begin{bmatrix} 0 & \frac{1}{5} \\ 0 & \frac{1}{50} \end{bmatrix}\boldsymbol{x}^{(k)} + \begin{bmatrix} \frac{3}{5} \\ \frac{49}{25} \end{bmatrix}, \quad k = 0, 1, 2, \ldots \end{aligned}$$

Selected iterates are shown in Table 9.4, starting with $\boldsymbol{x}^{(0)} = \boldsymbol{0}$. It is quite clear that the vector sequence $\{\boldsymbol{x}^{(k)}\}$ *is* convergent to the exact solution $[1, 2]$ after just five iterations.

That the method is convergent can be validated by inspecting the norm of $\boldsymbol{E}^{-1}\boldsymbol{F}$, that is the norm of $(\boldsymbol{D}+\boldsymbol{L})^{-1}\boldsymbol{U}$. By inspection the maximum row sum equals $1/5$. This value is less than 1 and so the convergence criterion (9.29) is satisfied.

Table 9.4 Gauss–Seidel vector sequence.

k	Iterate $x^{(k)}$	
0	0	0
1	0.6	1.96
2	0.992	1.9992
3	0.99984	1.999984
4	0.999997	2
5	1	2

Observations

(1) Gauss–Seidel iteration appears to converge more rapidly than Jacobi iteration, intuitively by virtue of using the latest elemental estimates at each step of the iteration process. By inspecting the element-by-element forms (9.35) and (9.39), no additional arithmetic is required by Gauss–Seidel. There are, however, cases in which Jacobi iteration does converge and Gauss–Seidel does not converge.

(2) Examples 9.24 and 9.25 highlight the 'looseness' of the norm approach to convergence. For Jacobi and *Gauss–Seidel, the norm is 1/5 (i.e. both converge). This does* not *distinguish the rates of convergence.*

DERIVE Experiment 9.4

Implementing Gauss–Seidel iteration uses the auxiliary functions defined in *DERIVE Experiment 9.3*. In addition, the iteration matrix $-(\boldsymbol{D}+\boldsymbol{L})^{-1}\boldsymbol{U}$ and vector $(\boldsymbol{D}+\boldsymbol{L})^{-1}\boldsymbol{b}$ are given by

```
g(a):=-(d(a)+l(a))^-1.u(a)
c(a,b):=(d(a)+l(a))^-1.b
```

and n steps of the iteration process are implemented with

```
gas(g,c,x,n):=iterates(g.x+c,x,vector(0,k,1,dim(x))‘,n)
```

where $\boldsymbol{x}^{(0)}=\boldsymbol{0}$. x represents the solution vector $\boldsymbol{x}$.

To reproduce Example 9.25, Author

```
a:=[[5,-1],[-1,10]]
b:=[3,19]‘
gas(g(a),c(a,b),[x,y],5)
```

and approX.

9.3.2.3 Successive over-relaxation

An improvement over Gauss–Seidel iteration is now proposed. By ‘adding and subtracting’ the term $x_i^{(k)}$ to the right-hand side of the element-by-element form (9.39), the Gauss–Seidel update formula can be written as

$$\begin{aligned} x_i^{(k+1)} &= x_i^{(k)} + \frac{1}{a_{ii}}\left[b_i - \sum_{j=1}^{i-1} a_{ij}x_j^{(k+1)} - \sum_{j=i}^{n} a_{ij}x_j^{(k)}\right] \\ &= x_i^{(k)} + \delta_i^{(k)}. \end{aligned} \tag{9.40}$$

Gauss–Seidel iteration can be viewed as adding a correction term to the estimate $x_i^{(k)}$ in order to produce the improved estimate $x_i^{(k+1)}$. For many classes of system of equations the correction term $\delta_i^{(k)}$ is *one-signed* throughout the iteration process and the sequence $\{x_i^{(k)}\}$ converges to the solution x_i monotonically from above or from below.

Example 9.26 In Figure 9.1(a) each correction term is shown to be positive and the element $x_i^{(k)}$ is converging from below to the exact value x_i.

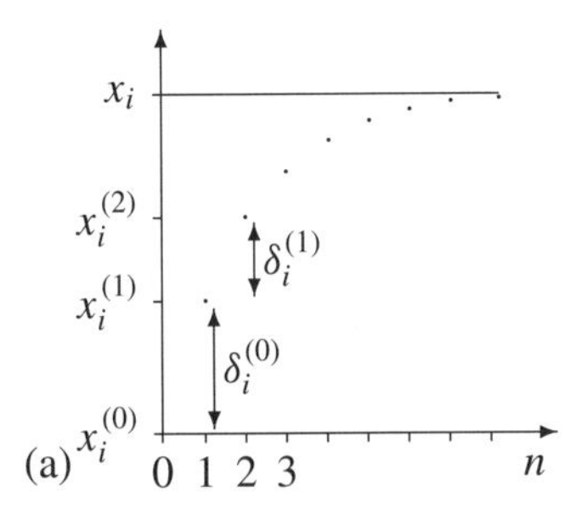

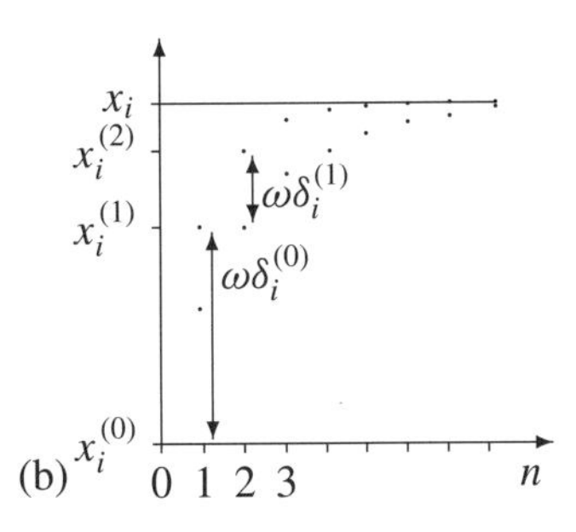

Figure 9.1 (a) Monotonic corrections. (b) Over-relaxed corrections.

Convergence to x_i could be *accelerated* if a *larger correction* were used at each step of the iteration process – not too large! For some $1 < \omega < 2$ the update formula (9.40) is replaced by[8]

$$x_i^{(k+1)} = x_i^{(k)} + \omega\delta_i^{(k)}. \tag{9.41}$$

If the value of ω is chosen carefully then the sequence $\{x_i^{(k)}\}$ might be expected to converge more rapdily to x_i, as illustrated by the SOR curve in Figure 9.1(b). Equation (9.41) is the *successive over-relaxation method*, or SOR for short.

The efficiency of the scheme in reducing the iterations required for convergence is dependent upon the value of the *acceleration parameter* ω. However, it is extremely difficult to determine the optimum value of ω, say ω_{opt}. One documented case where ω_{opt} can be determined is the following.

If all eigenvalues of the *Jacobi* point-iteration matrix $\boldsymbol{J} = -\boldsymbol{D}^{-1}(\boldsymbol{L}+\boldsymbol{U})$ are real and $0 < \rho(\boldsymbol{J}) < 1$ then[9]

$$\omega_{\text{opt}} = \frac{2}{1+\sqrt{1-\rho^2(\boldsymbol{J})}}. \tag{9.42}$$

If $\rho > 1$ then ω_{opt} is complex.[10]

Example 9.27 For Example 9.24, the Jacobi point-iteration matrix is

$$\boldsymbol{J} = \begin{bmatrix} 0 & \frac{1}{5} \\ \frac{1}{10} & 0 \end{bmatrix}.$$

[8] If $\omega = 1$ the Gauss–Seidel method is recovered.
[9] For a proof see [19], pp. 282–4.
[10] $\rho = \max|\lambda_i|$ is the *spectral radius*.

Its eigenvalues λ satisfy the equation $|-D^{-1}(L+U)-\lambda I|=0$, where I is the 2×2 identity matrix, that is $\lambda^2-\frac{1}{50}=0$. The solutions are $\lambda_{1,2}=\pm\frac{1}{5\sqrt{2}}$ which are real and of magnitude less than 1. Hence

$$\omega_{\text{opt}}=\frac{2}{1+\sqrt{1-\frac{1}{50}}}=\frac{10\sqrt{2}}{5\sqrt{2}+7}\approx 1.005.$$

The next step is to integrate the computed value of ω into the iteration process (9.41).

Example 9.28 Re-solve the 2×2 system

$$\begin{aligned} 5x \quad - y &= 3 \\ -x + 10y &= 19 \end{aligned}$$

The Jacobi iteration matrix satisfies the requirement $\rho(\boldsymbol{J})<1$, and $\omega_{\text{opt}}=1.005$ (see Example 9.27). The resulting SOR process is

$$\begin{aligned}\boldsymbol{x}^{(k+1)} &= \boldsymbol{x}^{(k)}+\omega\{(-(\boldsymbol{D}+\boldsymbol{L})^{-1}\boldsymbol{U}-\boldsymbol{I})\boldsymbol{x}^{(k)}+(\boldsymbol{D}+\boldsymbol{L})^{-1}\boldsymbol{b}\} \\ &= \boldsymbol{x}^{(k)}+1.005\left\{\begin{bmatrix}-1 & \frac{1}{5}\\ 0 & -\frac{49}{50}\end{bmatrix}\boldsymbol{x}^{(k)}+\begin{bmatrix}\frac{3}{5}\\ \frac{49}{25}\end{bmatrix}\right\}.\end{aligned}$$

Selected iterates are presented in Table 9.5, with the initial guess $\boldsymbol{x}^{(0)}=\boldsymbol{0}$. The vector sequence $\{\boldsymbol{x}^{(k)}\}$ converges to the exact solution [1, 2] after only five iterations. That the method is convergent can be validated by inspecting the norm of $\boldsymbol{E}^{-1}\boldsymbol{F}$, that is the norm of $\boldsymbol{I}-\omega((\boldsymbol{D}+\boldsymbol{L})^{-1}\boldsymbol{U}+\boldsymbol{I})$, or

$$\begin{bmatrix}-0.005 & 0.201\\ 0 & 0.0151\end{bmatrix}.$$

The maximum row sum equals 0.206 which is less than 1, and the convergence criterion (9.29) is again satisfied.

Table 9.5 SOR vector sequence.

k	Iterate $x^{(k)}$	
0	0	0
1	0.603	1.9698
2	0.995915	1.999544
3	0.999929	1.999993
4	0.999999	2
5	1	2

Not much has been gained over Gauss–Seidel iteration. The fact that ω_{opt} is *close* to 1 indicates that the Gauss–Seidel iteration is itself almost optimal for this 2×2 system!

Example 9.29 illustrates the effectiveness of SOR when ω_{opt} is genuinely distinct from 1.

Example 9.29 The 2×2 system

$$\begin{aligned} x \ + y &= 3 \\ x + 2y &= 5 \end{aligned}$$

has the solution $[1, 2]$ and associated matrices

$$\boldsymbol{A} = \begin{bmatrix} 1 & 1 \\ 1 & 2 \end{bmatrix}, \quad \boldsymbol{b} = \begin{bmatrix} 3 \\ 5 \end{bmatrix},$$

$$\boldsymbol{D} = \begin{bmatrix} 1 & 0 \\ 0 & 2 \end{bmatrix}, \quad \boldsymbol{L} = \begin{bmatrix} 0 & 0 \\ 1 & 0 \end{bmatrix}, \quad \boldsymbol{U} = \begin{bmatrix} 0 & 1 \\ 0 & 0 \end{bmatrix}.$$

The Jacobi iteration matrix is

$$\boldsymbol{J} = -\boldsymbol{D}^{-1}(\boldsymbol{L} + \boldsymbol{U}) = -\begin{bmatrix} 1 & 0 \\ 0 & \frac{1}{2} \end{bmatrix} \begin{bmatrix} 0 & 1 \\ 1 & 0 \end{bmatrix} = \begin{bmatrix} 0 & -1 \\ -\frac{1}{2} & 0 \end{bmatrix}$$

with eigenvalues given by $\lambda^2 - \frac{1}{2} = 0$, that is $\lambda = \pm\frac{1}{\sqrt{2}}$, and $\rho(J) = \frac{1}{\sqrt{2}} < 1$. The optimal SOR parameter is

$$\omega_{\text{opt}} = \frac{2}{1 + \sqrt{1 - \rho^2(J)}} = \frac{2}{1 + \sqrt{\frac{1}{2}}} \approx 1.17$$

The Jacobi process is

$$\boldsymbol{x}^{(k+1)} = \begin{bmatrix} 0 & -1 \\ -\frac{1}{2} & 0 \end{bmatrix} \boldsymbol{x}^{(k)} + \begin{bmatrix} 3 \\ \frac{5}{2} \end{bmatrix}.$$

The Gauss–Seidel process is

$$\boldsymbol{x}^{(k+1)} = \begin{bmatrix} 0 & -1 \\ 0 & \frac{1}{2} \end{bmatrix} \boldsymbol{x}^{(k)} + \begin{bmatrix} 3 \\ 1 \end{bmatrix}.$$

The SOR process is

$$\boldsymbol{x}^{(k+1)} = \boldsymbol{x}^{(k)} + 1.17 \left\{ \begin{bmatrix} -1 & -1 \\ 0 & -\frac{1}{2} \end{bmatrix} \boldsymbol{x}^{(k)} + \begin{bmatrix} 3 \\ 1 \end{bmatrix} \right\}.$$

Table 9.6 shows the solution sequence for each of these methods. The calculations are performed to 14 digits, and the table shows the results displayed to 5 digits (4 decimal places).

$$\rho(J) = \frac{1}{\sqrt{2}} \approx 0.707$$
$$\rho(G) = \frac{1}{2} = 0.5$$
$$\rho(S) = \sqrt{2} - 1 \approx 0.414$$

Table 9.6 Jacobi, Gauss–Seidel and SOR sequences.

k	**Jacobi**		**Gauss–Seidel**		**SOR**	
0	0	0	0	0	0	0
1	3.0000	2.5000	3.0000	1.0000	3.5100	1.1700
2	0.5000	1.0000	2.0000	1.5000	1.5444	1.6556
3	2.0000	2.2500	1.5000	1.7500	1.3105	1.8571
4	0.7500	1.5000	1.2500	1.8750	1.1145	1.9407
5	1.5000	2.1250	1.1250	1.9375	1.0499	1.9754
6	0.8750	1.7500	1.0625	1.9688	1.0203	1.9898
7	1.2500	2.0625	1.0312	1.9844	1.0085	1.9958
8	0.9375	1.8750	1.0156	1.9922	1.0035	1.9982
9	1.1250	2.0312	1.0078	1.9961	1.0015	1.9993
10	0.9688	1.9375	1.0039	1.9980	1.0006	1.9997
11	1.0625	2.0156	1.0200	1.9990	1.0003	1.9999
12	0.9844	1.9688	1.0010	1.9995	1.0001	1.9999
13	1.0312	2.0078	1.0005	1.9998	1.0000	2.0000
14	0.9922	1.9844	1.0002	1.9999		
15	1.0156	2.0039	1.0001	1.9999		
16	0.9961	1.9922	1.0001	2.0000		
17	1.0078	2.0020	1.0000	2.0000		
18	0.9980	1.9961				
19	1.0039	2.0010				
20	0.9990	1.9980				
21	1.0020	2.0005				
22	0.9995	1.9990				
23	1.0010	2.0002				
24	0.9998	1.9995				
25	1.0005	2.0001				
26	0.9999	1.9998				
27	1.0002	2.0001				
28	0.9999	1.9999				
29	1.0001	2.0000				
30	1.0000	1.9999				
31	1.0001	2.0000				
32	1.0000	2.0000				

DERIVE Experiment 9.5

To implement SOR iteration requires a few modifications to *DERIVE Experiment 9.4*. The identity matrix $\boldsymbol{I}$, iteration matrix $-(\boldsymbol{D}+\boldsymbol{L})^{-1}\boldsymbol{U}-\boldsymbol{I}$ and vector $(\boldsymbol{D}+\boldsymbol{L})^{-1}\boldsymbol{b}$ are given by

```
i(a):=vector(vector(if(k=j,1,0),j,1,dim(a)),k,1,dim(a))
s(a):=-(d(a)+l(a))^-1.u(a)-i(a)
c(a,b):=(d(a)+l(a))^-1.b
```

and n steps of the iteration process are implemented with

```
sor(s,c,w,x,n):=
   iterates(x+ws.x+wc,x,vector(0,k,1,dim(x))',n)
```

where $\boldsymbol{x}^{(0)}=\boldsymbol{0}$. `x` represents the solution vector $\boldsymbol{x}$.

To reproduce Example 9.29, Author

```
a:=[[1,1],[1,2]]
b:=[3,5]'
sor(s(a),c(a,b),1.17,[x,y],13)
```

and approX.

Summary

This chapter dealt with systematic solutions of the linear system $\boldsymbol{Ax}=\boldsymbol{b}$.

Direct methods (Section 9.2): simple cases were discussed in Sections 9.2.1 (diagonal systems) and 9.2.2 (triangular systems) leading to the back substitution algorithm (9.10). Section 9.2.3 discussed the reduction of a square system to triangular form using forward elimination (see Example 9.9) and the determinacy of a system was briefly mentioned in Section 9.2.3.1. Combining forward elimination and back substitution gave the Gauss elimination solution method (see Section 9.2.4). Several drawbacks were discussed, in particular loss of significance and its 'remedy' using partial pivoting (see Section 9.2.4.1).

Iterative methods (Section 9.3): The link between the system $\boldsymbol{Ax}=\boldsymbol{b}$, matrix iteration and vector sequences was introduced in Section 9.3.1. An error analysis was developed in Section 9.3.1.1 leading to the convergence criterion (9.29). Particular iterative schemes (Jacobi, Gauss–Seidel and SOR) were developed and compared in Sections 9.3.2.1, 9.3.2.2 and 9.3.2.3.

Exercises

9.1 Determine the operation count for solving the $n\times n$ diagonal system $\boldsymbol{Dx}=\boldsymbol{b}$ (see eqn (9.8)).

9.2 Use back substitution and/or forward substitution to solve the following triangular systems of equations.

$$\begin{bmatrix} -1 & 8 & -2 \\ 0 & 1 & 2 \\ 0 & 0 & -1 \end{bmatrix} \boldsymbol{x} = \begin{bmatrix} 0 \\ 0 \\ 1 \end{bmatrix}, \quad \begin{bmatrix} -1 & 0 & 0 \\ 2 & 1 & 0 \\ -2 & 8 & -1 \end{bmatrix} \boldsymbol{x} = \begin{bmatrix} 0 \\ 0 \\ 1 \end{bmatrix}.$$

9.3 Use forward elimination to reduce the matrix $\boldsymbol{A}$ to upper-triangular form $\boldsymbol{U}$. At each stage of the process insert the multipliers m_{ik} (see Algorithm 9.2) into the lower-triangular matrix $\boldsymbol{L}$ at row i, column k,

$$\boldsymbol{A} = \begin{bmatrix} 2 & 2 & 3 \\ 4 & 7 & 7 \\ -2 & 4 & 5 \end{bmatrix}, \quad \boldsymbol{L} = \begin{bmatrix} 1 & 0 & 0 \\ m_{21} & 1 & 0 \\ m_{31} & m_{32} & 1 \end{bmatrix}.$$

Compute the matrix product $\boldsymbol{LU}$ and comment on the result.

9.4 (a) Show that the following system has *no* solution:

$$\begin{aligned} x + 3y + 2z &= -1 \\ 2x + 2y \quad - z &= 2 \\ 3x + 5y \quad + z &= 0 \end{aligned}$$

(b) Show that the following system has a solution if and only if $a = 0$ or $a = 1/4$, and obtain the solution in each case.

$$\begin{aligned} x \quad + y \quad + z &= a^2 \\ 4x \quad - y \quad + z &= a \\ x \quad + y + 2z &= 1 \\ x + 6y + 5z &= 1 \end{aligned}$$

(c) Given the matrix

$$\boldsymbol{A} = \begin{bmatrix} 0 & 3 & 0 \\ 1 & 0 & 1 \\ 0 & 1 & 0 \end{bmatrix},$$

show that the system of equations $(\boldsymbol{A} - \lambda \boldsymbol{I})\boldsymbol{x} = \boldsymbol{0}$ has a non-zero solution if and only if $\lambda \in \{-2, 0, 2\}$. Find a form of the solution in each case. What have you found ?

(d) Show that the following system has an *infinity* of solutions and find the general form of these solutions.

$$\begin{aligned} x \quad + y \quad + 3z &= 1 \\ 3x + 2y \quad + z &= 5 \\ x + 2y + 11z &= -1 \end{aligned}$$

9.5 Use Gauss elimination to show that the following system has a unique solution, and find this solution.

$$\begin{bmatrix} 1 & 2 & 1 & 1 \\ 2 & 3 & 4 & 1 \\ 1 & 2 & 2 & 2 \\ 3 & 7 & -1 & -1 \end{bmatrix} \boldsymbol{x} = \begin{bmatrix} 2 \\ 6 \\ 2 \\ -8 \end{bmatrix}.$$

9.6 Solve the following system of equations using 2-digit floating-point arithmetic (a) by Gauss elimination without pivoting and (b) by Gauss elimination with partial pivoting.

$$\begin{aligned} 0.50x + 1.1y + 3.1z &= 6.0 \\ 2.0x + 4.5y + 0.4z &= 0.020 \\ 5.0x + 1.0y + 6.5z &= 1.0 \end{aligned}$$

Comment on your results.

9.7 Use the formulae derived for calculating the operation counts (see Section 9.2.4.3) to determine the size of the largest system of linear equations that could be solved during your lifetime, say 80 years at 365 days per year, (a) by Gauss elimination and (b) by Cramer's rule at 10^6 flops/s.

9.8 For the matrix $\boldsymbol{A}$ in Exercise 9.3, if $\boldsymbol{b} = [3 \;\; 1 \;\; -7]^{\mathrm{T}}$, solve the system $\boldsymbol{Ax} = \boldsymbol{b}$. Now, solve $\boldsymbol{Lc} = \boldsymbol{b}$ followed by $\boldsymbol{Ux} = \boldsymbol{c}$. Comment on the result, and prove it!

9.9 Solve the system

$$\begin{bmatrix} \varepsilon & 1 \\ 1 & 1 \end{bmatrix} \begin{bmatrix} x_1 \\ x_2 \end{bmatrix} = \begin{bmatrix} 1 \\ 2 \end{bmatrix}$$

using Gauss elimination without pivoting. Show that for any computing machine with a fixed word length, and for all sufficiently small $\varepsilon > 0$, the computer gives $x_2 = 1$ and then $x_1 = 0$. Show that the exact solution satisfies

$$\lim_{\varepsilon \to 0} x_1 = \lim_{\varepsilon \to 0} x_2 = 1$$

and comment. Repeat the solution process using Gauss elimination with partial pivoting, and comment.

9.10 Carry out four steps of Jacobi iteration to solve

$$\begin{bmatrix} -5 & 2 & 1 \\ 1 & -10 & 1 \\ 1 & 1 & -4 \end{bmatrix} \boldsymbol{x} = \begin{bmatrix} -3 \\ 27 \\ 4 \end{bmatrix}.$$

Using the 'norm' condition (9.29), check that the method converges and start iterating with $\boldsymbol{x}^{(0)} = [0 \;\; 0 \;\; 0]^{\mathrm{T}}$.

9.11 Show that the application of Gauss–Seidel iteration to the system

$$\begin{aligned} 2x \;+y &= 4 \\ x+2y &= 5 \end{aligned}$$

leads to the iteration

$$\boldsymbol{x}^{(k+1)} = \begin{bmatrix} 0 & -\frac{1}{2} \\ 0 & \frac{1}{4} \end{bmatrix} \boldsymbol{x}^{(k)} + \begin{bmatrix} 2 \\ \frac{3}{2} \end{bmatrix}.$$

in which $\boldsymbol{x}^{(k)} = [x^{(k)} \;\; y^{(k)}]^{\mathrm{T}}$ and k is the iteration number. Write this down in component form and deduce that the exact solution must be $x = 1$, $y = 2$.

Defining the *error* at iteration k as $\boldsymbol{e}^{(k)} = \boldsymbol{x} - \boldsymbol{x}^{(k)}$, where $\boldsymbol{x} = [x \;\; y]^{\mathrm{T}}$, show that

$$\boldsymbol{e}^{(k+1)} = \begin{bmatrix} 0 & -\frac{1}{2} \\ 0 & \frac{1}{4} \end{bmatrix} \boldsymbol{e}^{(k)}$$

and if $\boldsymbol{x}^{(0)} = \boldsymbol{0}$, determine how many iterations are required to reduce the error by a factor of at least 10.

9.12 The 'tightest' result on the convergence of iterative solvers states that if the *spectral radius* of the iteration matrix $-\boldsymbol{E}\boldsymbol{F}^{-1}$ is less than 1 then the solver will converge. The spectral radius equals the *eigenvalue* of largest magnitude. Apply this idea to the system of Exercise 9.10.

9.13 For the system

$$\begin{bmatrix} 1 & 2 & 1 \\ \frac{1}{8} & 1 & 1 \\ -1 & 4 & 1 \end{bmatrix} \boldsymbol{x} = \begin{bmatrix} 1 \\ 3 \\ 7 \end{bmatrix}$$

show that Jacobi converges whereas Gauss–Seidel does not! Use the idea of Exercise 9.12.

10 One-step methods for ordinary differential equations

In this chapter the numerical solution of *ordinary differential equations* (ODEs) is introduced. The subject is vast and entire text books have been devoted to treatments of the numerical techniques and associated analysis[1]. Here the discussion is restricted to the *one-step methods* and associated analysis (implementation, error, stability) applied to first-order ODEs.

10.1 Preliminaries

An *ordinary differential equation* (ODE) is an equation containing one or more derivatives of a function 'y of x'. The function name y may change (for example, velocity v, acceleration a, mass m), as may that of the *independent variable* x (for example, time t). The discussion here focuses on the first-order ODE

$$\frac{\mathrm{d}y}{\mathrm{d}x} = f(x, y) \tag{10.1}$$

where f is a known function of x and y. The derivative appearing on the left-hand side of eqn (10.1) is often written y'. The aim is to determine the function 'y of x' which satisfies eqn (10.1). If f is sufficiently simple then y may be found by a single integration.

[1] See [12, 17] for wide-ranging treatments of this subject.

Example 10.1 The rate of decrease of mass of a radioactive material is proportional to its mass m

$$\frac{\mathrm{d}m}{\mathrm{d}t} = -km$$

where k is the *decay constant* (the minus sign appears because the mass decreases with time t). The equation is *separable*, $\mathrm{d}m/m = -k\,\mathrm{d}t$, and can be integrated to give $\ln m = -kt + c^*$. c^* is a *constant of integration*. Taking the exponential of each side gives the solution $m(t) = c\mathrm{e}^{-kt}$ (where $c = \mathrm{e}^{c^*}$).

To evaluate the constant of integration appearing in the solution, an additional condition is required that specifies the value of the solution $y(x)$ for a particular value of x, and is of the form

$$y(x_0) = s. \tag{10.2}$$

Equation (10.2) is an *initial condition* and the combination of eqns (10.1) and (10.2) is termed an *initial value problem* (IVP).

Example 10.2 In Example 10.1, if the initial mass m is equal to 10, then the initial condition for the problem is $m(0) = 10$. Substitution into the solution $m(t) = c\mathrm{e}^{-kt}$ gives the equation $10 = c$ from which $m(t) = 10\mathrm{e}^{-kt}$ – the general solution (the value of k is a known material constant).

A difficulty in solving eqn (10.1) arises when the form of f prevents the integration from being performed (suitable primitives cannot be found).

10.1.1 Existence, uniqueness and Picard iteration

If the analytical process of finding a solution $y(x)$ is not feasible, it is still useful to know whether a solution *exists* and is *unique*. Existence serves to justify using a numerical method. Uniqueness is necessary so that once a solution is found we can be sure that it is the solution.

Example 10.3 The IVP $y' = -y$ $(x > 0)$, $y(0) = 2$ can be solved by integrating once to get $y(x) = c\mathrm{e}^{-x}$, followed by application of the initial condition $2 = c\mathrm{e}^0$ to give $c = 2$. The general solution is $y(x) = 2\mathrm{e}^{-x}$.

The IVP $y' = -y^2$ $(0 < x \leq 1)$, $y(0) = -2$ can be solved in a similar way to give $y(x) = 2/(2x - 1)$. An inspection of the graphs of the two solutions reveals the difficulty associated with this IVP. The solution is undefined at $x = \frac{1}{2}$, which lies in the domain of interest. In other words, this IVP does *not* have a solution.

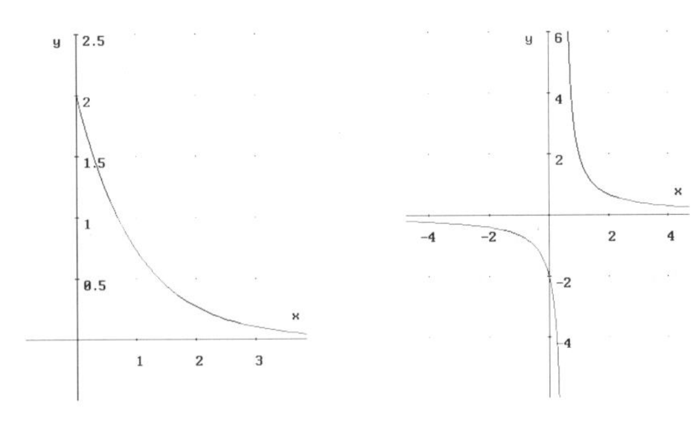

Figure 10.1 Graphs of the solutions $y(x) = 2e^{-x}$ and $y(x) = 2/(2x - 1)$.

Continuity of f with respect to x is *not* sufficient to ensure the existence of a solution, let alone its uniqueness! A *Lipschitz condition* is sufficient.

RUDOLF OTTO SIGISMUND LIPSCHITZ
(1832–1903)
German mathematician

WORK

Lipschitz worked on quadratic differential forms and mechanics. His work on the Hamilton–Jacobi method for integrating the equations of motion of a general dynamical system led to important applications in celestial mechanics.

Lipschitz is remembered for the 'Lipschitz condition', an inequality that guarantees a unique solution to the differential equation $y' = f(x,y)$. Peano gave an existence theorem for this differential equation, giving conditions which guarantee at least one solution.

Definition $f(x, y)$ satisfies a *Lipschitz condition* with respect to y if there exists a constant $L > 0$ such that

$$|f(x, y) - f(x, z)| \leq L|y - z|$$

for all $x \in [a, b]$. The *Lipschitz constant* L is *independent* of x.

Example 10.4 From Example 10.3, for the first IVP,

$$|f(x, y) - f(x, z)| = |-y + z| \leq |y - z|.$$

A Lipschitz constant exists, given by $L = 1$. For the second IVP

$$|f(x, y) - f(x, z)| = |-y^2 + z^2| \leq \left|\frac{\partial f}{\partial y}\right| |y - z| = |2y| |y - z|,$$

using the mean-value theorem. A value of y can always be found such that $2y$ is greater than any chosen L – the IVP does *not* satisfy a Lipschitz condition.

A condition for $y' = f(x, y)$, $y(x_0) = s$ to have a unique solution can now be stated.

Theorem Existence and uniqueness
The first-order IVP

$$y' = f(x, y), \quad x \in [x_0, b]; \quad y(x_0) = s, \quad x_0 \in [a, b],$$

has a unique solution $y(x)$ for $x_0 \leq x \leq b$ if

(1) $f(x, y)$ is continuous in x and

(2) $f(x, y)$ satisfies a Lipschitz condition with respect to y.

If both conditions are satisfied, there *exists a unique* solution to the IVP.

An outline proof, which constructs the solution $y(x)$, relies upon an integral form of the IVP

$$y(x) = s + \int_{x_0}^{x} f(x, y(x))\, dx.$$

A sequence of functions $\{y_m(x)\}$ is then defined recursively by

$$y_0(x) = s,$$

$$y_{m+1}(x) = s + \int_{x_0}^{x} f(x, y_m(x))\, dx, \quad m = 0, 1, \ldots$$

It can be shown that if conditions 1 and 2 of the Theorem are satisfied then

$$\lim_{m \to \infty} y_m(x) = y(x).$$

This approach is called *Picard iteration.*

CHARLES ÉMILE PICARD
(1856–1941)
French mathematician

LIFE

Picard was appointed lecturer at the University of Paris (1878) and professor at Toulouse (1879). In 1898 he was appointed professor at the Sorbonne in Paris.

He married Hermite's daughter. Picard's daughter and two sons were all killed in the First World War. His grandsons were wounded and captured in the Second World War.

WORK

He made his most important contributions in the field of analysis and analytic geometry. He used methods of successive approximation to show the existence of solutions of ordinary differential equations.

Building on work by Abel and Riemann, Picard's study of the integrals attached to algebraic surfaces and related topological questions developed into an important part of algebraic geometry. On this topic he published, with Georges Simart, *Théorie des fonctions algébriques de deux variables indépendantes* (1897, 1906). Picard also applied analysis to the study of elasticity, heat and electricity.

Example 10.5 The IVP $y' = -y$ $(0 < x \le 1)$, $y(0) = 2$ satisfies a Lipschitz condition,

$$|f(x, y) - f(x, z)| = |y - z|$$

with $L = 1$. Further, f is continuous in x (it is constant in x), so there exists a unique solution to the IVP. Picard iteration gives

$$\begin{aligned}
y_0(x) &= s && = 2 \\
y_1(x) &= 2 + \int_0^x (-2)\,\mathrm{d}x && = 2(1 - x) \\
y_2(x) &= 2 + \int_0^x [-2(1 - x)]\,\mathrm{d}x && = 2\left(1 - x + \frac{x^2}{2}\right) \\
y_3(x) &= 2 + \int_0^x \left[-2\left(1 - x + \frac{x^2}{2}\right)\right]\mathrm{d}x && = 2\left(1 - x + \frac{x^2}{2} - \frac{x^3}{6}\right)
\end{aligned}$$

and in general

$$y_m(x) = 2\left(1 - x + \frac{x^2}{2} - \frac{x^3}{6} + \cdots + (-1)^m \frac{x^m}{m!}\right).$$

$y_m(x)$ is the partial sum of the Maclaurin series for $2\mathrm{e}^{-x}$ (the analytic solution of the IVP).

10.1.2 Numerical solutions

If $f(x, y_m(x))$ cannot be integrated then Picard iteration fails and we resort to step-by-step numerical methods. The process is summarized as follows:

(1) The *solution domain* $x_0 \leq x \leq b$ is divided into N *cells* of equal size $h = (b - x_0)/N$, separated by *nodes* at $x_n = x_0 + nh$, $n = 0, \ldots, N$.

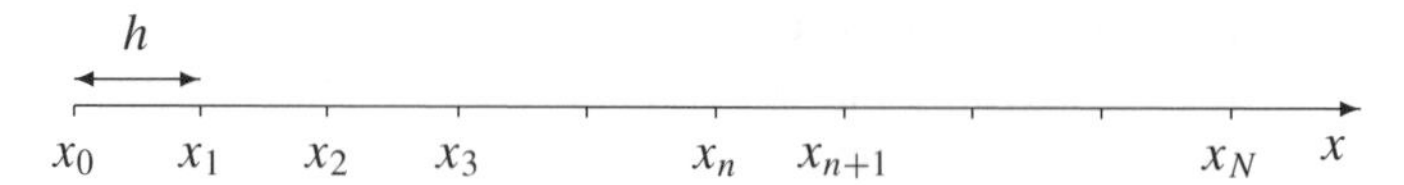

(2) Numerical estimates y_n to the exact solution $y(x_n)$ are sought at each node. The y_n are determined from a *recurrence relation* based upon the IVP $y' = f(x, y)$, $y(x_0) = s$.

Example 10.6 A simple recurrence relation is obtained by approximating y', at $x = x_n$, by the *forward difference* $(y_{n+1} - y_n)/h$ (see Section 7.4). Equation (10.1) is replaced (approximated) by

$$\frac{y_{n+1} - y_n}{h} = f(x_n, y_n)$$

which can be arranged to give

$$y_{n+1} = y_n + hf_n \qquad \textbf{(10.3)}$$

where $f_n = f(x_n, y_n)$. Equation (10.3) is an example of a *difference equation* and given $y_0 = s$ (obtained from the initial value) can be used to iteratively generate a sequence $\{y_n\}$.

s y_1 y_2 $y_3 = y_2 + hf(x_2, y_2)$

x_0 x_1 x_2 x_3 x_n x_{n+1} x_N x

10.2 Linear difference equations

Difference equations[2] are important for the development and analysis of numerical methods for ODEs.

[2] For those familiar with linear difference equations and their solutions, this section may be skipped. For those not familiar with the topic, it is recommended that the section is read. However, skipping the material should not preclude an understanding of later material in the chapter, but may impair the level of understanding. A full treatise on difference equations and their solutions may found in [13].

Let $\{y_n\}$ be a sequence of numbers defined over the field of integers. A *difference equation* expresses a relationship between several terms in the sequence $\{y_n\}$. y_n represents a *term* in the sequence, where n is the *index* (or position) of the term and y_n is the *value* of the term.

Example 10.7 The relationship described by

$$y_{n+1} = y_n^2 \quad \textbf{(10.4)}$$

is a difference equation and states that each term in the sequence $\{y_n\}$ equals the square of the previous term. The sequences 3, 9, 81, ... and 2, 4, 16, 256, ... both satisfy the relationship, as do the sequences 1, 1, 1, ... and 0.5, 0.25, 0.0625, ...

To determine the particular sequence that satisfies a given difference equation, one or more terms from the sequence must be *known*.

Example 10.8 If $\{y_n\}$ is generated by the recurrence $y_{n+1} = y_n^2$ and it is known that $y_0 = 2$, then the solution sequence is 2, 4, 16, 256, ...

Knowing the index of a particular term it may be possible to evaluate any term in the sequence as a function of the index.

Example 10.9 From Example 10.8, with $y_0 = 2$ we may use the recurrence relation (10.4) to obtain

Index n	0	1	2	3	4
Value y_n	2	4	16	256	65536

It is not difficult to verify that $y_n = 2^{2^n}$. Thus, having specified a term in the sequence and identified a pattern in the sequence, it has been possible to obtain an expression for the general term y_n as a function of the index n. Such an expression is termed a *solution*.

Equation (10.4) is an example of a *first-order difference equation* that relates two adjacent terms in a sequence $\{y_n\}$ and has the general form

$$F_n(y_n, y_{n+1}) = b_n \quad \textbf{(10.5)}$$

for all $n \in \aleph$. b_n is a known function n. If $b_n = 0$ the equation is *homogeneous*, otherwise it is *inhomogeneous*.

Example 10.10 Equation (10.4) is a first-order homogeneous difference equation, which can be written as

$$y_{n+1} - y_n^2 = 0. \tag{10.6}$$

The first-order difference equation

$$y_{n+1} = y_n^2 + n \tag{10.7}$$

is inhomogeneous with $F_n(y_n, y_{n+1}) = y_{n+1} - y_n^2$ and $b_n = n$.

For the difference equations considered here the function F_n will be a constant, that is the relation between two adjacent terms in the sequence $\{y_n\}$ will be identical (for example, $y_1 = y_0^2$, $y_2 = y_1^2$, $y_3 = y_2^2$, etc.).

A difference equation with only linear combinations of one or more terms of the sequence $\{y_n\}$ (and possibly an inhomogeneous term b_n) is *linear*.

Example 10.11 Equation (10.7) is non-linear – it contains the 'squared' term y_n^2. The equation $y_n - y_{n+1} = -2$ is linear since all components with terms from the sequence $\{y_n\}$ are linear. The equation $y_n y_{n+1} = 1$ is non-linear – it contains a product of terms.

A *solution* y_n of a difference equation is an expression involving functions of the index n, and possibly constants, that satisfies the difference equation. Substitution of a 'solution' into a difference equation results in an *identity*.

Example 10.12 A solution of the first-order linear inhomogeneous difference equation $y_n - y_{n+1} = -2$ is given by $y_n = 2n + c_1$, where c_1 is an arbitrary constant. This can be verified by substituting the solution into the left-hand side of the difference equation to obtain

$$2n + c_1 - 2(n+1) - c_1 = -2 \Rightarrow -2 = -2.$$

We shall shortly develop a general method for obtaining the solutions to linear difference equations, but first we consider how to determine the arbitrary constant(s) which may appear in a solution. The process is quite simple. For every arbitrary constant appearing in a solution, a term in the sequence $\{y_n\}$ must be specified.

Example 10.13 For the first-order equation of Example 10.12, one arbitrary constant appears in the solution. Specifying the term $y_0 = 3$ implies that $3 = 0 + c_1$, that is $c_1 = 3$. The *general solution* becomes $y_n = 2n + 3$.

10.2.1 Higher-order difference equations

A difference equation of *order m* relates $m+1$ consecutive terms in a sequence $\{y_n\}$ and may be written

$$F_n(y_n, y_{n+1}, \ldots, y_{n+m}) = b_n \tag{10.8}$$

for all $n \in \aleph$. The simplest example is the order m *linear homogeneous difference equation*

$$a_0 y_n + a_1 y_{n+1} + \cdots + a_m y_{n+m} = 0 \tag{10.9}$$

in which the a_i are constants. Solutions of eqn (10.9) take the form $y_n \propto \beta^n$ where β is a constant. Substituting this form into eqn (10.9) gives the algebraic equation

$$a_0\beta^n + a_1\beta^{n+1} + \cdots + a_m\beta_{n+m} = 0,$$

that is

$$\beta^n(a_0 + a_1\beta + \cdots + a_m\beta^m) = 0, \tag{10.10}$$

known as the *characteristic equation*. Equation (10.10) can be written

$$\beta^n p(\beta) = 0, \tag{10.11}$$

where $p(\beta) = a_0 + a_1\beta + \cdots + a_m\beta^m$ is the *characteristic polynomial*. If $\beta \neq 0$, eqn (10.11) has m roots, $\beta_1, \ldots, \beta_m$ (the m roots of the characteristic polynomial). If the roots are distinct, eqn (10.9) has the solution

$$y_n = c_1\beta_1^n + \cdots + c_m\beta_m^n. \tag{10.12}$$

β_i, $i = 1, \ldots, m$, denote the values of β such that β^n satisfies eqn (10.9) and each represents a *linearly independent* component of the solution. Linearity (the Principle of Superposition) permits the m components to be combined.

Example 10.14 Fibonacci[3] gave the following model for the population growth of rabbits, where r_n represents the number of pairs of rabbits after n breeding periods (see also Example 5.4):

$$r_{n+2} = r_{n+1} + r_n, \quad n = 0, 1, \ldots \tag{10.13}$$

This is a *second-order linear homogeneous equation* $r_{n+2} - r_{n+1} - r_n = 0$. Assuming the solution $r_n \propto \beta^n$ gives the (quadratic) characteristic equation $\beta^n(\beta^2 - \beta - 1) = 0$ which has two roots

$$\beta_1 = \frac{1-\sqrt{5}}{2}, \quad \beta_2 = \frac{1+\sqrt{5}}{2},$$

[3] Leonardo Fibonacci (known as Leonardo di Pisa) was an Italian mathematician who, in 1202, posed the following problem concerning the 'loose living' of rabbits: *'how many pairs of rabbits can be produced from a single pair in a year's time?'* The assumptions are (a) each fertile pair produces one new pair each month, (b) new pairs become fertile after one month, (c) rabbits never die.

the solution is then

$$r_n = c_1 \left(\frac{1 - \sqrt{5}}{2} \right)^n + c_2 \left(\frac{1 + \sqrt{5}}{2} \right)^n .$$

The solution to a difference equation of order m must contain m arbitrary constants, $c_1, \ldots, c_m$. To evaluate these constants, m terms in the solution sequence $\{y_n\}$ must be specified, say $y_0, \ldots, y_{m-1}$.

Example 10.15 In Example 10.14 the second-order equation led to two arbitrary constants in the solution, c_1 and c_2. The two specified terms $r_0 = 1$ and $r_1 = 1$ can be used to determine c_1 and c_2.

$$n = 0 : 1 = c_1 + c_2$$

$$n = 1 : 2 = (1 - \sqrt{5})c_1 + (1 + \sqrt{5})c_2$$

The solution of these linear equations is $c_1 = (\sqrt{5} - 1)/2\sqrt{5}$, $c_2 = (\sqrt{5} + 1)/2\sqrt{5}$. Hence, the *general solution* to the difference equation (10.13), subject to $r_0 = r_1 = 1$, is

$$r_n = \frac{1}{2^{n+1}\sqrt{5}}[(1 + \sqrt{5})^{n+1} - (1 - \sqrt{5})^{n+1}].$$

Returning to the solution (10.12) of the order m equation (10.9), we must consider the case of *repeated roots*. If, for example, β_1 is a double root ($\beta_1 = \beta_2$), the solution y_n will have just $m-1$ linearly independent solution components. An mth independent component is given by $n\beta_1^n$, and

$$y_n = c_1\beta_1^n + c_2 n\beta_1^n + c_3\beta_3^n + \cdots + c_m\beta_m^n. \tag{10.14}$$

Example 10.16 The third-order linear homogeneous equation $y_{n+3} - y_{n+2} - y_{n+1} + y_n = 0$ generates the cubic characteristic equation $\beta^n(\beta^3 - \beta^2 - \beta + 1) = 0$ (if $y_n \propto \beta^n$) with three non-trivial roots $\beta_{1,2,3} = -1, 1, 1$. The second root is repeated (a double root) and the solution is

$$y_n = c_1\beta_1^n + c_2\beta_2^n + c_3 n\beta_2^n = c_1(-1)^n + c_2 + c_3 n.$$

The *third-order equation* gives rise to *three* arbitrary constants, c_1, c_2, and c_3. Using three specified terms from the sequence $\{y_n\}$, such as $y_{-1} = 0$, $y_0 = 1$ and $y_1 = 2$, we obtain

$$n = -1 : 0 = -c_1 + c_2 - c_3$$

$$n = 0 : 1 = c_1 + c_2$$

$$n = 1 : 2 = -c_1 + c_2 + c_3$$

– three linear algebraic equations with solution $c_1 = 0$, $c_2 = c_3 = 1$. The general solution is

$$y_n = 1 + n.$$

A linear *inhomogeneous* difference equation of order m has the form

$$a_0 y_n + a_1 y_{n+1} + \cdots + a_m y_{n+m} = b_n \tag{10.15}$$

where b_n is a known function of n and the a_i are known constants. The general solution of eqn (10.15) is

$$y_n = y_n^{\mathrm{c}} + y_n^{\mathrm{p}} \tag{10.16}$$

where y_n^{c} is the *complementary solution* of the associated *homogeneous* difference equation and y_n^{p} is a *particular solution* of the *inhomogeneous* eqn (10.15). y_n^{p} is usually found either by inspection or by the *method of undetermined coefficients.*

10.2.1.1 Method of undetermined coefficients

Applied to difference equations, the method[4] of *undetermined coefficients* uses the assumption that the solution y_n^{p} has a form similar to the inhomogeneous term b_n. If b_n is a polynomial in n, say

$$\sum_{i=0}^{p} A_i n^i,$$

then y_n^{p} takes the polynomial form

$$\sum_{i=0}^{p} A_i^* n^i.$$

The undetermined coefficients A_i^* are found by substituting y_n^{p} into the inhomogeneous equation and comparing coefficients of powers of n.

Example 10.17 The first-order equation $y_n - 3y_{n+1} = -2n$ has the associated homogeneous equation $y_n - 3y_{n+1} = 0$. The characteristic equation is $\beta^n(1 - 3\beta) = 0$, with single root $\beta_1 = \frac{1}{3}$. The complementary solution is

$$y_n^{\mathrm{c}} = c_1 \left(\frac{1}{3}\right)^n.$$

The inhomogeneous term $b_n = -2n$ is a linear polynomial and y_n^{p} takes the form of a general linear polynomial (including the constant term), that is $y_n^{\mathrm{p}} = A_0^* + A_1^* n$. The difference equation becomes

$$A_0^* + A_1^* n - 3(A_0^* + A_1^*(n + 1)) = -2n$$

[4] See Section 8.3.4 for the use of undetermined coefficients as applied to numerical integration.

and comparing coefficients of $O(1)$ and $O(n)$ leads to a pair of linear algebraic equations, $-2A_0^* - 3A_1^* = 0$ and $-2A_1^* = -2$, from which $A_0^* = -\frac{3}{2}$ and $A_1^* = 1$. The solution is

$$y_n = y_n^c + y_n^p = c_1\left(\frac{1}{3}\right)^n - \frac{3}{2} + n.$$

Note At this point the arbitrary constant emanating from the homogeneous solution is still unknown. It is only at this stage of the solution process that specific terms of $\{y_n\}$ are used to enumerate arbitrary constants (in Example 10.17 this will be a single term).

Example 10.18 The second-order equation $-y_{n+2} + y_{n+1} + 2y_n = 1$ has the associated homogeneous equation $-y_{n+2} + y_{n+1} + 2y_n = 0$ with solution $y_n^c = c_1(-1)^n + c_2 2^n$. The inhomogeneous term b_n is a constant, and a constant particular solution $y_n^p = A_0^*$ is assumed. The inhomogeneous equation becomes $-A_0^* + A_0^* - 2A_0^* = 1$ from which $A_0^* = \frac{1}{2}$ and

$$y_n = y_n^c + y_n^p = c_1(-1)^n + c_2 2^n + \frac{1}{2}.$$

If $y_0 = 0$ and $y_1 = 2$ then $0 = c_1 + c_2 + \frac{1}{2}$ and $2 = -c_1 + 2c_2 + \frac{1}{2}$, from which $c_1 = -\frac{5}{6}$ and $c_2 = \frac{1}{3}$. The correct general solution is

$$y_n = -\frac{5}{6}(-1)^n + \frac{1}{3}2^n + \frac{1}{2}.$$

If, however, c_1 and c_2 are enumerated as soon as the complementary solution y_n^c is obtained, then $0 = c_1 + c_2$ and $2 = -c_1 + 2c_2$ with solution $c_1 = -\frac{2}{3}$, $c_2 = \frac{2}{3}$. The *incorrect* general solution is

$$y_n = -\frac{2}{3}(-1)^n + \frac{2}{3}2^n + \frac{1}{2}.$$

The difference between the solutions is shown in Figure 10.2.

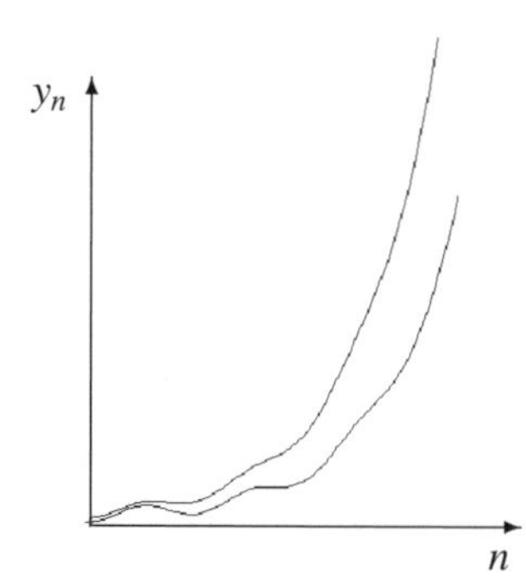

Figure 10.2 Correct (lower) and incorrect (upper) general solutions.

10.2.2 Difference inequalities

Difference inequalities are used in the error analysis of numerical methods (see Section 10.5.3). A *difference inequality* is a 'bounding' relation for one or more terms in a sequence $\{y_n\}$.

Example 10.19 The inequality $|y_n| \leq 1$ states that all terms in the sequence $\{y_n\}$ lie in the range -1 to $+1$.

The inequality $|y_n| \geq n^2$ states that n^2 is a lower bound on y_n.

$3|y_{n+1}| \leq |y_n|$ states that the magnitude of each term in the sequence $\{y_n\}$ is no greater than $\frac{1}{3}$ times the magnitude of the previous term.

The approach to solving a difference inequality for the sequence $\{y_n\}$ is first to solve an associated difference equation for a sequence $\{z_n\}$ and then to identify a relation between y_n and z_n.

Example 10.20 Given $|y_0| = s$, $s \geq 0$, and the inequality $3|y_{n+1}| \leq |y_n|$, an upper bound is required for the general term $|y_n|$.

(1) The first step consists of solving the associated first-order difference equation $3z_{n+1} = z_n$ subject to the initial condition $z_0 = |y_0| = s$. The equation is linear and homogeneous with solution $z_n = c_1(1/3)^n$. The inital condition gives $c_1 = s$, and hence the general solution is $z_n = s(1/3)^n$. z_n is non-negative.

(2) The second step is to identify a relation between y_n and z_n (typically by induction). We are ultimately looking to bound $|y_n|$ above, which suggests that we should aim to 'prove' that $|y_n| \leq z_n$. The hypothesis is true for $n = 0$, since $|y_0| \leq s = z_0$. Assuming that the hypothesis is true for integers $k = 0, \ldots, n$, that is $|y_k| \leq z_k$ for $k = 0, \ldots, n$, then for $k = n + 1$ the difference inequality/equation gives

$$3|y_{n+1}| \leq |y_n| \leq z_n = 3z_{n+1},$$

that is $|y_n| \leq z_n$ implies $|y_{n+1}| \leq z_{n+1}$. Thus, $|y_n| \leq s(1/3)^n$.

10.3 Taylor series method

Once again Taylor's theorem surfaces. From the discussion in Section 10.1.2 we need to develop a means of evaluating the solution $y(x)$ at a neighbouring point $x = x_{n+1}$ given information on the function y at $x = x_n$. From Taylor's theorem

$$\begin{aligned} y(x_{n+1}) &= y(x_n + h) \\ &= y(x_n) + h'(x_n) + \frac{h^2}{2}y''(x_n) + \frac{h^3}{6}y'''(x_n) + \cdots \end{aligned} \quad \textbf{(10.17)}$$

This formula represents a relationship between two successive values of an arbitrary function 'y of x'. We require to 'link' the formula to the problem at hand, that of solving eqn (10.1).

From eqn (10.1), $y' = y^{(1)} = f$. On differentiating eqn (10.1),

$$\frac{\mathrm{d}^2 y}{\mathrm{d}x^2} = y'' = y^{(2)} = \frac{\mathrm{d}}{\mathrm{d}x} f.$$

The right-hand side represents a *total derivative*, and since y is itself a function of x the *function of a function rule* is applied, giving

$$y^{(2)} = \frac{\mathrm{d}}{\mathrm{d}x} f = \frac{\partial f}{\partial x} + \frac{\partial f}{\partial y}\frac{\mathrm{d}y}{\mathrm{d}x} = f^{(1)}.$$

The differentiation can be repeated to generate the relation

$$y^{(r+1)} = f^{(r)}$$

where the *Taylor derivatives* $f^{(r)}$ are defined recursively by

$$f^{(r)} = \frac{\partial}{\partial x} f^{(r-1)} + \frac{\partial}{\partial y} f^{(r-1)} \frac{\mathrm{d}y}{\mathrm{d}x} = f_x^{(r-1)} + f^{(r-1)} y'.$$

Hence, the original Taylor series (10.17) developed for y can be written

$$\begin{aligned} y(x_{n+1}) &= y(x_n + h) \\ &= y(x_n) + h f^{(0)}(x_n, y(x_n)) + \frac{h^2}{2} f^{(1)}(x_n, y(x_n)) \\ &\quad + \frac{h^3}{6} f^{(2)}(x_n, y(x_n)) + \cdots \end{aligned} \tag{10.18}$$

where $f^{(0)} \equiv f$.

Example 10.21 If $f(x, y) = y$ then

$$\begin{aligned} f^{(1)}(x, y) &= f_x^{(0)} + f_y^{(0)} f^{(0)} = 0 + 1 \times y = y, \\ f^{(2)}(x, y) &= f_x^{(1)} + f_y^{(1)} f^{(0)} = 0 + 1 \times y = y, \end{aligned}$$

and in general $f^{(r)}(x, y) = y$. Hence

$$\begin{aligned} y(x_{n+1}) &= y(x_n) + h y(x_n) + \frac{h^2}{2} y(x_n) + \frac{h^3}{6} y(x_n) + \cdots \\ &= \left(1 + h + \frac{h^2}{2} + \frac{h^3}{6} + \cdots\right) y(x_n) \\ &= \mathrm{e}^h y(x_n). \end{aligned}$$

If $y(x_0) = s$ it is not difficult to show that the linear first-order homogeneous difference equation has the solution $y(x_n) = s\mathrm{e}^{x_n - x_0}$, which is precisely equal to the exact solution at $x = x_n$.

The series on the right-hand side of eqn (10.18) has an infinite number of terms in order to preserve the equality, and is not a practical formula for evaluating $y(x_{n+1})$. In practice all terms up to and including that involving h^p are included, that is

$$\begin{aligned} y(x_{n+1}) &= y(x_n + h) \\ &= y(x_n) + hf^{(0)}(x_n, y(x_n)) + \frac{h^2}{2} f^{(1)}(x_n, y(x_n)) \\ &\quad + \cdots + \frac{h^p}{p!} f^{(p-1)}(x_n, y(x_n)) + h^{p+1} R_{p+1}(\xi_n), \end{aligned} \tag{10.19}$$

where $R_{p+1}(\xi_n)$, $x_n < \xi_n < x_n + h$, is the *remainder term*, is approximated (replaced) by the *truncated series*

$$y_{n+1} = y_n + hf^{(0)}(x_n, y_n) + \frac{h^2}{2} f^{(1)}(x_n, y_n) + \cdots + \frac{h^p}{p!} f^{(p-1)}(x_n, y_n). \tag{10.20}$$

Equation (10.20) is the *Taylor series method* of order p. y_n is taken to be an estimate to the exact value $y(x_n)$.

Example 10.22 For Example 10.21, the Taylor series method of order p gives the recurrence relation

$$\begin{aligned} y_{n+1} &= y_n + hy_n + \frac{h^2}{2} y_n + \cdots + \frac{h^p}{p!} y_n \\ &= \left(1 + h + \frac{h^2}{2} + \cdots + \frac{h^p}{p!}\right) y_n \\ &= \beta y_n. \end{aligned}$$

If $y_0 = s$, the difference equation has the solution $y_n = s\beta^n$. As $p \to \infty$, $\beta \to \mathrm{e}^h$ and $\beta^n \to \mathrm{e}^{nh}$, that is $y_n \to y(x_n)$.

The Taylor derivatives $f^{(r)}$ rapidly become extremely difficult to obtain for all but the most simple functions f.

Example 10.23 The second Taylor derivative $f^{(2)}$ is given by

$$\begin{aligned} f^{(2)} &= f_x^{(1)} + f_y^{(1)} f \\ &= f_{xx} + 2f_{xy} f + f_{yy} f^2 + (f_x + f_y f) f_y. \end{aligned}$$

However, for $p = 1$ the relatively simple *Euler's method* is obtained;

$$y_0 = s; \quad y_{n+1} = y_n + hf_n, \quad n = 0, 1, \ldots, \tag{10.21}$$

described as a *Taylor series method of order 1*.

LEONHARD EULER
(1707–83)
Swiss mathematician

LIFE

Euler's father sent him to the University of Basel to prepare for the ministry. Geometry soon became his favourite subject and he obtained his father's consent to change to mathematics after Johann Bernoulli had used his persuasion.

He joined the St Petersburg Academy of Science (1727) two years after it was founded by Catherine I (wife of Peter the Great). Euler was a medical lieutenant in the Russian navy (1727–30). In St Petersburg he lived with Daniel Bernoulli, became professor of physics at the academy (1730) and professor of mathematics (1733). He married and had 13 children, of whom 5 survived. He claimed that he made some of his greatest discoveries while holding a baby on his arm with other children playing round his feet.

In 1741, at the invitation of Frederick the Great, Euler joined the Berlin Academy of Science, where he remained for 25 years, still receiving part of his salary from Russia. During this time he wrote over 200 articles, three books on mathematical analysis, and a popular scientific publication *Letters to a Princess of Germany* (1768–72).

In 1766 Euler returned to Russia (he had argued with Frederick over academic freedom). Euler lost the sight of his right eye at the age of 31 and soon after his return to St Petersburg he became almost entirely blind after a cataract operation. Because of his remarkable memory he was able to continue with his work on optics, algebra and lunar motion. After 1765 (when Euler was 58) he produced almost half his works despite being totally blind! After his death (1783) the St Petersburg Academy continued to publish his unpublished work for nearly 50 more years. He was the most prolific writer on mathematics of all time. His complete works contains 886 books and papers.

WORK

The publication of many articles and his book *Mechanica* (1736–37), which presented Newtonian dynamics in the form of mathematical analysis for the first time, started Euler on the way to major mathematical work.

In number theory he did much work with Goldbach (he stated the prime number theorem and the law of biquadratic reciprocity). He integrated Leibniz's differential calculus and Newton's method of fluxions into mathematical analysis.

He proposed the notations $f(x)$ (1734), e for the base of natural logs (1727), i for the square root of -1 (1777), π for pi, Σ for summation (1755) etc. He introduced beta and gamma functions, integrating factors for differential equations etc.

He studied continuum mechanics, lunar theory with Clairaut, the three-body problem, elasticity, acoustics, the wave theory of light, hydraulics, music etc. He laid the foundation of analytical mechanics, especially in his *Theory of the Motions of Rigid Bodies*

(1765), and worked on the catenary, the epicycloid, the hypocycloid, the tricuspoid and the trident of Newton.

Example 10.24 The IVP $y' = -y + x + 1$ $(0 < x \leq 1)$, $y(0) = 1$, is to be solved by Euler's method. The first step is to construct the difference equation,

$$y_{n+1} = y_n + hf_n = y_n + h(-y_n + x_n + 1) = (1 - h)y_n + h(1 + x_n).$$

With $h = 0.1$ and $x_n = 0.1n$,

$$y_{n+1} = 0.9y_n + 0.1 + 0.01n, \quad n = 0, 1, \ldots$$

The result of implementing this relation is shown below (where $y(x) = e^{-x} + x$, $\Delta = |y(x_n) - y_n|$ and $\delta = \Delta/|y(x_n)|$).

n	x_n	$y(x_n)$	y_n	Δ	δ
0	0.0	1.0	1.0	0	0
1	0.1	1.00484	1.0	0.00484	0.0048
2	0.2	1.01873	1.01	0.00873	0.0086
3	0.3	1.04082	1.029	0.01182	0.0114
4	0.4	1.07032	1.0561	0.01422	0.0133
5	0.5	1.10653	1.09049	0.01604	0.0145
6	0.6	1.14881	1.13144	0.01737	0.0151
7	0.7	1.19659	1.17830	0.01829	0.0153
8	0.8	1.24933	1.23047	0.01886	0.0151
9	0.9	1.30657	1.28742	0.01915	0.0147
10	1.0	1.36788	1.34868	0.01920	0.0140

Observations *The behaviour of the numerical solution y_n clearly resembles the exact solution $y(x_n)$. The absolute error is bounded, adding approximately 0.005 per step. The relative error is bounded at approximately 0.015, indicating about 1 significant digit of accuracy in the numerical solution. We can say that Euler's method appears to provide an 'acceptable' solution procedure.*

Note An alternative approach to generating the numerical estimates shown in column 'y_n' is to solve the original difference equation $y_{n+1} = (1-h)y_n + h(1+x_n)$ subject to $y_0 = 1$. The homogeneous equation $y_{n+1} = (1-h)y_n$ has solution $y_n^c = c_1(1-h)^n$. The inhomogeneous term $h(1 + x_n)$ can be written $h(1 + nh)$ and is linear in n. Assuming $y_n^p = A_0^* + A_1^* n$, the difference equation gives $A_0^* + A_1^*(1+n) = (1-h)(A_0^* + A_1^* n) + h + nh^2$. Comparing coefficients of $O(1)$ and $O(n)$ results in $hA_0^* + A_1^* = h$ and $hA_1^* = h^2$ with solution $A_0^* = 0$, $A_1^* = h$ and $y_n = c_1(1-h)^n + nh$. Enforcing $y_0 = 1$ gives $c_1 = 1$ and hence

$$y_n = (1 - h)^n + nh.$$

For $h = 0.1$, the sequence $\{y_n\}$ is generated by $y_n = 0.9^n + 0.1n$. It is a simple matter to evaluate the expression for suitable values of n.

DERIVE Experiment 10.1
Numerical ODE solvers can be implemented in *DERIVE* with the intrinsic function `iterates`. Euler's method is specified by

```
euler(f,x,y,x0,y0,h,n):=iterates(
   [xn+h,yn+hlim(f,[x,y],[xn,yn]),[xn,yn],[x0,y0],n)
```

To reproduce the results of Example 10.24, `Author` and `approX` the expression `euler(-y+x+1,x,y,0,1,.1,10)`.

10.3.1 Practical estimation of order

It has been stated that Euler's method is first-order (it agrees with the Taylor series expansion for y up to the linear term). In practice, what does order mean?

Order is related to the behaviour of the error as the step size h is refined. Intuitively we expect the error to decrease. Given a numerical estimate $y(h)$ to y based upon a step size h, we can write

$$y = y(h) + ah^p. \tag{10.22}$$

In other words, the error $y - y(h)$ behaves as a power of h, where a is independent of h. The power p is the *order of the method* and indicates the rate at which the error decreases as h is refined. If a and p are known then, given $y(h)$, y can be calculated!

With two further estimates to y, $y(h/2)$ and $y(h/4)$ based upon refined step sizes $h/2$ and $h/4$, we may write

$$y = y(h/2) + a\,(h/2)^p \tag{10.23}$$

$$y = y(h/4) + a\,(h/4)^p \tag{10.24}$$

Forming (10.22)–(10.23) and (10.23)–(10.24) results in

$$0 = y(h) - y(h/2) + ah^p\left(1 - \frac{1}{2^p}\right) \tag{10.25}$$

$$0 = y(h/2) - y(h/4) + a\frac{h^p}{2^p}\left(1 - \frac{1}{2^p}\right) \tag{10.26}$$

Dividing (10.25) by (10.26) produces

$$\frac{y(h) - y(h/2)}{y(h/2) - y(h/4)} = 2^p$$

from which the order p is obtained,

$$p = \frac{\log\left(\frac{y(h)-y(h/2)}{y(h/2)-y(h/4)}\right)}{\log 2}. \tag{10.27}$$

Example 10.25 In Example 10.24 Euler's method gave the estimate $y_6 = 1.13144$ to the exact value $y(0.6) = 1.14881$ using a step size $h = 0.1$. With step sizes $h = 0.05$ and $h = 0.025$, the estimates $y_{12} = 1.14036$ and $y_{24} = 1.14464$ are obtained. Applying eqn (10.27) gives the order

$$p = \frac{\log\left(\frac{1.13144-1.14036}{1.14036-1.14464}\right)}{\log 2} \approx 1.059.$$

This value is close to 1, that is $e_n = |y(x_n) - y_n| \simeq h^1$ – the method is first order.

The following statements are equivalent:

- Order of a method
- Order of Taylor series up to which the method agrees
- Rate, as a power of step size h, at which error decreases with h.

10.4 Runge–Kutta methods

The *Taylor series method* of Section 10.3 provides the formal definition of a 'step-by-step' numerical method for solving the IVP (10.1)–(10.2). Several useful observations can be made.

(1) The error $e_n = |y(x_n) - y_n|$ can be reduced by *refining the step size h* (see Examples 10.24 and 10.25 for Euler's method). This is inefficient – for Euler's method h must be reduced by a factor of $\log 10/\log 2$ (≈ 3.32) for each extra decimal place of accuracy in the estimate y_n.

(2) The error can be reduced by *increasing the order p* of the Taylor expansion (for $0 < h < 1$). This requires higher-order Taylor derivatives $f^{(r)}$ which, as seen, are likely to be extremely complicated.

We now explore a class of methods that agree with the first $p + 1$ terms of the Taylor series (10.18) using function values only (that is, without having to construct $f^{(r)}$). These are the *Runge–Kutta methods*.

CARLE DAVID TOLME RUNGE
(1856–1927)
German mathematician and scientist

LIFE

After leaving school at 19, Runge spent 6 months with his mother visiting the cultural centres of Italy. On his return to Germany he enrolled at the University of Munich to study literature. However, after 6 weeks of the course he changed to mathematics and physics. Runge attended courses with Max Planck and they became close friends. In 1877 both went to Berlin but Runge turned to pure mathematics after attending Weierstrass's lectures. His doctoral dissertation (1880) dealt with differential geometry. After taking his secondary school teachers examinations he returned to Berlin where he was influenced by Kronecker.

Runge was always a fit and active man and on his 70th birthday he entertained his grandchildren with handstands. A few months later he had a heart attack and died.

WORK

Runge worked on a procedure for the numerical solution of algebraic equations in which the roots were expressed as infinite series of rational functions of the coefficients. He published little at that stage but after visiting Mittag-Leffler in Stockholm in September 1884 he produced a large number of papers in Mittag-Leffler's journal *Acta Mathematica* (1885).

Runge obtained a chair at Hanover in 1886 and remained there for 18 years. Within a year he had moved away from pure mathematics to study the wavelengths of the spectral lines of elements other than hydrogen.

Runge did a great deal of experimental work and published a great quantity of results. He succeeded in arranging the spectral lines of helium in two spectral series and, until 1897, this was thought to be evidence that hydrogen was a mixture of two elements.

In 1904 Klein persuaded Göttingen to offer Runge a chair of Applied Mathematics, a post which Runge held until he retired in 1925.

MARTIN WILHELM KUTTA
(1867–1944)
Polish mathematician

LIFE

Kutta studied at Breslau (1885–90) and Munich (1891–94), later becoming an assistant to von Dyck. During this period he spent the year 1898–99 in England at the University of Cambridge. Kutta held posts at Munich, Jena and Aachen. He became professor at Stuttgart (1911) and remained there until he retired in 1935.

WORK

He is best known for the Runge–Kutta method (1901) for solving ordinary differential equations and for the Zhukovskii–Kutta aerofoil. Runge presented Kutta's methods.

Let the solution sequence $\{y_n\}$ be defined by

$$\begin{aligned} y_0 &= y(x_0) = s, \\ y_{n+1} &= y_n + h(\alpha_1 k_1 + \alpha_2 k_2), \quad n = 0, 1, \ldots \end{aligned} \tag{10.28}$$

where $k_1 = f(x_n, y_n) = f_n$ and $k_2 = f(x_n + h\lambda_2, y_n + h\mu_2 k_1)$. α_1, α_2, λ_2 and μ_2 are constants to be determined such that the first three terms of the Taylor expansion of (10.18) agree with the first three terms of the Taylor expansion (10.18). To expand eqn (10.28), Taylor's theorem in two variables is required, the first few terms of which are

$$f(x + h\lambda, y + h\mu k) = f(x, y) + h\left(\lambda\frac{\partial}{\partial x} + \mu k\frac{\partial}{\partial y}\right) f(x, y) + \cdots \tag{10.29}$$

For $p = 2$, the second-order Taylor series method (10.20) is

$$y_{n+1} = y_n + hf_n + \frac{h^2}{2}\left(\frac{\partial f_n}{\partial x} + \frac{\partial f_n}{\partial y} f_n\right). \tag{10.30}$$

Applying (10.29) to eqn (10.28)

$$\begin{aligned} y_{n+1} &= y_n + h(\alpha_1 k_1 + \alpha_2 k_2) \\ &= y_n + h[\alpha_1 f_n + \alpha_2 f(x_n + h\lambda_2, y_n + h\mu_2 k_1)] \\ &= y_n + h[\alpha_1 f_n + \alpha_2 f(x_n + h\lambda_2, y_n + h\mu_2 f_n)] \\ &= y_n + h\left[\alpha_1 f_n + \alpha_2\left\{f_n + h\lambda_2\frac{\partial f_n}{\partial x} + h\mu_2 f_n\frac{\partial f_n}{\partial y} + O(h^2)\right\}\right] \\ &= y_n + h(\alpha_1 + \alpha_2) f_n + h^2\left(\alpha_2\lambda_2\frac{\partial f_n}{\partial x} + \alpha_2\mu_2\frac{\partial f_n}{\partial y} f_n\right) + O(h^3). \end{aligned} \tag{10.31}$$

On comparing powers of h in eqns (10.30) and (10.31), a system of three linear equations is obtained,

$$\begin{aligned} O(h) : \alpha_1 + \alpha_2 &= 1 \\ (f_n)_x\, O(h^2) : \quad \lambda_2\alpha_2 &= \tfrac{1}{2} \\ (f_n)_y\, f_n\, O(h^2) : \quad \mu_2\alpha_2 &= \tfrac{1}{2} \end{aligned} \tag{10.32}$$

The four unknowns α_1, α_2, λ_2 and μ_2 imply an infinity of solutions (see Section 9.2.3.1) in which one unknown takes an arbitrary value. Two common solutions are given.

(a) *Simple Runge–Kutta method*

With $\alpha_1 = \frac{1}{2}$ ($\alpha_2 = \frac{1}{2}$, $\lambda_2 = \mu_2 = 1$), eqn (10.28) gives

$$y_{n+1} = y_n + \frac{h}{2}[f(x_n, y_n) + f(x_n + h, y_n + hf(x_n, y_n))]. \qquad \textbf{(10.33)}$$

(b) *Modified Euler's method*

With $\alpha_1 = 0$ ($\alpha_2 = 1$, $\lambda_2 = \mu_2 = \frac{1}{2}$), eqn (10.28) gives

$$y_{n+1} = y_n + hf\left(x_n + \frac{h}{2}, y_n + \frac{h}{2}f(x_n, y_n)\right). \qquad \textbf{(10.34)}$$

Example 10.26 The IVP $y' = -y + x + 1$ ($0 < x \leq 1$), $y(0) = 1$ is solved with the Runge–Kutta methods (10.33) and (10.34). The first step is to construct the difference equations. For the simple Runge–Kutta method:

$$\begin{aligned} y_{n+1} &= y_n + \frac{h}{2}[f(x_n, y_n) + f(x_n + h, y_n + hf(x_n, y_n))] \\ &= y_n + \frac{h}{2}[-y_n + x_n + 1 - y_n - hf_n + x_n + h + 1] \\ &= y_n + \frac{h}{2}[-2y_n + 2x_n + 2 - h(-y_n + x_n + 1) + h] \\ &= \left(1 - h + \frac{h^2}{2}\right) y_n + \frac{h}{2}[(2 - h)x_n + 2]. \end{aligned}$$

For the modified Euler's method:

$$\begin{aligned} y_{n+1} &= y_n + hf\left(x_n + \frac{h}{2}, y_n + \frac{h}{2}f(x_n, y_n)\right) \\ &= y_n + h\left(-y_n - \frac{h}{2}f_n + x_n + \frac{h}{2} + 1\right) \\ &= y_n + h\left[-y_n - \frac{h}{2}(-y_n + x_n + 1) + x_n + \frac{h}{2} + 1\right] \\ &= \left(1 - h + \frac{h^2}{2}\right) y_n + \frac{h}{2}[(2 - h)x_n + 2]. \end{aligned}$$

Note Both second-order methods give the same recurrence relation. This occurs if $f(x, y)$ is at most linear in x and y. For non-linear dependence of f on x and/or y, each second-order method will generate a different recurrence relation.

With $h = 0.1$ and $x_n = 0.1n$, the recurrence relation becomes

$$y_{n+1} = 0.905y_n + 0.0095n + 0.1.$$

The result of implementing the relation is shown below, and compared with the Euler solution (see Example 10.24).

			Euler			Runge–Kutta		
n	x_n	$y(x_n)$	y_n	Δ	δ	y_n	Δ	δ
0	0.0	1.0	1.0	0	0	1.0	0	0
1	0.1	1.00484	1.0	0.00484	0.0048	1.005	0.00016	0.00016
2	0.2	1.01873	1.01	0.00873	0.0086	1.01903	0.00029	0.00029
3	0.3	1.04082	1.029	0.01182	0.0114	1.04122	0.00040	0.00038
4	0.4	1.07032	1.0561	0.01422	0.0133	1.07080	0.00048	0.00045
5	0.5	1.10653	1.09049	0.01604	0.0145	1.10708	0.00055	0.00049
6	0.6	1.14881	1.13144	0.01737	0.0151	1.14940	0.00059	0.00052
7	0.7	1.19659	1.17830	0.01829	0.0153	1.19721	0.00062	0.00052
8	0.8	1.24933	1.23047	0.01886	0.0151	1.24998	0.00065	0.00052
9	0.9	1.30657	1.28742	0.01915	0.0147	1.30723	0.00066	0.00050
10	1.0	1.36788	1.34868	0.01920	0.0140	1.36854	0.00066	0.00048

Observations *The behaviour of the Runge–Kutta solution y_n closely resembles the exact solution $y(x_n)$. The absolute error is bounded, adding approximately 0.001 per step. The relative error is bounded at approximately 0.0005 indicating about three significant digits of accuracy in the numerical solution, which is approximately two orders of magnitude more accurate than Euler's method. Note the requirement for* nested function *evaluations.*

DERIVE Experiment 10.2

The second-order Runge–Kutta methods are slightly more difficult than Euler's method to implement in *DERIVE* because of the nested function evaluations. Several auxiliary functions are needed. For the modified Euler's method `Author`

```
me2(f,x,y,xn,yn,k1,h):=lim(f,[x,y],[xn+h/2,yn+hk1/2])
me1(f,x,y,xn,xy,h):=me2(f,x,y,xn,yn,lim(f,[x,y],[xn,yn]),h)
```

`me2` computes $f(x_n + \frac{h}{2}, y_n + \frac{h}{2}k_1)$ and `me1` passes the correct value of k_1 $(= f(x_n, y_n))$ to `me2`. To simulate the iterative process `Author`

```
meuler(f,x,y,x0,y0,h,n):=iterates(
   [xn+h,yn+hme1(f,x,y,xn,yn,h)],[xn,yn],[x0,y0],n)
```

To reproduce the results of Example 10.26, `Author` and `approX` the expression `meuler(-y+x+1,x,y,0,1,.1,10)`.

The corresponding functions for the simple Runge–Kutta method are

```
sk2(f,x,y,xn,yn,k1,h):=k1+lim(f,[x,y],[xn+h,yn+hk1])
sk1(f,x,y,xn,xy,h):=sk2(f,x,y,xn,yn,lim(f,[x,y],[xn,yn]),h)
srk(f,x,y,x0,y0,h,n):=iterates(
   [xn+h,yn+hsk1(f,x,y,xn,yn,h)/2],[xn,yn],[x0,y0],n)
```

10.4.1 Classical Runge–Kutta method

Discussion

The *classical RK method* is similar to the second-order methods except that eqn (10.28) now contains two additional terms

$$y_{n+1} = y_n + h(\alpha_1 k_1 + \alpha_2 k_2 + \alpha_3 k_3 + \alpha_4 k_4),$$

and is forced to agree with the first five terms of the Taylor series method (10.18). The resulting formula is

$$\begin{aligned} y_0 &= s, \\ y_{n+1} &= y_n + \frac{h}{6}(k_1 + 2k_2 + 2k_3 + k_4), \quad n = 0, 1, \ldots \end{aligned} \tag{10.35}$$

where

$$\begin{aligned} k_1 &= f(x_n, y_n), \\ k_2 &= f\left(x_n + \frac{h}{2}, y_n + \frac{h}{2}k_1\right), \\ k_3 &= f\left(x_n + \frac{h}{2}, y_n + \frac{h}{2}k_2\right), \\ k_4 &= f(x_n + h, y_n + hk_3). \end{aligned}$$

This method is fourth order.

DERIVE Experiment 10.3

The fourth-order Runge–Kutta method is implemented using auxiliary functions to simulate k_1, k_2, k_3 and k_4. Author

```
kk4(f,x,y,xn,yn,k1,k2,k3,h):=
   k1+2(k2+k3)+lim(f,[x,y],[xn+h,yn+hk3])
kk3(f,x,y,xn,yn,k1,k2,h):=kk4(f,x,y,xn,yn,k1,k2,
   lim(f,[x,y],[xn+h/2,yn+hk2/2]),h)
```

```
kk2(f,x,y,xn,yn,k1,h):=kk3(f,x,y,xn,yn,k1,
   lim(f,[x,y],[xn+h/2,yn+hk1/2]),h)
kk1(f,x,y,xn,xy,h):=kk2(f,x,y,xn,yn,lim(f,[x,y],[xn,yn]),h)
```

To define the iterative process, Author

```
rk4(f,x,y,x0,y0,h,n):=iterates(
   [xn+h,yn+hkk1(f,x,y,xn,yn,h)/6],[xn,yn],[x0,y0],n)
```

To solve the IVP of Example 10.26, `Author` and `approX` the expression `rk4(-y+x+1,x,y,0,1,.1,10)`.

10.5 Error analysis

Having developed, implemented and observed the behaviour of several one-step numerical methods for solving IVPs, we arrive at the point where some 'serious' analysis is required! The performance of the numerical methods has been couched in terms relating the observed behaviour of the numerical solution to that of the exact solution. More mathematical rigour is needed.

Error analysis concerns the conditions for which $y_n \simeq y(x_n)$, that is $e_n = |y(x_n) - y_n| \to 0$. In other words, what conditions must be satisfied such that the numerical solution y_n can be made as close as desired to the exact solution $y(x_n)$ at a *fixed* point $x = x_n$? This is a difficult question to answer and the approach adopted here is to split the response into two parts – enforcing consistency and a Lipschitz condition.

10.5.1 Local truncation error and consistency

One-step methods for solving $y' = f(x, y)$ take the form

$$y_{n+1} = y_n + h\phi(x_n, y_n; h) \tag{10.36}$$

where $\phi(x, y; h)$ is the *increment function*. Table 10.1 shows the increment function for several of the methods developed to date.

Table 10.1 Increment functions.

Method	Increment function $\phi(x,y;h)$
Euler	$f(x,y)$
Simple Runge–Kutta	$[f(x,y) + f(x+h,y+hf(x,y))]/2$
Modified Euler	$f(x+h/2,y+hf(x,y)/2)$

If $h \neq 0$, eqn (10.36) can be arranged as

$$\frac{y_{n+1} - y_n}{h} = \phi(x_n, y_n; h) \Rightarrow \frac{y_{n+1} - y_n}{h} - \phi(x_n, y_n; h) = 0. \quad \textbf{(10.37)}$$

The aim is to quantify how well the replacement (10.37) imitates the ODE

$$y'(x) = f(x, y(x)) \Rightarrow y'(x) - f(x, y(x)) = 0, \quad \textbf{(10.38)}$$

at $x = x_n$. Replacing $y'(x)$ by $(y(x+h) - y(x))/h$ and $f(x, y(x))$ by $\phi(x, y(x); h)$, the *local truncation error* (LTE) is defined by

$$t(x; h) = \frac{y(x+h) - y(x)}{h} - \phi(x, y(x); h) \quad \textbf{(10.39)}$$

where $x \in [x_0, b]$ and $h > 0$. If the replacement (10.37) were exact then $t(x; h)$ would be zero. This is rarely the case, but we should expect that as h is refined so the 'accuracy' of the replacement improves. Equation (10.37) is a *consistent replacement* to the ODE $y' = f(x, y)$ if

$$\lim_{h \to 0} t(x; h) = 0, \quad x \in [x_0, b]. \quad \textbf{(10.40)}$$

By definition

$$\lim_{h \to 0} \frac{y(x+h) - y(x)}{h} = y'(x),$$

and for a consistent replacement

$$\lim_{h \to 0} \phi(x, y; h) = f(x, y).$$

Further, the derived method (10.36) is *consistent of order p* if there exists an $N \geq 0$, $h_0 > 0$ and integer p such that

$$\sup_{x_0 \leq x \leq b} |t(x; h)| \leq N h^p, \quad 0 < h < h_0. \quad \textbf{(10.41)}$$

Example 10.27 For Euler's method $\phi(x, y; h) \equiv f(x, y)$. The LTE is

$$\begin{aligned} t(x; h) &= \frac{y(x+h) - y(x)}{h} - f(x, y(x)) \\ &= \frac{1}{h}\left[y + hy' + \frac{h^2}{2}y'' + O(h^3) - y\right] - f \\ &= y' + \frac{h}{2}y''(\xi) - f \\ &= Nh \end{aligned}$$

where $x < \xi < x + h$ and

$$N \geq \max_{x \in [x_0, b]} \left| \frac{1}{2} y''(x) \right|.$$

Euler's method is consistent of order 1.

All Taylor series methods and Runge–Kutta methods using $p + 1$ terms are consistent of order p.

10.5.2 Local discretisation error

On arranging eqn (10.39) we can write

$$y(x + h) = y(x) + h\phi(x, y; h) + ht(x; h). \qquad \textbf{(10.42)}$$

In other words, if in addition to h times the increment function, h times the LTE is added to the exact value $y(x)$, then the exact value $y(x + h)$ is recovered. This is similar to the one-step method (10.36), which provides *estimates*, except that the term $ht(x; h)$ is included.

$$ht(x; h) \qquad \textbf{(10.43)}$$

is the *local discretisation error* (LDE) and measures the additional (local) error incurred by the solution process (10.36) when advancing the numerical solution over the interval $[x_n, x_{n+1}]$.

Example 10.28 For Euler's method (see Example 10.27) the discretisation error (per step) is bounded by

$$|ht(x; h)| \leq h^2 N$$

where $N = \frac{1}{2} \max |y''(x)|$. The solution of the IVP in Example 10.24 is $y(x) = e^{-x} + x$ from which $y''(x) = e^{-x}$. On the interval [0, 1], $N = \frac{1}{2}$ and with $h = 0.1$, $|ht(x; h)| \leq 0.005$ (this is a good estimate of the additional error per step in column Δ of Example 10.24).

Summary *The LTE is a measure of how well the difference replacement (10.37) imitates the ODE and the LDE is a measure of the error incurred by the numerical solution (10.36) over* one *interval.*

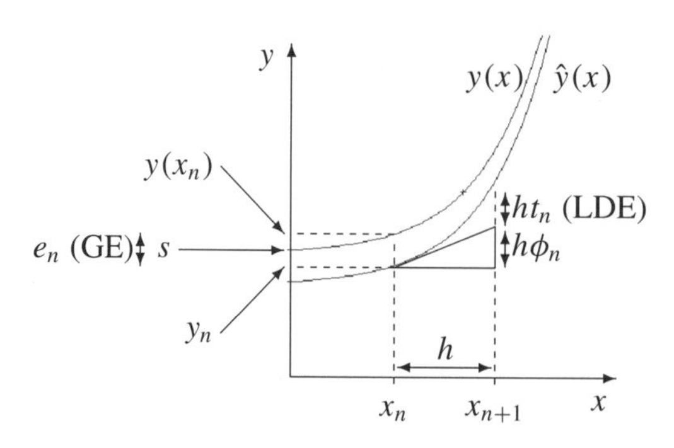

Figure 10.3 Errors. $\hat{y}(x)$ is the exact solution with initial value y_n.

10.5.3 Global error and convergence

Sections 10.5.1 to 10.5.2 dealt with the error on integrating the ODE $y' = f(x, y)$ over *one* interval of size h. The analysis implied that the difference replacement is more accurate for a refined mesh size h – consistency.

Here we assess the accuracy of the estimate y_n as an approximation to $y(x_n)$, for any fixed $x = x_n$ as $h \to 0$. We are concerned with the *global error* (GE) defined by

$$e_n = |y(x_n) - y_n|. \tag{10.44}$$

e_n is non-negative and eqn (10.44) begs the question of how to add together all the local LDEs (see Figure 10.3).

Definition Convergence

If

$$\lim_{\substack{h\to 0\\ x=x_n\,(\text{fixed})}} y_n = y(x_n), \text{ for all } x \in [x_0, b],$$

the one-step method $y_{n+1} = y_n + h\phi(x_n, y_n; h)$ is *convergent*, that is

$$\lim_{\substack{h\to 0\\ x=x_n\,(\text{fixed})}} e_n = 0. \tag{10.45}$$

This definition is due to Dalquist.

A sensible definition of convergence is *not* obtained by considering $\lim_{h\to 0}$ for n fixed. In this case, as h is refined, the point $x_n = x_0 + nh$ would no longer remain fixed, and would ultimately tend to x_0 at which $y_0 = y(x_0)$ (by this definition the error would always be zero!).

The following theorem provides the conditions for which e_n has an upper bound, and gives a formula for the bound from which convergence naturally follows – it is a 'practical' theorem.

Theorem Convergence of one-step methods
The IVP

$$y(x_0) = 0; \quad y' = f(x, y), \quad a \le x_0 \le x \le b,$$

with solution $y(x)$, is replaced by the one-step method

$$y_0 = s; \quad y_{n+1} = y_n + h\phi(x_n, y_n; h), \quad n \ge 0.$$

If

(A) the method is *consistent of order* p, i.e. $|t(x; h)| \le Nh^p$, and
(B) ϕ satisfies the *Lipschitz condition* $|\phi(x, y; h) - \phi(x, z; h)| \le L_\phi |y - z|$ for all $x \in [x_0, b]$, $y, z \in \Re$,

then the global error is bounded by

$$e_n \le \left(\frac{e^{(x_n - x_0)L_\phi} - 1}{L_\phi} \right) Nh^p, \quad L_\phi > 0. \tag{10.46}$$

In the case where $L_\phi = 0$, $e_n \le (x_n - x_0)Nh^p$ (see Exercise 10.16). Such a method is *order* p. e_n approaches zero and the method is convergent.

Proof We show that if conditions (A) and (B) are satisfied then formula (10.46) holds and, consequently, the method is convergent (e_n satisfies condition (10.45)). The one-step method

$$y_{n+1} = y_n + h\phi(x_n, y_n; h) \tag{10.47}$$

is obtained by ignoring the LDE in the Taylor series

$$y(x_{n+1}) = y(x_n) + h\phi(x_n, y(x_n); h) + ht(x_n; h). \tag{10.48}$$

Subtracting eqn (10.47) from eqn (10.48) gives

$$y(x_{n+1}) - y_{n+1} = y(x_n) - y_n + h[\phi(x_n, y(x_n); h) - \phi(x_n, y_n; h)] + ht(x_n; h)$$

and taking the modulus of each side (using the triangle inequality) leads to

$$e_{n+1} \le e_n + h|\phi(x_n, y(x_n); h) - \phi(x_n, y_n; h)| + h|t(x_n; h)|.$$

Using 'consistency (A) and Lipschitz on ϕ (B)'

$$\begin{aligned} e_{n+1} &\le e_n + hL_\phi |y(x_n) - y_n| + Nh^{p+1} \\ &\le (1 + hL_\phi)e_n + Nh^{p+1}, \end{aligned} \tag{10.49}$$

in which h and L_ϕ are *non-negative*. Equation (10.49) is a *difference inequality* (see Section 10.2.2) and the solution requires an associated difference *equation* whose solution ξ_n provides an upper bound to e_n.

We assert that e_n is bounded above by ξ_n where ξ_n satisfies

$$\xi_{n+1} = (1 + hL_\phi)\xi_n + Nh^{p+1}, \quad \xi_0 = e_0. \tag{10.50}$$

By induction, the statement is true for $n = 0$ ($e_0 \le \xi_0$ is given). We now assert that $e_n \le \xi_n$ is true for integers $k = 0, \ldots, n$, and show that this implies that the result is also true for $k = n + 1$. From the inequality (10.49) and equation (10.50)

$$e_{n+1} \le (1 + hL_\phi)e_n + Nh^{p+1} \le (1 + hL_\phi)\xi_n + Nh^{p+1} = \xi_{n+1}.$$

The next step is to solve eqn (10.50). It is a linear first-order inhomogeneous equation, with associated homogeneous equation $\xi_{n+1} - (1 + L_\phi)\xi_n = 0$. The form $\xi_n \propto \beta^n$ gives rise to the characteristic equation $\beta - 1 - hL_\phi = 0$, and the complementary solution is

$$\xi_n^c = c_1(1 + hL_\phi)^n.$$

The inhomogeneous term in the equation is a constant and the particular solution takes the form k. Equation (10.50) becomes $k = (1 + hL_\phi)k + Nh^{p+1}$ from which $k = -Nh^p/L_\phi$. The solution is then

$$\xi_n = c_1(1 + hL_\phi)^n - \frac{Nh^p}{L_\phi}.$$

Using the initial condition $\xi_0 = e_0 = 0$ (assuming that $y_0 = y(x_0)$) we obtain $c_1 = Nh^p/L_\phi$. The general solution is then

$$\xi_n = \frac{Nh^p}{L_\phi}[(1 + hL_\phi)^n - 1].$$

For any $x \ge 0$, $\mathrm{e}^x = \sum_{j=0}^{\infty} \frac{x^j}{j!} \ge 1 + x$. Here $x = hL_\phi$ and so

$$\xi_n \le \frac{Nh^p}{L_\phi}[(\mathrm{e}^{hL_\phi})^n - 1] = \frac{Nh^p}{L_\phi}[\mathrm{e}^{nhL_\phi} - 1].$$

Since $nh = x_n - x_0$, the final result is obtained:

$$\mathrm{e}_n \le \xi_n \le \left(\frac{\mathrm{e}^{(x_n - x_0)L_\phi} - 1}{L_\phi}\right) Nh^p \tag{10.51}$$

(see Exercise 10.16 for the case $L_\phi = 0$). An in-depth illustration is now discussed, showing how the concepts of local truncation error (consistency) and 'Lipschitz' imply convergence as applied to a one-step method.

Example 10.29 The simple Runge–Kutta method

$$y_{n+1} = y_n + \frac{h}{2}[f(x_n, y_n) + f(x_n + h, y_n + hf(x_n, y_n))].$$

(1) *Consistency*: The first step is to ensure that the method is based on a consistent replacement of the ODE $y' = f(x, y)$, established via the LTE. With $y = y(x)$ and $f = f(x, y(x))$ at $x = x_n$,

$$
\begin{aligned}
t(x; h) &= \frac{y(x+h) - y(x)}{h} - \frac{1}{2}[f(x, y) + f(x+h, y + hf(x, y))] \\
&= \frac{1}{h}\left[y + hy' + \frac{h^2}{2}y'' + \frac{h^3}{6}y''' + \cdots - y\right] \\
&\quad - \frac{1}{2}\left[f + \left\{f + hf^{(1)} + \frac{h^2}{2}(f_{xx} + 2ff_{xy} + f^2 f_{yy})\right\} + \cdots\right] \\
&= \frac{h^2}{12}[2y''' - 3(f_{xx} + 2ff_{xy} + f^2 f_{yy})]_{x=\xi} \\
&= \frac{h^2}{12}[-y''' + 3y'' f_y]_{x=\xi}
\end{aligned}
$$

where $x_n < \xi < x_n + h$, $\eta = y(\xi)$. If $N = \frac{1}{12}\max|-y''' + 3y'' f_y|$ then $|t(x; h)| \le Nh^2$ and the method is consistent of order 2.

(2) *Lipschitz*: The second step is to show that ϕ satisfies a Lipschitz condition, *given* that f does ($|f(x, y) - f(x, z)| \le L|y - z|$).

$$
\begin{aligned}
&|\phi(x, y; h) - \phi(x, z; h)| \\
&= \frac{1}{2}|f(x, y) + f(x+h, y + hf(x, y)) \\
&\quad - f(x, z) - f(x+h, z + hf(x, z))| \\
&\le \frac{1}{2}[|f(x, y) - f(x, z)| \\
&\qquad + |f(x+h, y + hf(x, y)) - f(x+h, z + hf(x, z))|] \\
&\le \frac{1}{2}[L|y - z| + L|y + hf(x, y) - z - hf(x, z)|] \\
&\le \frac{L}{2}[|y - z| + |y - z| + h|f(x, y) - f(x, z)|] \\
&\le \frac{L}{2}[2|y - z| + hL|y - z|] \\
&= L\left[1 + \frac{hL}{2}\right]|y - z| \\
&= L_\phi|y - z|.
\end{aligned}
$$

A Lipschitz constant L_ϕ exists (the method *is* convergent of order 2).

Summary *For the one-step method* $y_0 = s$, $y_{n+1} = y_n + h\phi(x_n, y_n; h)$ *applied to the IVP* $y(x_0) = s$, $y' = f(x, y)$,

Consistency	+	Lipschitz on ϕ	$\Rightarrow$	Convergence
$\lvert t(x; h)\rvert \le Nh^p$		$\lvert\phi(x, y) - \phi(x, z)\rvert \le L_\phi \lvert y - z\rvert$		$\lim_{h\to 0} e_n = 0$

10.5.4 Rounding errors

A startling conclusion may be deduced from Section 10.5.3. It appears that consistency and convergence are synonymous, and by reducing the LTE (refining h) an arbitrarily small error e_n can be achieved! This, however, neglects the effect of *rounding errors* in the iterative process $y_{n+1} = y_n + h\phi_n$.

If h is refined, an increasing number of steps are required to numerically integrate the solution to a specified fixed point x. This implies more floating-point arithmetic operations and hence more rounding errors. In place of computing a sequence $\{y_n\}$ using the numerical process

$$y_{n+1} = y_n + h\phi(x_n, y_n; h),$$

the sequence $\{z_n\}$ is computed from the relation

$$z_{n+1} = z_n + h\phi(x_n, z_n; h) + \varepsilon_n \tag{10.52}$$

where $\varepsilon_n > 0$ is the rounding error on storing intermediate values to *finite machine precision*. Subtracting eqn (10.52) from the Taylor series (10.48)

$$y(x_{n+1}) - z_{n+1} = y(x_n) - z_n + h[\phi(x_n, y(x_n); h) - \phi(x_n, z_n; h)] + ht(x_n; h) + \varepsilon_n$$

and taking the modulus of each side (utilizing the triangle inequality)

$$e_{n+1} \le e_n + h|\phi(x_n, y(x_n); h) - \phi(x_n, z_n; h)| + h|t(x_n; h)| + \varepsilon_n.$$

The global error is now defined as $e_n = |y(x_n) - z_n|$. Using consistency (condition (A)) and the Lipschitz condition on ϕ (condition (B))

$$\begin{aligned} e_{n+1} &\le e_n + hL_\phi|y(x_n) - z_n| + Nh^{p+1} + \varepsilon_n \\ &= (1 + hL_\phi)e_n + Nh^{p+1} + \varepsilon_n \\ &= (1 + hL_\phi)e_n + h\left(Nh^p + \frac{\varepsilon_n}{h}\right). \end{aligned} \tag{10.53}$$

This is eqn (10.49) with the term Nh^p replaced by $Nh^p + \varepsilon_n/h$. The upper bound (10.51) for e_n is now replaced by

$$e_n \le \left(\frac{e^{(x_n - x_0)L_\phi} - 1}{L_\phi}\right)\left(Nh^p + \frac{\varepsilon}{h}\right), \tag{10.54}$$

in which $\varepsilon_n \le \varepsilon$. The consequence of this 'small' change is enormous! While the LTE term Nh^p decreases as h is refined, the rounding term ε/h increases (see Section 7.6 on errors in numerical differentiation). In other words, there is an optimum value of h, say h_{opt}, below which the rounding error dominates the truncation error (Exercise 10.21 deals with the determination of h_{opt}).

10.6 Numerical stability

The error analysis of Section 10.5 focused on the conditions under which the LTE, $t(x;h)$, the LDE, $ht(x;h)$, and the GE $e_n = |y(x_n) - y_n|$ (including rounding errors) were 'small'. Each error measures the 'goodness of fit' of the difference model and numerical solution to the ODE and exact solution.

IVP/solution			Numerical model/solution
$y(x_0)$	$= s$		$y_0 = s$
$y'(x)$	$= f(x, y(x))$	LTE	$\frac{y(x+h)-y(x)}{h} = \phi(x, y(x); h)$
$y(x_n)$	: One step	LDE	$y_n \to y_{n+1}$
	: n steps	GE	$e_n = \lvert y(x_n) - y_n \rvert$, $x = x_n$ fixed

Here we focus upon errors *inherent* in the recurrence relation used to generate the sequence $\{y_n\}$. We are less concerned with the relationship between $y(x_n)$ and y_n (the theoretical numerical solution) than with the conditions which ensure that the computed numerical solution, say y_n^*, is close to y_n.

10.6.1 Absolute stability

Example 10.30 The solution of the IVP $y(0) = \frac{1}{6}$, $y' = -60y + 10x$ $(x > 0)$ is $y(x) = \frac{1}{6}(x + e^{-60x})$, a rapidly decreasing function for $0 \le x \le 0.05$ followed by linear behaviour (e^{-60x} is negligble for $x > 0.05$) (see Figure 10.4).

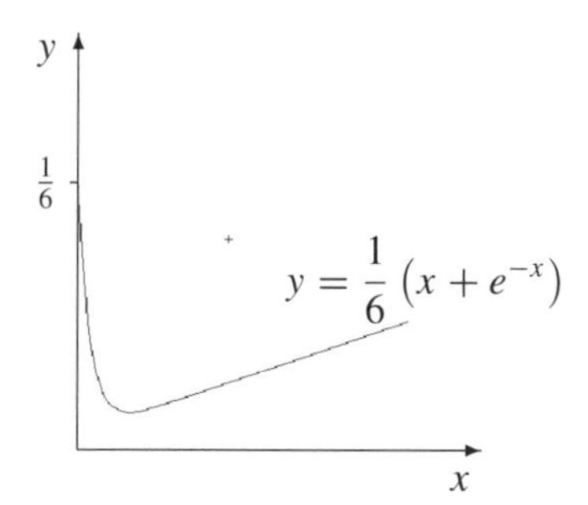

Figure 10.4 A rapidly decaying solution component.

The second-order simple Runge–Kutta method generates

$$\begin{aligned} y_{n+1} &= y_n + \frac{h}{2}[f(x_n, y_n) + f(x_n + h, y_n + hf(x_n, y_n))] \\ &= (1 - 60h + 1800h^2)y_n + \sigma_s(x_n; h) \end{aligned} \tag{10.55}$$

and the second-order *trapezoidal method* generates

$$\begin{aligned} y_{n+1} &= y_n + \frac{h}{2}[f(x_n, y_n) + f(x_n + h, y_{n+1})] \\ &= \left(\frac{1 - 30h}{1 + 30h}\right) y_n + \sigma_t(x_n; h). \end{aligned} \tag{10.56}$$

σ_s and σ_t are known functions of the independent variable x and step size h. Table 10.2 shows the results of implementing the methods (10.55) and (10.56) starting at $x_0 = 1$ with the analytical solution $y_0 = y(1) = \frac{1}{6}(1 + e^{-60})$.

Table 10.2 One-step instability.

n	x_n	$y(x_n)$ Analytical	y_n Simple RK	y_n Trapezoidal
0	1.0	0.167	0.167	0.167
1	1.1	0.183	0.187	0.183
2	1.2	0.200	0.247	0.200
3	1.3	0.217	0.827	0.217
4	1.4	0.233	8.168	0.233

What is happening to the SRK solution? Both methods generate first-order difference equations when applied to the first-order ODE. They are of the same order, so we might expect similar numerical solutions. Further, both methods start at $x_0 = 1$, which is well into the linear region of the analytical solution – the methods should cope with this (almost) exactly!

To analyse the behaviour of the simple Runge–Kutta method (see Example 10.30) and to serve as a focus for analysis, the test problem

$$y(x_0) = s; \; y' = \lambda y, \quad x_0 < x \leq b, \tag{10.57}$$

is introduced, where $\Re(\lambda) < 0$, with the *single* solution

$$y(x) = e^{\lambda(x - x_0)}, \quad x \geq x_0. \tag{10.58}$$

λ is real and negative or complex with negative real part. The solution (10.58) represents exponential decay. All one-step methods applied to the test problem generate a first-order difference equation of the form

$$y_{n+1} = p(\omega) y_n \tag{10.59}$$

where p is a polynomial function of $\omega = h\lambda$. Inhomogeneous terms in the test problem, say $y' = \lambda y + \mu$, where μ is a known function of x, cause the difference equation to take the modified form

$$y_{n+1} = p(\omega)y_n + q(x_n; h). \qquad \textbf{(10.60)}$$

Equations (10.59) and (10.60) are linear and first-order with solutions

$$y_{n+1} = s[p(\omega)]^n \quad \text{and} \quad y_{n+1} = (s - r(x_0; h))[p(\omega)]^n + r(x_n; h).$$

For a decaying solution the modulus of $p(\omega)$ must be less than 1,

$$|p(\omega)| < 1. \qquad \textbf{(10.61)}$$

A numerical method satisfying condition (10.61) is *absolutely stable*, or simply *A-stable*.

Example 10.31 For the test problem, the simple Runge–Kutta method generates the difference equation $y_{n+1} = p(\omega)y_n$ where $p(\omega) = 1 + \omega + \frac{1}{2}\omega^2$. Condition (10.61) requires

$$-1 < 1 + \omega + \frac{1}{2}\omega^2 < 1.$$

The left-hand inequality $\omega^2 + 2\omega + 4 > 0$ is true for all ω ($p(\omega)$ has a turning point at $\omega = -1$, which is a minimum since the second-derivative is positive, at which its value is $1/2$). The right-hand inequality $\omega^2 + 2\omega < 0$ ($\omega(\omega + 2) < 0$) is true for $-2 < \omega < 0 \Rightarrow 0 < -\omega < 2 \Rightarrow 0 < h < -2/\lambda$.

h must lie within a specific domain for the method to be A-stable. For Example 10.30 the value of λ is -60 and for A-stability h should lie within the interval $(0, 1/30)$. The value used was $h = 0.1$, which lies outside the region of A-stability (see Figure 10.5(a)).

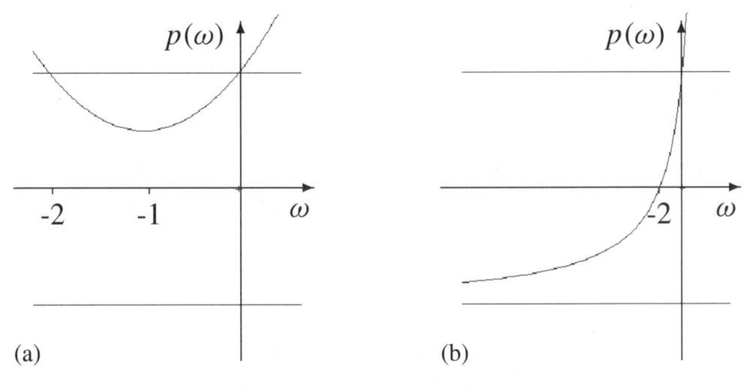

Figure 10.5 Absolute stability intervals for (a) the simple Runge–Kutta ($-2 < \omega < 0$) and (b) the trapezoidal ($\omega < 0$) methods.

The trapezoidal method applied to the test problem generates the difference equation

$$y_{n+1} = \left(\frac{1 + \frac{\omega}{2}}{1 - \frac{\omega}{2}} \right) y_n.$$

Since $\lambda < 0$ ($\omega < 0$), $1 - \omega/2$ has magnitude greater than $1 + \omega/2$ and

$$|p(\omega)| = \left| \frac{2 + \omega}{2 - \omega} \right| < 1$$

for all negative ω, that is for all $h > 0$. The method is A-stable (see Figure 10.5(b)).

10.6.2 L-stability and error propagation

If the error $\varepsilon_0 = y_0 - y_0^*$ is introduced at the beginning of an iterative solution process, in place of $\{y_n\}$ being computed from

$$y_{n+1} = p(\omega) y_n$$

the sequence $\{y_n^*\}$ is computed from

$$y_{n+1}^* = p(\omega) y_n^*.$$

Subtracting these equations gives the error formula

$$\varepsilon_{n+1} = p(\omega) \varepsilon_n. \tag{10.62}$$

Errors decrease, that is $|\varepsilon_{n+1}| < |\varepsilon_n|$, if $|p(\omega)| < 1$. In other words, the condition for A-stability is equivalent to ensuring that errors (introduced at any stage of the iterative process) will decrease.

The solution of eqn (10.62) is $\varepsilon_n = \varepsilon_0 [p(\omega)]^n$. For large and negative ω (large and negative λ) the numerical solution $y_n = s[p(\omega)]^n$ will be small for most $n > 0$ (the solution has a very high rate of decay) and the errors may dominate the solution. In order that the error ε_n does not dominate the solution y_n^* we require

$$\lim_{\Re(\omega) \to -\infty} |p(\omega)| = 0. \tag{10.63}$$

A numerical method satisfying (10.63) is *L-stable*. No Runge–Kutta method is L-stable since $p(\omega)$ will be a polynomial in ω of degree at least 1.

Summary *To précis the discussion of stability several points are highlighted with reference to the numerical solution of the ODE* $y' = f(x, y)$.

(1) A-stability*: to ensure a decaying solution,* $|p(\omega)| < 1$. *This may require finding a suitable interval* $0 < h < h_A$.

(2) L-stability*: if necessary, determine an interval* $0 < h < h_L$ *such that* $|p(\omega)| \to 0$ ($|\beta_1(\omega)| \to 0$) *as* $\Re(\omega) \to -\infty$.

A step size h is selected such that $h = \min\{h_A, h_L\}$.

Note A third type of instability may occur, called *parasitic instability*, when a first-order ODE is replaced by a difference equation of higher order. This introduces independent spurious solution components that cannot imitate the decaying solution (10.58). If these spurious (parasitic) components dominate, the numerical solution will deteriorate. Euler's method and all Runge–Kutta methods are *P-stable* since they are one-step methods and generate difference equations of order 1 (for the first-order ODE $y' = f(x, y)$ discussed in this chapter)[5].

10.6.3 Complex λ

Finally, we briefly consider the question of numerical stability when the decay constant is complex (with negative real part), that is $\lambda = a + ib$ with $a < 0$ and $\mathrm{i}^2 = -1$. The solution of the test problem can be written

$$y(x) = s\mathrm{e}^{\lambda(x-x_0)} = s\mathrm{e}^{(a+\mathrm{i}b)(x-x_0)}. \tag{10.64}$$

The real part, $\mathrm{e}^{a(x-x_0)}$, causes decay (since $a < 0$) and the complex part, $\mathrm{e}^{\mathrm{i}b(x-x_0)} = \cos(b(x - x_0)) + \mathrm{i}\sin(b(x - x_0))$, imposes an oscillatory structure upon the decay.

To indicate the approach to A-stability for the complex case, we consider an illustration based upon Euler's method.

Example 10.32 Applied to the test problem $y' = \lambda y$, Euler's method gives the first-order difference equation $y_{n+1} - (1 + \omega)y_n = 0$. The solution $y_n \propto \beta^n$ generates the characteristic equation $\beta - (1 + \omega) = 0$ (at this point we normally solve the equation for β as a function of ω in order to determine the general form of y_n).

The aim here is to find regions in the complex ω-plane, where $\omega = u + \mathrm{i}v$, such that $|\beta| < 1$. For real ω this means solving two inequalities, $-1 < \beta$ and $\beta < 1$. For complex ω the approach is to let β be a general complex number with modulus 1 and then trace the locus of points in the complex ω-plane such that $|\beta| = 1$.

A complex number with unit modulus is $\beta = \mathrm{e}^{\mathrm{i}\theta}$, where θ varies. Substitution into the characteristic equation gives $\mathrm{e}^{\mathrm{i}\theta} - (1 + \omega) = 0$, that is $\mathrm{e}^{\mathrm{i}\theta} - (1 + u + \mathrm{i}v) = 0$. Expanding $\mathrm{e}^{\mathrm{i}\theta}$, $\cos\theta + \mathrm{i}\sin\theta - (1 + u + \mathrm{i}v) = 0$, and comparing real and imaginary parts gives the formulae

$$\begin{aligned} u &= \cos\theta - 1 \\ v &= \sin\theta. \end{aligned}$$

These equations represent the totality of (u, v) points in the complex ω-plane for which $|\beta| = 1$. The locus of these points traces out a curve that separates the region of A-stability from the unstable region. A little manipulation results in

$$(1 + u)^2 + v^2 = \cos^2\theta + \sin^2\theta = 1$$

which is the equation of a circle of radius 1 and centre $(u, v) = (-1, 0)$ (see Figure 10.6).

[5] A discussion of parasitic instability can be found in [18], Section 13.11.

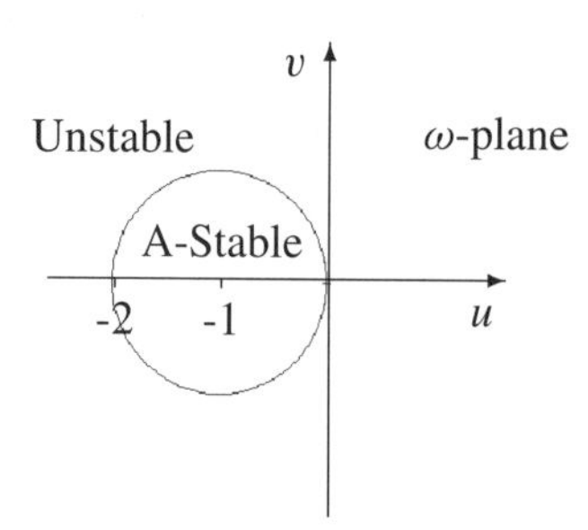

Figure 10.6 A-stable region for Euler's method with complex λ.

The final step is to determine which part of the ω-plane is A-stable. We simply pick a point (which specifies ω) and solve the characteristic equation for β. The point $(-1, 0)$ lies inside the circle and the characteristic equation is $\beta - (1 + (-1 + i0)) = 0$, that is $\beta = 0$. This value has modulus less than 1 and the inner region is A-stable.

The relationships between solutions presented at the beginning of Section 10.6 can now be updated to include the stability analyses.

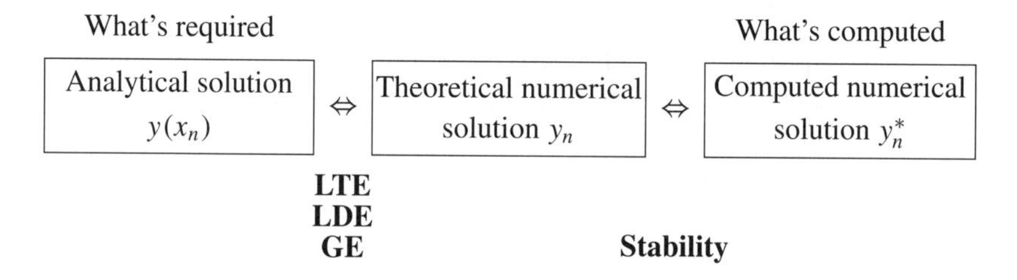

To have a reliable numerical method for solving ODEs it is essential that accuracy (LTE, LDE, GE) *and* insensitivity to errors (stability) are established so that the computed numerical solution y_n^* is 'close to' the required analytical solution $y(x_n)$. Table 10.3 summarises the properties of the five methods highlighted in this chapter.

Table 10.3 Properties of selected numerical ODE solvers.

Method		Order	Stability A	Stability L
Euler	Eqn (10.21)	1	$-2 < \omega < 0$	No
Simple RK	Eqn (10.33)	2	$-2 < \omega < 0$	No
Modified Euler	Eqn (10.34)	2	$-2 < \omega < 0$	No
Classical RK	Eqn (10.35)	4	$-2.785 < \omega < 0$	No
Trapezoidal	Ex 10.30	2	Yes ($\forall \omega < 0$)	No

Summary

This chapter dealt with the development, implementation and analysis of numerical methods for solving the IVP $y(x_0) = s$, $y' = f(x, y)$. In Section 10.1 several preliminary concepts of ODEs were introduced, including existence and uniqueness (Section 10.1.1) and numerical solutions (Section 10.1.2), followed by the solution of linear difference equations (Section 10.2) and inequalities (Section 10.2.2). Taylor series methods were discussed in Section 10.3, and their impracticality exposed (except for Euler's method (10.21)), and the concept of order and its estimation were given in Section 10.3.1. The imitation of Taylor series methods by Runge–Kutta methods was developed in Section 10.4 and several specific Runge–Kutta methods were described; simple Runge–Kutta (10.33), modified Euler (10.34) and the classical fourth-order method (10.35).

Section 10.5 dealt with the analysis of errors; local truncation error and consistency (Section 10.5.1), local discretisation error (Section 10.5.2) and global error and convergnce (Section 10.5.3). The effect of rounding errors was assessed in Section 10.5.4. In Section 10.6 the idea of numerical stability was developed with specific reference to A-stability (Section 10.6.1) and L-stability (Section 10.6.2). The chapter concluded with a summary of the one-step methods developed (see Table 10.3).

Exercises

If you have access to *DERIVE* you are urged to `Plot` the various analytical and numerical solutions developed in these exercises.

10.1 Show that the following IVPs have a unique solution.

(a) $y(0) = 2; \quad \dfrac{dy}{dx} = 3xy, \; 0 < x \leq 1.$

(b) $y(x_0) = s; \quad \dfrac{dy}{dx} = \cos 2y,$ on any interval containing x_0.

10.2 Determine the value(s) of $\varepsilon \geq 0$ for which the following IVP has a unique solution.

$$y(0) = s; \quad \frac{dy}{dx} = \frac{1}{\varepsilon + y^2}, \; x > 0.$$

10.3 Show that the following IVP has a unique solution and find the first three terms of the solution series using Picard iteration.

$$y(0) = 1; \quad y' = y\cos x, \; x > 0.$$

10.4 Find the general solution of the following linear difference equations.

(a) $y_{n+1} + ky_n = 0$, constant k.

(b) $y_{n+2} - 3y_{n+1} + y_n = 0$.
(c) $y_{n+2} + 2y_{n+1} + y_n = 0$.
(d) $-2y_{n+1} + y_n = n - 2$.
(e) $y_{n+2} - 4y_{n+1} + 3y_n = 2^n$, $y_0 = 0$, $y_1 = 1$.

10.5 In the absence of players, the alien population A_n on the screen of a video game doubles every 4 seconds, $A_n = 2A_{n-1}$. The game ends when 10^7 aliens are on the screen. Solve the difference equation with $A_0 = 10$ to determine how long the game takes to finish.

10.6 Rework Exercise 5.13 on pollution by solving the difference equations (as opposed to performing the iteration).

10.7 Given $|y_0| = s$, solve the following first-order difference inequalities. State any restrictions on s.

(a) $|y_{n+1}| \leq \frac{1}{2}|y_n|$.
(b) $|y_{n+1}| \leq 3|y_n| - 2$.

10.8 Show that the following IVP has a unique solution.

$$y(0) = 1; \quad \frac{dy}{dx} = \cos y, \ 0 < x \leq 1.$$

Determine the difference equation corresponding to the Taylor series method of order 2 and use it with $h = 0.1$ to estimate the solution at $x = 0.2$.

10.9 Apply Euler's method (Taylor series method of order 1) to the IVP

$$y(0) = \frac{1}{4}; \quad \frac{dy}{dx} = y - 1, \ 0 < x \leq \frac{1}{2}.$$

Solve the difference equation to obtain an explicit formula for y_n and use the formula to obtain three estimates to $y(0.5)$ with step sizes $h = 0.1$, $h = 0.05$ and $h = 0.025$ (you could use the function `euler` (see *DERIVE* Experiment 10.1) to check your three estimates by iteration). Apply extrapolation to these estimates to determine the approximate order of the method.

`Plot` the first two *DERIVE* solutions (with `Options State Discrete` set) and compare them with the exact solution $y(x) = \frac{1}{4}(4 - 3e^x)$.

10.10 Show that the Runge–Kutta method of order 1 is Euler's method.

10.11 Use the simple Runge–Kutta method with $h = 0.2$ to solve the IVP

$$y(0) = \frac{1}{2}; \quad \frac{dy}{dx} = x^2 - y, \ 0 < x \leq 0.6.$$

Find an approximation to $y(0.6)$ and compare your answer with the exact solution $y(x) = -\frac{3}{2}e^{-x} + x^2 - 2x + 2$ at $x = 0.6$. Use extrapolation to estimate the order of the method (you will find the function `srk` of *DERIVE* Experiment 10.2 useful).

10.12 Obtain the second-order Runge–Kutta method from eqn (10.32) in which $\alpha_1 = \frac{1}{4}$ (*Heun's method*). Obtain the difference equation generated by applying the method to the linear ODE in Example 10.26 and comment upon your answer.

10.13 Repeat Exercise 10.11 using (a) the modified Euler's method and (b) Heun's method (see Exercise 10.12).

10.14 Apply Euler's method and the second-order Runge–Kutta methods (simple Runge–Kutta, modified Euler's and Heun (see Exercise 10.12)) to the IVP $y(0) = 1$, $y' = x^2 + y^2$ $(x > 0)$. Obtain estimates to $y(0.2)$ using a step size of $h = 0.1$. The *DERIVE* functions `euler`, `meuler`, `srk` and `heun` will be helpful!

10.15 Apply the classical fourth-order Runge–Kutta method to the IVP of Exercise 10.8 with $h = 0.2$ to approximate $y(1)$. If you are using *DERIVE* set `Options Precision Digits` to 10. Estimate the order of the method.

10.16 Using L'Hôpital's rule, take the limit as $L_\phi \to 0$ of eqn (10.51) to obtain an upper bound on e_n when the Lipschitz constant equals zero.

10.17 Show that the modified Euler's method gives a consistent second-order replacement when applied to the differential equation $y' = f(x, y)$ (see Examples 10.27 and 10.29).

10.18 Apply the simple Runge–Kutta method to the IVP $y(0) = 1$, $y' = -y+x^2+1$ $(0 < x \leq 1)$ with $h = 0.2$. Write down the *observed* error at the nodes $x_1, \ldots, x_5$ given that $y(x) = -2e^{-x} + x^2 - 2x + 3$. Determine an expression for the local discretisation error and verify that it is comparable with the observed *local* errors.

10.19 Apply Euler's method to the IVP

$$y(0) = 0; \quad \frac{\mathrm{d}y}{\mathrm{d}x} = x^2 + x - y, \; 0 < x \leq 1,$$

with $h = 0.2$, to obtain a numerical estimate to $y(1)$.

Find values for N, p and L_ϕ such that $|t(x; h)| \leq Nh^p$ and $|\phi(x, y; h) - \phi(x, z; h) \leq L_\phi|y - z|$. Hence determine an *upper* bound for the global error (see eqn (10.51)) and verify that the value does provide an upper bound for the *observed* error. Note that $y(x) = x^2 - x + 1 - e^{-x}$.

10.20 Applied to the ODE $y' = f(x, y)$, show, for each of the second-order Runge–Kutta methods (modified Euler's and Heun's) that the Lipschitz constant is

$$L_\phi = \left(1 + \frac{h}{2}L\right) L$$

where L is a Lipschitz constant for $f(x, y)$ (see Example 10.29).

10.21 Use the ideas of Section 7.6 to show that the combined effect of truncation and rounding errors in eqn (10.54) is minimized when

$$h = h_{\text{opt}} = \sqrt[p+1]{\frac{\varepsilon}{pN}}.$$

For the IVP $y(0) = 1$, $y' = -y$ $(x > 0)$, obtain the values of h_{opt} for Euler's method, the simple Runge–Kutta method and the modified Euler's method if $\varepsilon = \frac{1}{2} \times 10^{-6}$ (typical of a PC). Appropriate expressions for N are

$$\frac{1}{2}\max|y''|, \quad \frac{1}{24}\max|y''' + 3y''f_y|, \quad \frac{1}{12}\max|-y''' + 3y''f_y|.$$

10.22 Repeat Exercise 10.21 for the IVP $y(0) = 0$, $y' = x^2 + x - y$ $(0 < x \leq 2)$, which has the exact solution $y(x) = x^2 - x + 1 - e^{-x}$. This time include Heun's method for which $N = \frac{1}{6}\max|y''f_y|$.

10.23 Is Euler's method L-stable ? Justify your answer.

Set up a *DERIVE* Experiment to empirically test your answer. Use the function euler to solve $y(0) = 1$, $y' = \lambda y$ for λ large and negative, and compare with the exact solution $y(x) = e^{\lambda x}$.

10.24 From Example 10.31, the second-order Runge–Kutta methods are deemed A-stable if $-1 < 1+\omega+\frac{\omega^2}{2} < 1$ (for negative real ω). For complex ω we need to solve (and plot) the equation $p = 1 + \omega + \frac{\omega^2}{2}$ where p is replaced by the general complex number $e^{i\theta}$ of unit magnitude and $\omega = u+iv$. What does the A-stable region look like?

Author p=1+w+w^2/2 and Manage Substitute p with exp(<alt>it) and w with u+<Alt>iv. Take the real and imaginary parts of the resulting equation (re#* and im#*, where * is the expression number). Using the fact that $\cos^2 t + \sin^2 t = 1$, create an expression involving just u and v, and Plot it.

10.25 The one-step scheme $y_{n+1} = y_n + hf_{n+1}$ is the *backward Euler's method* (note the change from f_n (Euler's method) to f_{n+1}).

(a) Obtain the difference equation on applying the method to the IVP of Example 10.24. With $h = 0.1$ use the recurrence relation to estimate $y(x)$, $x \in \{0.1, 0.2, \ldots, 1\}$ and compare the behaviour of the numerical solution with Euler's method. You may wish to use *DERIVE* to implement the method.
Repeat your calculations using $h = 0.05$ and $h = 0.025$ and make an informed guess as to the order of the method. Confirm your guess by using the extrapolation procedure of Section 10.3.1.

(b) Perform a convergence analysis (see Example 10.29) to confirm the theoretical order of the method.

(c) Analyse the A- and L-stability of the method for negative real λ. Repeat the *DERIVE* Experiment of Exercise 10.23 for this method.
Repeat the A-stability analysis for complex λ and sketch the locus of the A-stable

region. Compare and contrast with the analogous results for Euler's method (see Example 10.32).

10.26 One type of instability not discussed in the text is that due to an *ill-conditioned problem*. Small errors/changes in the statement of the IVP lead to gross changes in the numerical solution. Consider the non-linear IVP $y(0) = s$, $y' = xy(y-1)$ $(x > 0)$ which has the analytical solution

$$y(x) = \frac{s}{s + (1-s)e^x}, \quad x \geq 0.$$

Define the *DERIVE* expression `y(x,s):=s/(s+(1-s)expx)` and create the vector `vector(y(x,s),s,.5,1.5,.1)`. `approX` and `Plot` the result and comment on the picture.

10.27 Set up a *DERIVE* Experiment to solve the IVP $y(0) = 1$, $y' = y^2$ $(0 < x \leq 2)$ using both Euler's method and the modified Euler's method. Describe the results and explain them!

Bibliography

[1] Abramowitz M. and Stegun I. A. (eds.) (1972). *Handbook of Mathematical Functions*. New York: Dover.

[2] Baker G. A. Jr and Graves-Morris P. R. (1996). *Padé Approximants*, 2nd edn. Cambridge: Cambridge University Press.

[3] Barnett S. (1990). *Matrices: Methods and Applications*. Oxford: Clarendon Press.

[4] Barnett S. (1995). *Some Modern Applications of Mathematics*. London: Ellis Horwood.

[5] Bronson R. (1988). *Schaum's Outline of Matrix Operations*. New York: McGraw-Hill.

[6] Bruce J. W., Gilpin P. J. and Rippon P. J. (1990). *Microcomputers and Mathematics*. Cambridge: Cambridge University Press.

[7] Burden R. L., Faires J. D. and Reynolds A. C. (1981). *Numerical Analysis*, 2nd edn. Boston: Prindle, Weber & Schmidt.

[8] Davis P. J. and Rabinowitz P. (1975). *Methods of Numerical Integration*. New York: Academic Press.

[9] *Derive User Manual – Ver. 3* (1994). 7th edn. Honolulu Soft Warehouse.

[10] Golub G. H. and Van Loan C. F. (1989). *Matrix Computations*. Baltimore: Johns Hopkins University Press.

[11] Graham E., Berry J. S. and Watkins A. J. P. (eds.) (1997). *Mathematical Activities with DERIVE*. Bromley: Chartwell-Bratt.

[12] Hall G. and Watt J. M. (eds.) (1976). *Modern Numerical Methods for Ordinary Differential Equations*. Oxford: Clarendon Press.

[13] Kelly W. G. (1991). *Difference Equations*. New York: Academic Press.

[14] Kutzler B. (1994). *Mathematics on the PC: Introduction to Derive*. Hagenberg: Soft Warehouse.

[15] Kutzler B. (1996). *Introduction to Derive for Windows*. Hagenberg: Soft Warehouse.

[16] O'Connor J. J. and Robertson E. F. *MacTutor History of Mathematics Archive*. University of St Andrews. `http://www-groups.dcs.st-and.ac.uk/~history/`

[17] Ortega J. M. and Poole W. G. Jr. (1981). *Numerical Methods for Differential Equations*. Marshfield MA: Pitman.

[18] Phillips G. M. and Taylor P. J. (1996). *Theory and Applications of Numerical Analysis*, 2nd edn. London: Academic Press.

[19] Smith G. D. (1985). *Numerical Solution of Partial Differential Equations*, 3rd edn. Oxford: Clarendon Press.

A Answers to exercises

Chapter 1

1.1 $p_0(x) = e$, $p_1(x) = ex$, $p_2(x) = \frac{e}{2}(x^2+1)$.

1.2 Use eqn (1.4) to obtain expressions for $p_3(x)$, $p_3'(x)$, $p_3''(x)$ and $p_3'''(x)$, and evaluate these expressions at $x = x_0$.

1.3 Taylor polynomial p_n is

$$f(x_0) + (x - x_0)f'(x_0) + \cdots + \frac{(x-x_0)^k}{k!} f^{(k)}(x_0) + \cdots + \frac{(x-x_0)^n}{n!} f^{(n)}(x_0).$$

If $f(x) = \sum_{i=0}^{n} a_i (x-x_0)^i$, $f^{(k)}(x) = \sum_{i=k}^{n} k \ldots (k-i+1) a_i (x-x_0)^{i-k}$ and $f^{(k)}(x_0) = k! a_k$. Hence

$$\begin{aligned} p_n(x) &= a_0 + (x-x_0)a_1 + \cdots \\ &\quad + \frac{(x-x_0)^k}{k!} k! a_k + \cdots \\ &\quad + \frac{(x-x_0)^n}{n!} n! a_n \\ &= f(x). \end{aligned}$$

1.4 $\sin x = x - \frac{x^3}{6} + \frac{x^5}{120} + O(x^7)$, $\cos x = 1 - \frac{x^2}{2} + \frac{x^4}{24} + O(x^6)$. $\cos(a+h) = \cos a \cos h - \sin a \sin h$.

1.5 $\tan^{-1} x = x - \frac{x^3}{3} + \frac{x^5}{5} - \frac{x^7}{7} + \cdots$. $|R_n(x)| = |\frac{x^{n+1}}{n+1}|$. For 6 d. p., $|R_n(x)| \le \frac{1}{2} \times 10^{-6}$. $x = \frac{2}{5}$ and $x = \frac{3}{7}$ require $n = 12.03$ and $n = 13.01$. With $n = 13$, $\tan^{-1} x \approx x - \frac{x^3}{3} + \frac{x^5}{5} - \frac{x^7}{7} + \frac{x^9}{9} - \frac{x^{11}}{11} + \frac{x^{13}}{13}$. $\tan^{-1}\frac{2}{5} \approx 0.380506$, $\tan^{-1}\frac{3}{7} \approx 0.404891$, $\pi = 4(\tan^{-1}\frac{2}{5} + \tan^{-1}\frac{3}{7}) \approx 3.14159$

1.6 $e^x = 1 + x + \frac{x^2}{2!} + \frac{x^3}{3!} + \frac{x^4}{4!} + \frac{x^5}{5!} + \frac{x^6}{6!} + \cdots$

$e^{-x} = 1 - x + \frac{x^2}{2!} - \frac{x^3}{3!} + \frac{x^4}{4!} - \frac{x^5}{5!} + \frac{x^6}{6!} + \cdots$
$\cosh x = \frac{e^x + e^{-x}}{2}$.

$$\frac{1}{(n+2)!} \le 10^{-4} \Rightarrow (n+2)! \ge 10000 \Rightarrow n+2 \ge 8 \Rightarrow n \ge 6.$$

$$\int_0^x \cosh\sqrt{t}\,dt \approx \int_0^1 \left[1 + \frac{t}{2!} + \frac{t^2}{4!} + \frac{t^3}{6!}\right] dt = x + \frac{x^2}{2 \times 2!} + \frac{x^3}{3 \times 4!} + \frac{x^4}{4 \times 6!}.$$

$$\text{At } x = 0.5\text{, value is } \begin{array}{r} 0.5 \\ +\ 0.0625 \\ +\ 0.001736 \\ +\ 0.0000217 \\ \hline =\ 0.\mathbf{5642}577 \end{array}$$

1.7 $\tan x = x + \frac{x^3}{3} + O(x^5)$, $\frac{\tan x}{x} = 1 + \frac{x^2}{3} + O(x^4)$. Limit: 1.

1.8 Maclaurin series for $\cos x$. `Author` `taylor (cosx,x,0,6)` and `Simplify`

1.9 Cubic Taylor polynomial is $p_3(x) = \frac{5 + 15x - 5x^2 + x^3}{16}$.

$$|R_3(x)| = \frac{(x-1)^4}{4!} \max_{x<\xi<1} \left|\frac{15}{16\xi^{7/2}}\right| = \frac{5(x-1)^4}{128 x^{7/2}}.$$

x	$p_3(x)$	$\lvert\sqrt{x} - p_3(x)\rvert$	$\lvert R_3(x)\rvert$
0	0.3125	0.3125	∞
0.5	0.710937	3.83068×10^{-3}	2.76213×10^{-2}
0.75	0.866210	1.85516×10^{-4}	4.17643×10^{-4}
0.875	0.935424	1.04911×10^{-5}	1.52185×10^{-5}
0.9375	0.968246	7.24917×10^{-7}	7.47103×10^{-7}

The error bound improves (*less* pessimistic) as $x \to x_0$.

1.10 $R_{4,0} = x - x^2 + 2x^3 - 6x^4$,
$R_{1,3} = \frac{x}{1+x-x^2+3x^3}$, no solution.

1.11 Power series: $1 + x + \frac{x^2}{2}$,
$1 + x + \frac{x^2}{2} + \frac{x^3}{6} + \frac{x^4}{24}$.

$$R_{1,1} = \frac{1 + \frac{x}{2}}{1 - \frac{x}{2}},\ R_{2,2} = \frac{1 + \frac{x}{2} + \frac{x^2}{12}}{1 - \frac{x}{2} + \frac{x^2}{12}}.$$

Chapter 2

2.1 $10, 37, \frac{99}{128}$

2.2 $(.00011)_2$, .00625, .00625

2.3 $.145 \times 10^2 (m = 3, n = 2)$,
$-.14636 \times 10^2 (m = 5, n = 2)$,
$.93 \times 10^{-2} (m = 2, n = -2)$, $.16 \times 10^{-1}$ $(m = 2, n = -1)$

2.4 $(.01100)_2$, .025, .0625

2.5
(i) 1228.8675, .353, $.287 \times 10^{-3}$, [1228.5154, 1229.2205], 1229.
(ii) 465.43822, .050005, $.107 \times 10^{-3}$, [465.388215, 465.488225], 465.
(iii) 125.6045602, .050000245, $.398 \times 10^{-3}$, [125.554, 125.654], 126.
(iv) 87579.8722, .5025, $.574 \times 10^{-5}$, [87579.3697, 87580.3747], 87580.

2.6 -1.2059, .0279, $.231 \times 10^{-1}$.

2.7
(i) $\frac{x-y}{\sqrt{x}+\sqrt{y}}$
(ii) Use $\sin(a + b) - \sin(a - b) = 2\cos a \cos b$ with $x = a + b$ and $y = b - a$ to give

$$\sin x - \sin y = 2\cos\frac{x+y}{2}\cos\frac{x-y}{2}.$$

(iii) $\ln(\frac{x}{y})$
(iv) $2\cos(\alpha + \frac{x}{2})\sin\frac{\alpha}{2}$
(v) $\frac{\alpha}{x+\sqrt{x^2-\alpha}}$

2.8 $x = \{-.4001, -.00015\}$, $\Delta = \{.1 \times 10^{-3}, .5 \times 10^{-4}\}$, $\delta = \{.25 \times 10^{-3}, .25\}$.
Cancellation error $b \simeq \sqrt{b^2 - 4ac}$
$.8 \times 10^{-4} y^2 + .4002y + 1 = 0$,
$y \in \{-5001, -1.875\}$,
$x_{\text{small}} = 1/y_{\text{large}} = -.0002$

2.9 `iterates([(1-i)/n,n-1],[i,n],[0,15],15)`, 0.632120

2.10 $\Delta_{n+1} = 3\Delta_n + \Delta_b$,
$\Delta_n = 3^n\Delta_0 + \Delta_b \sum_{i=0}^{n-1} 3^i$.

$$\frac{3^n}{2} \times 10^{-2} + \frac{1}{2} \times 10^{-2} \sum_{i=0}^{n-1} 3^i$$

$$= \frac{1}{2} \times 10^{-1}, \text{ i.e. } \sum_{i=0}^{n} 3^i = 10,$$

i.e. $n \approx 1.77$ (use $n = 2$).

2.11 `vector(taylor(ln(1+x),x,0,n),n,0,10)`
At -0.5, $-0.693 \to -8.2 \times 10^{-5}$, improves with n
At 0, exact (by Maclaurin series definition)
At 0.5, $0.405 \to 3.05 \times 10^{-5}$, improves with n (alternating sign)
At 1, $0.093 \to 0.048$ (slow convergence)
At 1.5, slow decrease then increase with n – *not* convergent.

2.12 `vector(taylor(exp-x,x,0,n),n,0,20)`. Values increase before decreasing with n ($n = 20$ gives 0.00674272). Reason is $\pm$ large numbers
Use $e^{-x} = 1/e^x$. $n = 15$ gives 0.006739

2.13 Use $x^*_{n+1} = 2.5x^*_n - x^*_{n-1}$ and $x^*_{n+1} = 0.5x^*_n$.

n	x_n	x^*_n	
0	0.6667	0.6667	0.6667
1	0.3333	0.3333	0.3334
2	0.1667	0.1666	0.1667
3	0.08333	0.0832	0.08335
4	0.04166	0.0414	0.04168
5	0.02083	0.0203	0.02084
6	0.01042	0.00935	0.01042
7	0.00520	0.003075	0.00521
8	0.00260	−0.001662	0.002605
9	0.00130	−0.00723	0.001303
10	0.0006510	−0.01641	0.0006515

True seq. $x_n = \frac{2}{3}(\frac{1}{3})^n$.
1st seq. $x^*_n = (\frac{2}{3} + \varepsilon)(\frac{1}{2})^n + \varepsilon 2^n$,
$|x_n - x^*_n| \simeq \varepsilon 2^n$ – exponential.
2nd seq. $x^*_n = (\frac{2}{3} + \varepsilon)(\frac{1}{2})^n$,
$|x_n - x^*_n| \simeq \varepsilon(\frac{1}{2})^n \leq \varepsilon n$ – linear.

Chapter 3

3.1 $x^4 - 4x^3 + 4x^2 + 2$

3.2 (a) $e := e * x/n$, (b) $e := (\frac{n-2}{n})(\frac{x-1}{x+2})^2 e$

3.3 0.0925285, 8. 0.0914446319590, 0.0914544425580. Subtractive cancellation.

3.4 Extract coefficients of f_0 and f_1 from the first form.

3.5 (a) $p_1(x) = \frac{11+x}{6}$, (b) Apply (3.12) and use properties of logs.

3.6 0.622, true value 0.643, observed error 0.021, error bound (3.16) 1.511 – pessimistic.

3.7 1.1230, true value 1.1447, observed error 0.0217, error bound (3.16) 0.0278 – quite 'tight'.

3.8 $g(x) = (x - x_0)(x - x_1)$. Max. at $g'(x) = 0 \Rightarrow x = (x_0 + x_1)/2 = x_m$ – mid-point. $g(x_m)\frac{M_2}{2} = \frac{h^2 M_2}{8}$. $M_2 = 1$ so $\frac{h^2}{8} \leq \frac{1}{2} \times 10^{-4} \Rightarrow h \approx 0.028$ (use $h = 0.025$ or $h = 0.02$).

3.9 Expand determinant by first row.

3.10 $p_2(x) = x^2 + 1$.

3.11 Error $e_n(x) \propto f^{(n+1)}$.
$f \in P_n \Rightarrow f^{(n+1)}(x) \equiv 0 \Rightarrow e_n(x) \equiv 0 \Rightarrow f(x) - p_n(x) \equiv 0 \Rightarrow p_n(x) \equiv f(x)$.
$n + 1 = 10 \Rightarrow \deg(p_n) \leq 9$. $1 - x^2$ passes through data.

3.12 $f''(x) = \cos x$, $M_2 = 1$, $\frac{h^2}{8} \leq \frac{10^{-4}}{2} \Rightarrow h \leq 0.02$.

3.13 Divided-difference table

−2	5		
0	1	−2	
1	2	1	1

Polynomial

$$\begin{aligned} p_2(x) &= f_0 + (x - x_0)(f[x_0, x_1] \\ &\quad + (x - x_1) f[x_0, x_1, x_2]) \\ &= 5 + (x + 2)(-2 + x) \\ &= x^2 + 1. \end{aligned}$$

3.14 Divided-difference table

−2	−15			
0	1	7		
1	−3	−4	−4	
2	−7	−4	0	1
5	41	16	5	1

Constant third differences imply cubic polynomial.

$$\begin{aligned} p_2(x) &= -15 + (x + 2) \\ &\quad (8 + x(-4 + (x - 1))) \\ &= x^3 - 3x^2 - 2x + 1. \end{aligned}$$

3.15 Divided difference table

−2	**8**		
−1	4	**−4**	
0	2	−2	**2**
1	2	0	2
2	**4**	**2**	**2**
3	**8**	**4**	**2**

3.16 Substitute $x = x_0 + sh$, $x_i = x_0 + ih$ into eqn (3.51).
True for $k = 0$: $f[x_0] = f_0$. Assume true for integers $0, \dots, k$. For $k + 1$ use divided-difference recurrence relation

$$f[x_0, \dots, x_{k+1}] = \frac{f[x_1, \dots, x_{k+1}] - f[x_0, \dots, x_k]}{x_{k+1} - x_0}.$$

3.17 Set $s = (x - x_n)/h$ and use relation (3.45)

$$\begin{aligned} f(x) &= f(x_n + sh) \\ &= E^s f(x_n) \\ &= (1 - \nabla)^{-s} f(x_n) \\ &= \left[1 + s\nabla + \frac{s(s+1)}{2!}\nabla^2 \right. \\ &\quad \left. + \frac{s(s+1)(s+2)}{3!}\nabla^3 + \cdots \right] f(x_n) \\ &= \sum_{k=0}^{\infty} (-1)^k \binom{-s}{k} \nabla^k f_n. \end{aligned}$$

Truncate the result at $k = n$.

3.18 (a) Difference table

0.0	0.55		
0.2	0.82	0.27	
0.4	1.15	0.33	0.06
0.6	1.54	0.39	0.06
0.8	1.99	0.45	0.06
1.0	2.50	0.51	0.06

(b) $p_f(s) = 0.55 + 0.24s + 0.03s^2$,
$p_b(s) = 2.5 + 0.54s + 0.03s^2$.
(c) $s = 5x$, $s = 5x - 5$,
$p_f(x) = p_b(x) = 0.55 + 1.2x + 0.75x^2$.

3.19 Difference table

1	1			
2	5	4		
3	14	9	5	
4	30	16	7	2
5	55	25	9	2
6	91	36	11	2
7	140	49	13	2
8	204	64	15	2

Constant third differences imply cubic polynomial.
$p_3(s) = (6 + 13s + 9s^2 + 2s^3)/6$, $s = (n - n_0)/h = n - 1$, $p_3(n) = n(n+1)(2n+1)/6$ – well-known result!

3.20 $p_1(x) = f_0 + (f_1 - f_0)x/h$, $h(f_0 + f_1)/2$ – trapezium rule.
$p_2(x) = p_1(x) + (f_2 - 2f_1 + f_0)x(x - h)/2h^2$, $h(f_0 + 4f_1 + f_2)/3$ – Simpson's rule.

3.21 `Author` and `Plot s(7)` – data. `Author vector(p(n,x,z),n,3,7)` – computes p_n to $n + 1$ data points. Should *all* be cubic since data lies on cubic curve. `Simplify` the expression with `Digits` set to 10, 20, 40, 45, 46 and 47. It is only with 47 digits that all 5 polynomials equal the required cubic.

Chapter 4

4.1 Use data

x	0	$\frac{1}{4}$	$\frac{1}{2}$	$\frac{3}{4}$	1
x^2	0	$\frac{1}{16}$	$\frac{1}{4}$	$\frac{9}{16}$	1

and eqn (4.1).

$$p_{1,1}(x) = \frac{x}{4},\ p_{1,2}(x) = \frac{3x}{4} - \frac{1}{8},$$

$$p_{1,3}(x) = \frac{5x}{4} - \frac{3}{8},\ p_{1,4}(x) = \frac{7x}{4} - \frac{3}{4}.$$

4.2 Natural spline – solve $\begin{bmatrix} 1 & 0 \\ 0 & \pi \end{bmatrix}\begin{bmatrix} b_1 \\ b_2 \end{bmatrix} = \begin{bmatrix} g_1 \\ g_2 \end{bmatrix}$,

$g_1 = 0$, $g_2 = 3\left(\frac{\nabla f_2}{h_2} - \frac{\nabla f_1}{h_1}\right) = -\frac{12(\sqrt{2}-1)}{\pi}$.

Coefficients:

i	a_i	b_i	c_i
1	−0.241	0	−0.214
2	−0.637	−0.504	0.214

4.3 Solve system $\begin{bmatrix}1 & 0 & 0\\ 0 & 12 & 3\\ 0 & 3 & 12\end{bmatrix}\begin{bmatrix}b_1\\ b_2\\ b_3\end{bmatrix} = \begin{bmatrix}0\\ 14\\ 50\end{bmatrix}$.

Spline coefficients are

Interval	a_i	b_i	c_i
$1: 0 \le x \le \frac{1}{3}$	$\frac{1}{45}$	0	$\frac{2}{15}$
$2: \frac{1}{3} \le x \le \frac{2}{3}$	$\frac{1}{15}$	$\frac{2}{15}$	4
$3: \frac{2}{3} \le x \le 1$	$\frac{67}{45}$	$\frac{62}{15}$	$-\frac{62}{15}$

Interp. poly.: $p_n(3) = 2x^3 - \frac{11x^2}{9} + \frac{2x}{9}$. Figure A.1 shows $x^4 - S_3(x)$ and $x^4 - p_3(x)$. Interp. poly. is better, except close to $x = 0$ (the spline enforces $S''(1) = 0$; actual value is 12).

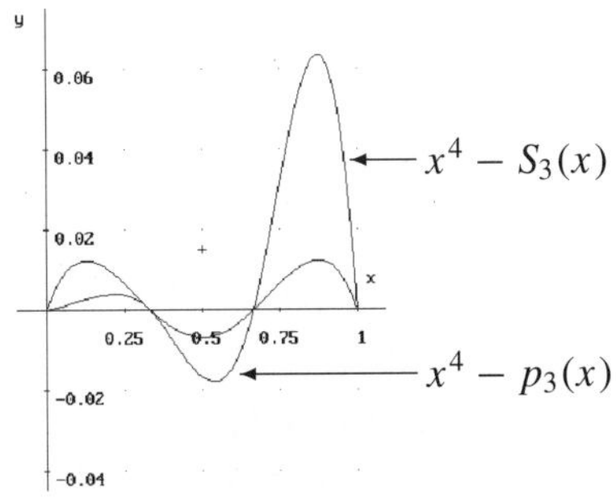

Figure A.1 Errors: Spline vs. interpolating polynomial.

DERIVE spline: Author

```
y:=vector([z,z^4],z,0,
1,1/3)
u:=nco(b(y),y)
s3:=ns3(u,z,x(y))
```

Interpolating polynomial: Author

```
fit([x,r+mz+nz^2+tz^3],y)
```

4.4 Author

```
y:=[[0,0],[.1,.5],[.2,3.35],
   [.3,3.3],[.4,1.65],[.5,1.6],
   [.6,1.6],[.7,1.6],[.8,1.6],
   [.9,.6],[1,0]]
u:=nco(b(y),y)
s3:=ns3(u,z,x(y))
```

approX and Plot this last expression. Plot the data set y. Use fit to obtain the interpolating polynomial.

4.5 Interpolation (see Figure A.2(a)):

Degree	Polynomial
1	$p_1(x) = \frac{1}{26}$
2	$p_2(x) = 1 - \frac{25}{26}x^2$
3	$p_3(x) = \frac{259}{884} - \frac{225}{884}x^2$
4	$p_4(x) = 1 - \frac{3225}{754}x^2 + \frac{1250}{377}x^4$

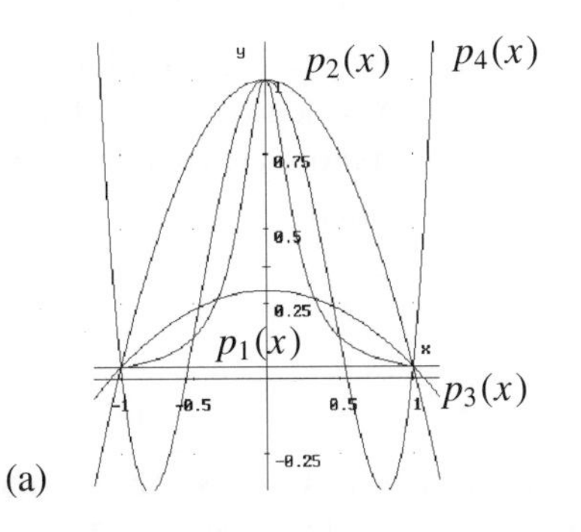

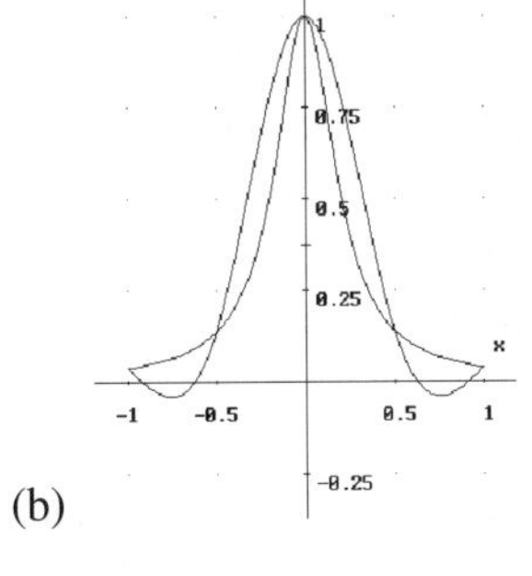

Figure A.2 (a) Interpolating $1/(1 + 25x^2)$. (b) Spline approximation.

Spline system:

$$\begin{bmatrix} 1 & 0 & 0 & 0 \\ 0 & 4 & 1 & 0 \\ 0 & 1 & 4 & 1 \\ 0 & 0 & 1 & 4 \end{bmatrix} \begin{bmatrix} b_1 \\ b_2 \\ b_3 \\ b_4 \end{bmatrix} = \begin{bmatrix} 0 \\ -\frac{7350}{377} \\ -\frac{600}{29} \\ -\frac{7350}{377} \end{bmatrix}.$$

Spline coefficients:

Interval	a_i	b_i	c_i
1: $-1 \le x \le -\frac{1}{2}$	−0.483	0	2.728
2: $-\frac{1}{2} \le x \le 0$	1.563	4.092	−7.541
3: $0 \le x \le \frac{1}{2}$	0	−7.219	7.541
4: $\frac{1}{2} \le x \le 1$	−1.563	4.092	−2.728

Reasonable agreement (see Figure A.2(b)).

4.6 Coefficients:

i	a_i	b_i	c_i
1	−0.5	0	1.5
2	4	4.5	−1.5

$s_{3,1} = 2-0.5x+1.5x^3$, $s_{3,2} = 3+4(x-1) +4.5(x-1)^2 - 1.5(x-1)^3$.

Not equal to $f(x) = 2 + x^3$ since $f''(2) \neq 0$, i.e. right-hand condition $s_3''(x_n) = 0$ is not correct. Requires $s_3''(2) = f''(2)$.

4.7 $y = \frac{1529}{610} - \frac{119x}{122} = 2.507 - 0.975x$, $\sum e_i^2 = 0.668$.

4.8 (a) Using eqn (4.16),

$$y = a + bx = (\bar{f} - \bar{x}b) + bx = \bar{f} + b(x - \bar{x}). \; x = \bar{x} \Rightarrow y = \bar{f}.$$

(b) $y = a$. Sum of squares of errors is $E(a) = \sum_{i=0}^n (f_i - a)^2$.

Differentiate and set to zero

$$\frac{dE}{da} = -2\sum_{i=0}^{n}(f_i - a) = 0 \Rightarrow a\sum_{i=0}^{n} = \sum_{i=0}^{n} f_i \Rightarrow a = \frac{1}{n+1}\sum_{i=0}^{n} f_i = \bar{f}.$$

4.9 $x = a + by$. Normal equations are

$$(n+1)a + \sum_{i=0}^{n} f_i b = \sum_{i=0}^{n} x_i$$
$$\sum_{i=0}^{n} f_i a + \sum_{i=0}^{n} f_i^2 b = \sum_{i=0}^{n} f_i x_i$$

with solution

$$b = \frac{\bar{f}\,\bar{x} - \bar{fx}}{(\bar{f})^2 - \bar{f^2}}, \quad a = \bar{x} - \bar{f}\,b.$$

For Exercise 4.7, $x = 1.431 - 0.477y \Rightarrow y = 3.000 - 2.097x$.

4.10 $M = \ln m = \ln a + bt = A + bt$. Add row $\ln m$ to data (see Table A.1 below). Find $M = A + bt = 1.489 - 0.00176t$. $m = e^A e^{bt} = 4.433e^{-0.00176t}$. Half-life: $e^{-0.00176t} = \frac{1}{2} \Rightarrow t \approx 394$ days.

4.11 Define error $E(a, b, c) = \sum_{i=0}^n (f_i - a - bx_i - cx_i^2)^2$ and solve

$$\frac{\partial E}{\partial a} = \frac{\partial E}{\partial b} = \frac{\partial E}{\partial c} = 0.$$

$y = 2.936 - 1.571x - 0.714x^2$.

4.12 $\log y = \log a + k \log x$. Let $Y = \log y$, $X = \log x$, so $Y = A + kX$. Add appropriate columns to data (using $\log_{10}$)

Table A.1 See answer 4.10

t	0	5	23	50	94	145	180	200
m	5	4.88	4.46	3.90	3.14	2.44	2.05	1.88
$M = \ln m$	1.61	1.59	1.50	1.36	1.14	0.89	0.72	0.63

y	x		
Pressure	Flowrate	X	Y
70	7100	3.851	1.845
112	8900	3.949	2.049
175	11000	4.041	2.243
280	13500	4.130	2.447
420	17300	4.238	2.623

$Y = -6.205 + 2.090X$,
$y = 6.24 \times 10^{-7} x^{2.09}$.

Chapter 5

5.1 (a) Differences 1, 2, 4, 8, **16**, next term 31.
(b) Differences 1, 2, 3, 4, **5**, next term 16.
(c) As (b), next term 15.
(d) Reciprocals, next term $\frac{1}{5}$.
(e) Inverse powers, next term $\frac{1}{25}$.
(f) Location of first letter of months in alphabet, next term 1.
(g) Sum of previous two terms, next term 13.

5.2 Fibonacci sequence: $x_0 = 1, x_1 = 1$.

k	0	1	2	3	4	5	6
x_k	1	1	2	3	5	8	13
$\frac{x_{k+1}}{x_k}$	1	2	1.5	1.667	1.6	1.625	1.615
	7	8	9	10	11	12	13
	21	34	55	89	144	233	377
	1.619	1.618	1.618	1.618	1.618	1.618	

Limiting value ≈ 1.618 ($= (1+\sqrt{2})/2$ – Golden Ratio).
Lucas sequence: $x_0 = 1, x_1 = 3$.

k	0	1	2	3	4	5	6	7
x_k	1	3	4	7	11	18	29	47
$\frac{x_{k+1}}{x_k}$	3	1.333	1.75	1.571	1.636	1.611	1.621	1.617

Same limiting value!
'Tribonacci' sequence: $x_0 = 1, x_1 = 1, x_2 = 1$.

k	0	1	2	3	4	5	6	7	8
x_k	1	1	1	3	5	9	17	31	57
$\frac{x_{k+1}}{x_k}$	1	1	3	1.667	1.8	1.889	1.823	1.839	

Different limiting value.

5.3 Substitute soln. into eqn. RHS:

$$
\begin{aligned}
&ax_n + b \\
&= a\left[x_0 a^n + b\left(\frac{1-a^{n+1}}{1-a}\right)\right] + b \\
&= x_0 a^{n+1} + b\left[a\left(\frac{1-a^n}{1-a}\right) + 1\right] \\
&= x_0 a^{n+1} + b\left[\frac{a(1-a^n) + 1 - a}{1-a}\right] \\
&= x_0 a^{n+1} + b\left(\frac{1-a^{n+1}}{1-a}\right) = x_{n+1}.
\end{aligned}
$$

5.4 Using $M_{n+1} = 7 + \left\langle \frac{M_n}{3} \right\rangle$, for $1 \le M_0 \le 11$,

M_0	1	2	3	4	5	6	7	8	9	10	11
M_1	8	8	8	9	9	9	10	10	10	11	11
M_2	10	10	10	10	10	10	11	11	11		
M_3	11	11	11	11	11	11					

Max. moves required is 3. For $11 \le M_0 \le 21$,

$$M_n \ge 11 \Rightarrow M_{n+1} = 7 + \left\langle \frac{M_n}{3} \right\rangle \ge 7 + \left\langle \frac{11}{3} \right\rangle = 7 + 4 = 11.$$

M_0	11	12	13	14	15	16	17	18	19	20	21
M_1	11	11	12	12	12	13	13	13	14	14	14
M_2			11	11	11	12	12	12	12	12	12
M_3						11	11	11	11	11	11

5.5 Möbius sequence: $ad - bc = 0 \Rightarrow ad = bc \Rightarrow a/b = c/d$, and

$$x_{n+1} = \frac{ax_n + b}{cx_n + d} = \frac{b}{d}\left[\frac{\frac{a}{b}x_n + 1}{\frac{c}{d}x_n + 1}\right] = \frac{b}{d}\left[\frac{\frac{a}{b}x_n + 1}{\frac{a}{b}x_n + 1}\right] = \frac{b}{d}$$

– a constant.

5.6 Use the function `mobi` (see *DERIVE Experiment 5.4*). Repeat simulations of Figure 5.3 with $x_0 = 0$. Try $a = b = c = d = 1$, then take each parameter equal to 0 in turn. Make up your own experiments!

5.7 3-cycle if $\lambda \simeq 3.83$

n	x_n	n	x_n
0	0.5000	7	0.9574
1	0.9575	8	0.1561
2	0.1559	9	0.5046
3	0.5039	10	0.9574
4	0.9574	11	**0.1561**
5	0.1561	12	**0.5046**
6	0.5044	13	**0.9574**

After 13 iterations, sequence reaches 3-cycle: 0.1561, 0.5046, 0.9574. Author `iterates(3.83x(1-x),x,.5,13)` and `approX`

5.8 $x_{n+1} = (1 + r/100)x_n$. If $r = 6\%$ then $x_{n+1} = 1.06x_n$.
$x_0 = 1500$, $x_1 = 1.06x_0 = 1590$,
$x_2 = 1.06x_1 = (1.06)^2 x_0 = 1685.4$,
$x_3 = 1.06x_2 = (1.06)^3 x_0 = 1786.52$.
Middle-year cheque accrues interest for 2 years, $y_2 = (1.06)^2 \times 1500 = 1685.40$.
Final-year cheque for 1 year, $z_1 = 1.06 \times 1500 = 1590$. Total 5061.92.
Monthly rate is 6%/12 (0.5%).
$x_{n+1} = 1.005x_n \Rightarrow x_n = (1.005)^n x_0$ (n measures months).
First-year cheque $n = 36$, $x_{36} = (1.005)^{36} \times 1500 = 1795.02$, Middle-year cheque $n = 24$, $y_{24} = (1.005)^{24} \times 1500 = 1690.74$.
Final-year cheque $n = 12$, $z_{12} = (1.005)^{12} \times 1500 = 1592.52$.
Total 5078.28.

5.9 $i = r/100m$. Apply relation 3 times; $w_1 = (1+i)w_0 = (1+i)C$, $w_2 = (1+i)w_1 = (1+i)^2C$, $w_3 = (1+i)w_2 = (1+i)^3C$.
Pattern is $w_n = (1 + i)^n C$. For $r = 10\%$, $C = 20\,000$

Compound	$m = n$	w_n
Yearly	1	22 000
6-monthly	2	22 050
Quarterly	4	22 076.26
Monthly	12	22 094.26
Weekly	52	22 101.30
Daily	365	22 103.12

Wealth increases as 'compound' period reduces. Limiting value

$$w = C \lim_{m\to\infty} \left(1 + \frac{r}{100m}\right)^m = Ce^{rm/100}.$$

Author `w(c,r,m):=c(1+r/(100m))^m` and `approX vector(w(20000,10,m),m,[1,2,4,12,52,365])`. Simplify `lim(w(c,r,m),m,inf)`. Plot the `vector` and the limiting expression!

5.10 $a = (V - 7)/(V - 2)$, $b = 7q$, and $c_{n+1} = ac_n + b$ where $c_0 = Vq$.

$$\begin{aligned} c_1 &= ac_0 + b \\ c_2 &= ac_1 + b = a^2c_0 + (a + 1)b \\ c_3 &= ac_2 + b = a^3c_0 + (a^2 + a + 1)b \\ c_n &= a^nc_0 + (a^{n-1} + a^{n-2} + \cdots + a^2 \\ &\quad + a + 1)b = a^nc_0 + b\sum_{i=0}^{n-1} a^i, \end{aligned}$$

Geometric series $s_m = \alpha + \alpha r + \alpha r^2 + \cdots + \alpha r^{m-1} = \frac{\alpha(1-r^m)}{1-r}$ with $\alpha = 1$, $r = a$ and $m = n$. Hence

$$\begin{aligned} c_n &= a^nc_0 + b\left(\frac{1-a^n}{1-a}\right) \\ &= \left(c_0 - \frac{b}{1-a}\right)a^n + \frac{b}{1-a}. \end{aligned}$$

$1 - a = 5/(V - 2)$,
$b/(1 - a) = \frac{7}{5}q(V - 2)$,

$$c_n = \frac{2q}{5}(7 - V)\left[\frac{V - 7}{V - 2}\right]^n + \frac{7q(V - 2)}{5}.$$

Limiting concentration: $\frac{7q(V-2)}{5V}$. Max. as $V \to \infty$, i.e. $1.4q$. Weeks 1–3, $c_{n+1} = c_n + 2q$. Week 4,

$$c_4 = \left(\frac{V - 22}{V - 2}\right) c_3 + 22q = \left(\frac{V - 22}{V - 2}\right) [c_0 + 6q] + 22q.$$

4-weekly cycle:
$M_{n+1} = \left(\frac{V-22}{V-2}\right) [M_n + 6q] + 22q$.

$$M_n = M_0\alpha^n + \beta\left(\frac{1 - \alpha^n}{1 - \alpha}\right),$$

$\alpha = (V - 22)/(V - 2)$, $\beta = 4q(7V - 44)/(V - 2)$. Limiting concentration $(\beta/[(1 - \alpha)V])$ is $\frac{q(7V-44)}{5V}$. Max. as $V \to \infty$ is $1.4q$.

5.11 Iterates (see Figure A.3):

n	0	1	2	3	4	5	6	···	23	24
I_n	2	3	3.4	3.61	3.739	3.824	3.880		4.000	4.000

Monotonic growth to a steady state! If only this were possible!

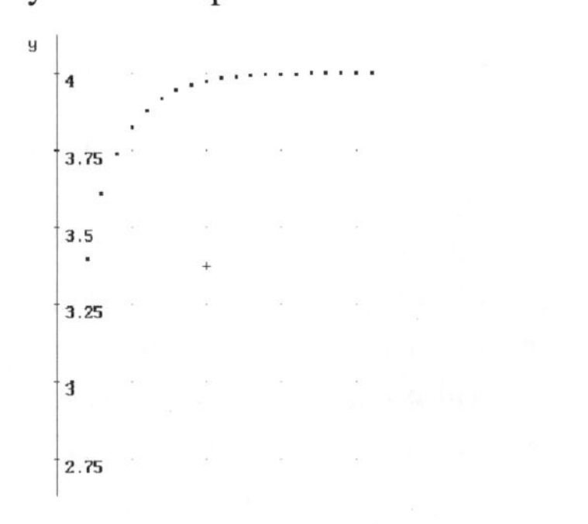

Figure A.3 The ideal economy?

5.12 Price declines if supply increases. Supply in any year is proportional to price in previous year. Price equation:

$$\begin{aligned} p_n &= a - bs_n = a - b(cp_{n-1}) \\ &\Rightarrow p_{n+1} = a - bcp_n \\ p_1 &= a - bcp_0 \\ p_2 &= a - bcp_1 = a(1 - bc) + (bc)^2 p_0 \\ p_3 &= a - bcp_2 = a(1 - bc \\ &\quad + (bc)^2) - (bc)^3 p_0 \end{aligned}$$

General $p_n = p_0(-bc)^n + \frac{a}{1+bc}[1-(-bc)^n]$, 'settles' if $|bc| < 1$.

5.13 In one year 38% of Erie's water replaced (pollution reduced to 62% of previous value); $E_{n+1} = 0.62E_n$.

Ontario receives 38% of Erie's pollution from previous year ($O_n \to O_n + 0.38E_n$). 13% of its water is replaced by fresh water (reducing pollution by 13% of previous value); $O_{n+1} = O_n + 0.38E_n - 0.13O_n = 0.38E_n + 0.87O_n$.

$$\begin{aligned} e_1 &= 0.62e_0 \\ e_2 &= 0.62e_1 = (0.62)^2 e_0 \\ e_3 &= 0.62e_2 = (0.62)^3 e_0 \end{aligned}$$

General: $e_n = (0.62)^n e_0$. Pollution at 10% when $e_n = 0.1$, $(0.62)^n = 0.1 \Rightarrow n = \log 0.1/\log 0.62 = 4.82$ ($\approx$ 5 years).

$E_0 = 3O_0$. Set $E_0 = 3$, $O_0 = 1$ (implement the coupled recurrence relations)

n	O_n	n	O_n	n	O_n
0	1.0000				
1	2.0100	11	1.1779	21	0.2983
2	2.4555	12	1.0307	22	0.2596
3	2.5745	13	0.9004	23	0.2259
4	2.5115	14	0.7857	24	0.1965
5	2.3535	15	0.6849	25	0.1710
6	2.1520	16	0.5968	26	0.1488
7	1.9370	17	0.5197	27	0.1294
8	1.7253	18	0.4525	28	0.1126
9	1.5259	19	0.3939	29	**0.0980**
10	1.3430	20	0.3428		

Pollution increases, then tails off to 10% after approx. 29 years.

5.14 $B_{n+1} = 0.2A_n + (0.9 - \alpha)B_n$ (put in matrix iteration). With $\alpha = 0.9$ (all remaining B felled), number still rises because of high A numbers. Using initial values of Example 5.7, $B_4(\alpha = 0) = 3784$, $B_4(\alpha = 0.9) = 2883$.

$A_{n+1} = 5.42A_n + 15B_n + 2C_n$ (new A equals old A plus seeding rate times 0.85 old A, plus B and C seedings). $B_{n+1} = 0.17A_n + (0.9 - \alpha)B_n$ (A trees coming through now equals 0.2×0.85). With $\alpha = 0.9$, numbers still rise, $B_4 = 1973$.

Chapter 6

6.1 $f(x) = 1 - 2x + \cos x$ is *continuous* on $[0, 1]$. $f(0) = 2$, $f(1) = \cos(1) - 1$ and $f(x)$ changes sign on $[0, 1]$ – a *root* exists.

Require error $(b - a)/2^{n+1} \le \frac{1}{2} \times 10^{-3}$. $a = 0, b = 1$,

$$\frac{1}{2^{n+1}} \le \frac{1}{2} \times 10^{-3} \Rightarrow 2^n \ge 10^3 \Rightarrow n \ge \frac{3 \log 10}{\log 2} = 9.97$$

At least 10 iterations (root is 0.510973).

6.2 $f(x) = x^2 + x - 1$ is continuous, $f(0)f(1) = -1$, $f(x)$ changes sign on $[0, 1]$ – a root exists. One d.p.,

$$\frac{1}{2^{n+1}} \le \frac{1}{2} \times 10^{-1} \Rightarrow 2^n \ge 10 \Rightarrow n \ge \frac{\log 10}{\log 2} = 3.32$$

At least 4 iterations: $\alpha = 0.6$ (to 1 d.p.).

n	x_0	x_1	α^*	$f(\alpha^*)$
0	1	2	1.5	−0.25
1	1.5	2	1.75	0.3125
2	1.5	1.75	1.625	0.015625
3	1.5	1.625	1.5625	−0.121094
4	1.5625	1.625	1.59375	−0.053711

6.3 With $[a, b] = [0, 2]$ (For $x = 0$ the goat cannot graze any of the field. For $x = 2$ the goat can graze the whole field, and more! Hence, the required value must lie somewhere between) Table A.2 shows the first few iterations.

Table A.2 Bisection results for the 'tethered goat' problem.

n	x_0	c	x_1	Max. error	$\lvert f(c)\rvert$
0	0	1	2	1	0.839
1	0	0.500	1.000	0.5	0.461
2	0.500	0.750	1.000	0.25	0.057
3	0.750	0.875	1.000	0.125	0.285
4	0.750	0.813	0.875	0.0625	0.096
5	0.750	0.781	0.813	0.03125	0.016
6	0.750	0.766	0.781	0.015625	0.022
7	0.766	0.773	0.781	0.007813	0.003
8	0.773	0.777	0.781	0.003906	0.006
9	0.773	0.775	0.777	0.001953	0.002
10	0.773	0.774	0.775	0.000977	0.001

6.4 1 processor requires $n \ge k\frac{\log 10}{\log 2}$ iters. for k d.p. p processors require $m \ge k\frac{\log 10}{\log(p+1)}$ iters. Speed-up

$$S_p = \frac{T_1}{T_p} \simeq \frac{n}{m} = \frac{\log(p+1)}{\log 2} = \frac{\log_2(p+1)}{\log_2 2} = \log_2(p+1).$$

p	1	2	3	4	5	6	7	8	9	10
S_p	1	1.58	2	2.32	2.58	2.81	3	3.17	3.32	3.46

S_p is non-linear function of p (to double S_p requires virtually squaring p). Values not attained in practice as communication costs have been ignored (at each step the processor inspecting the bracketing interval must broadcast (send) the new interval to the other $p - 1$ processors).

6.5 Write equation as $f(x) = 0$: $C - \frac{P}{i}[1 - (1 + i)^{-n}] = 0, i = i_{\text{month}}$.

Initial interval [0.001, 0.01] gives $f(0.001) = -66586.15$, $f(0.01) = 7274.055$. Bisection with 3 d.p. and $n = 300$

	−	+		
n	i_0	i_1	α^*	$f(\alpha^*)$
0	0.001	0.01	0.0055	−16033.61
1	0.0055	0.01	0.00775	−2335.715
2	0.00755	0.01	0.008875	2875.363
3	0.00755	0.008875	0.0083125	

$i\%_{\text{annual}} = 0.0083125 \times 100 \times 12 = 9.97\%$.

6.6 $S - \frac{P}{i}[(1+i)^n - 1] = 0$, $i = i_{\text{quarter}}$. Initial interval [0.051, 0.06] gives $f(0.051) = 59.13794$, $f(0.06) = -18.29273$. Bisection with 3 d.p. and $n = 12$

	+	−		
n	i_0	i_1	α^*	$f(\alpha^*)$
0	0.051	0.06	0.0555	21.00195
1	0.0555	0.06	0.05775	1.502808
2	0.05775	0.06	0.058875	−8.358033
3	0.05775	0.058875	0.0583125	−3.42041

$i\%_{\text{annual}} = 0.0583125 \times 100 \times 4 = 23.33\%$.

6.7 Use eqn (3.12) with $x_0 = a$, $x_1 = b$, $f_0 = f(a)$, $f_1 = f(b)$ to give

$$y = f(a) + \left(\frac{f(b) - f(a)}{b - a}\right)(x - a).$$

Equate RHS to zero: $c = a - f(a)\left(\frac{b-a}{f(b)-f(a)}\right)$.

	+	−				
n	x_0	x_1	f_0	f_1	c	$f(c)$
0	0	1	2	−0.4597	0.8131	0.06105
1	0.8131	1	0.06105	−0.4597	0.8350	0.001178
2	0.8350	1	0.001178	−0.45971	0.8354	0.81×10^{-5}

Just 2 iters. required.

6.8 (a) $\frac{(b-a)}{2^{n+1}} \le \frac{10^{-k}}{2}$, $a = 0, b = 1, k = 4$ gives $n \ge 13.29$. 14 iters.
(b) $x_{n+1} = \frac{1}{5}(2^{x_n} + 1)$.

n	0	1	2	3
x_n	0.1	0.4144	0.4665	0.4764
n	4	5	6	7
x_n	0.4782	0.4786	0.4787	0.4787

7 iterations.

6.9 $g(x) = \frac{2^x+1}{5}$, $\lambda = -g'(x)$, $\lambda = -g'(0.1) = \frac{-2^x \ln 2}{5}\Big|_{x=0.1} = -0.1486$. Iterative scheme $x_{n+1} = G(x_n)$ gives

n	0	1	2	3	4
x_n	0.1	0.4692	0.4782	0.4787	0.4787

4 iterations.

6.10 For the sequence of exercise 6.8(b), apply the formula

$$x^*_{n+1} = x_{n+1} - \frac{(x_{n+1} - x_n)^2}{x_{n+1} - 2x_n + x_{n-1}}$$

n	0	1	2	3	4
x_n	0.1	0.4144	0.4665	0.4764	0.4782
x^*_n	–	–	0.4768	0.4787	0.4787

5 iterations.

6.11 Use the functions `aitken` and `fixpt` from *DERIVE Experiment 6.5*. `Author` and `approX`

```
aitken(fixpt(<Alt>p/2-
.9sinx,x,1,10))
```

n	0	1	2	3	4	5	6
x_n	1.0000	2.3281	2.2248	2.2851	2.2508	2.2706	2.2593
x^*_n	–	–	2.2323	2.2629	2.2632	2.2634	2.2534

Convergence in 6 iters. (to 4 d.p.).

6.12 $V = iR \Rightarrow 7 = i(50) + 2i^{\frac{2}{3}} \Rightarrow i = (7 - 2i^{\frac{2}{3}})/50$.

n	0	1	2	3	4
i_n	0	0.14	0.138	0.139	0.139

6.13 Remove subscripts, multiply by $c - ax^2$ to give $x(c - ax^2) = bx^2 + 2cx$. Divide by x (admitting $x \neq 0$) $ax^2 + bx + c = 0$. *Exactly* second order: show $g'(\alpha) = 0$, $g''(\alpha) \neq 0$ ($e_{n+1} \propto e_n^2$).

$$g(x) = \frac{bx^2 + 2cx}{c - ax^2},$$

$$g'(x) = 2c\frac{ax^2 + bx + c}{(c - ax^2)^2}.$$

At root α, $ax^2 + bx + c$ is identically zero and $g'(\alpha) = 0$ if $c - a\alpha^2 \neq 0$. Further differentiation (use $\mathcal{DERIVE}$!) gives

$$g''(x) = 2c\frac{2ax(ax^2 + bx + c) + ax(bx + 4c) + bc}{(c - ax^2)^3}$$

– non-zero at a root of the original quadratic.
$c = 0$: eqn is $ax^2 + bx = 0$ with roots $x = 0$, $x = -\frac{b}{a}$. Iterative scheme gives root *immediately*.

$$x_{n+1} = \frac{bx_n^2}{-ax_n^2} = -\frac{b}{a}.$$

6.14 $g(x) = cx^{1-p}$, $g'(x) = (1 - p)cx^{-p}$. At a root α, $x = c^{\frac{1}{p}}$,

$$g'(\alpha) = (1 - p)c\left(c^{\frac{1}{p}}\right)^{-p} = 1 - p.$$

$|g'(\alpha)| \geq 1$ if $p \geq 2$ (method not convergent).

6.15 See Figure A.4. Slope of a^x must be less than or equal to 1 for $0 \leq x \leq z$ where $x = z$ is intersection (root) of x and a^x. Set slope equal to 1 at $x = z$, $a^z \ln a = 1 \Rightarrow z = -\ln(\ln a)/\ln a$. For x and a^x to meet, $a^z \leq z$. Solving $a^z = z \Rightarrow \mathrm{e} = 1/\ln a \Rightarrow a = \mathrm{e}^{1/\mathrm{e}}$. For this case $x = a^x \Rightarrow x = \mathrm{e}\ln x \Rightarrow x = \mathrm{e}$. Exponential sequence is $x_{n+1} = \mathrm{e}^{x_n/\mathrm{e}}$.

n	0	1	2	3	4	5	6	7	8	9
x_n	0	1	1.445	1.702	1.870	1.990	2.079	2.149	2.205	2.251

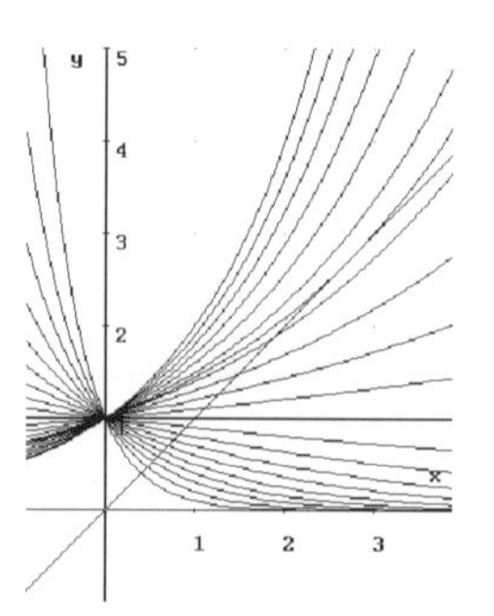

Figure A.4 The equation $x = a^x$, $a = 0.1(0.1)2$.

Slow! As $x_n \to \mathrm{e}$ so slope of g ($\mathrm{e}^{x/\mathrm{e}}$) approaches 1. Newton–Raphson gives 2.71759 afer 10 iters.

6.16 (a) $f(x) = 2^x - 5x + 1$, $f'(x) = 2^x \ln 2 - 5$, $x_{n+1} = \frac{(x_n \ln 2 - 1)2^{x_n} - 1}{2^{x_n}\ln 2 - 5}$.

n	0	1	2	3
x_n	0.1	0.4692	0.4787	0.4787

3 iters., 4 d.p.

(b) $f(x) = x^p - c$, $f'(x) = px^{p-1}$, $x_{n+1} = \frac{(p-1)x_n^p + c}{px_n^{p-1}}$.

(c) $x_{n+1} = (x_n^2 + c)/2x_n$. If $x_n < 0$, $x_n^2 + c > 0$ and $2x_n < 0$, so $(x_n^2 + c)/2x_n < 0$. Subsequent term is negative – method converges to *negative* square root.

(d) $f(x) = 1/x - c$, $f'(x) = -1/x^2$, $x_{n+1} = x_n(2 - cx_n)$. Limiting value gives *reciprocal* of c ($x_\infty = 1/c$). Reciprocation implies division – there is no division in this iterative process!

(e) $f(x) = \sin x$, $f'(x) = \cos x$, $x_{n+1} = x_n - \tan x_n$,

$$x_{n+2} = x_{n+1} - \tan x_{n+1} = x_n - \tan x_n - \tan(x_n - \tan x_n).$$

$x_{n+2} = x_n = x \Rightarrow \tan x = -\tan(x - \tan x)$. tan is odd function, $\tan x = \tan(\tan x - x) \Rightarrow x = \tan x - x \Rightarrow \tan x = 2x$. The root of this equation in the vicinity of π is $x = 1.165561185\ldots$ – a non-terminating decimal. This cannot be stored exactly on a PC and the critical point can never be reached. This is an instance where computer rounding error has a beneficial effect!

(f) Roots are $x = 0$, $x = 1 - 1/\beta$. $f(x) = x - \beta x(1-x)$, $f'(x) = 1 - \beta(1-2x)$, $x_{n+1} = \frac{\beta x_n^2}{1-\beta(1-2x_n)}$. See Table A.3 below. Converges for all β to $1 - 1/\beta$ (to zero when $\beta = 0.5$). Note the slow convergence when $\beta = 1$. Any ideas why?

(g) $f(x) \simeq p_1(x) = f(x_0) + (x - x_0)f'(x_0)$ and

$$p_1(x) = 0 \Rightarrow f(x_0) + (x - x_0)$$

$$f'(x_0) = 0 = 0 \Rightarrow x = x_0 - \frac{f(x_0)}{f'(x_0)}.$$

The Newton–Raphson formula – Taylor series are everywhere!

6.17 Use the `secant` function defined in *DERIVE Experiment 6.8*. With $R = 5$ and $A = 1.2$, the equation is $\theta - \sin\theta - 0.096 = 0$. Author and approX

```
secant(x-sinx-.096,x,0,
<Alt>p/2,10)
```

assuming $\theta_0 = 0$ and $\theta_1 = \frac{\pi}{2}$.

n	0	1	2	3	4	5
θ_n	0	1.5708	0.2642	0.4781	1.5946	0.6290
	6	7	8	9	10	
	0.7254	0.8794	0.8367	0.8417	0.8419	

$\theta \approx 0.842$ rads, or $48.2°$.

6.18 Roots:

$f(x)$	$x - f(x)/f'(x)$	x_{10}	Root
$\sin x$	$x - \tan x$	6.4×10^{-63}	$x = 0$
$\frac{\sin x}{x}$	$\frac{x(x\cos x - 2\sin x)}{x\cos x - \sin x}$	6.28318	$x = 2\pi$
$\frac{\sin x}{x(x-2\pi)}$		−298.451	$x = -95\pi$

6.19 Quadratic: $r_2(x) = \frac{q_3(x)}{x-2} = x^2 - 7x + 12$.

$x_0 = 0.5$ k	2	1	0
a_k	1	−7	12
b_k	1	−6.5	$8.75 = r_2(0.5)$
d_k	1	−6	$= r_2'(0.5)$

$$x_1 = x_0 - r_2(x_0)/r_2'(x_0) = 1.958$$

Table A.3 See Answer 6.16(f)

β	**0.5**	**1.0**	**1.5**	**2.0**	**2.5**	**3.0**	**3.5**	**4.0**
x_0	0.5	0.5	0.5	0.5	0.5	0.5	0.5	0.5
x_1	0.25	0.25	0.375	0.5	0.625	0.75	0.875	1
x_2	0.013	0.125	0.338		0.601	0.675	0.739	0.8
x_3	0.0002	0.063	0.333		0.6	0.667	0.715	0.753
x_4	$< 10^{-7}$	0.031	0.333			0.667	0.714	0.75
x_5	$< 10^{-15}$	0.016						
x_6	$< 10^{-30}$	0.008						
x_7		0.004						
x_8		0.002						
x_9		0.001						
x_{10}	0	0.000						

$x_1 = 1.958$ k	2	1	0
b_k	1	−5.042	$2.128 = r_2(1.958)$
d_k	1	−3.084	$= r_2'(1.958)$

$$x_2 = x_1 - r_2(x_1)/r_2'(x_1) = 2.648$$

$x_2 = 2.648$ k	2	1	0
b_k	1	−4.352	$0.476 = r_2(2.648)$
d_k	1	−1.704	$= r_2'(2.648)$

$$x_3 = x_2 - r_2(x_2)/r_2'(x_2) = 2.927$$

$x_3 = 2.927$ k	2	1	0
b_k	1	−4.073	$0.078 = r_2(2.927)$
d_k	1	−1.146	$= r_2'(2.927)$

$$x_4 = x_3 - r_2(x_3)/r_2'(x_3) = 2.995$$

$x_4 = 2.995$ k	2	1	0
b_k	1	−4.005	$0.005 = r_2(2.995)$
d_k	1	−1.01	$= r_2'(2.995)$

$$x_5 = x_4 - r_2(x_4)/r_2'(x_4) = 3.000$$

$x_5 = 3.000$ k	2	1	0
b_k	1	−4.000	$0.000 = r_2(3.000)$

Root: 3. $s_1(x) = \frac{r_2(x)}{x-3} = x - 4$. Root: 4.

6.20 (a) For copper $\alpha = \frac{550}{250} = 2.2$.
For polymer $\alpha = \frac{120}{250} = 0.48$.
Use Newton–Raphson for λ:

n	0	1	2	3	4	5
Cu λ_n	1	0.746	0.553	0.463	0.446	0.446
Poly λ_n	1	0.861	0.816	0.812	0.812	0.812

$S_{\text{Cu}}(2) = 1.26$, $S_{\text{Poly}}(2) = 2.30$.

(b) Simpson's rule: $h = \frac{R}{2}$

$$\frac{R}{6}\left[1 + 4\cos^2\frac{R}{2} + \cos^2 R\right] - \frac{R}{2}$$

$$= 0 \Rightarrow 4\cos^2\frac{R}{2} + \cos^2 R - 2 = 0.$$

Newton-Raphson, with $R_0 = 1$.

n	0	1	2	3	4
R_n	1	1.52947	1.57	1.57079	1.57079

Iteration could be avoided here. A little use of trig. identities gives the original equation as $\cos^2 R +$ $2\cos R = 0$.

6.21 Roots 1 to 8: unchanged to 1 d. p. 9th, 20th roots: 8.9 and 20.8. Roots 10 to 19: 5 complex conjugate pairs!
An example of ill-conditioning for multiple-root problems (see [17] p.135 for more information).

Chapter 7

7.1 Bookwork! Use the definitions in Table 7.2.

7.2

t	0	4	8	12	16	20
h	0	0.84	3.53	8.41	15.97	27.00
v eqn (7.9)	0.21	0.6725	1.22	1.89	2.7575	
a eqn (7.9)	0.1156	0.1369	0.1675	0.2169		
a eqn (7.14)		0.1156	0.1369	0.1675	0.2169	

Accelerations: $h \to v \to a$ and $h \to a$ are same (time is different). Second-order rule is difference of two first-order rules.

$$\frac{\delta^2 f_i}{h^2} = \frac{f_{i+1} - 2f_i + f_{i-1}}{h^2} = \frac{1}{h}\left[\frac{f_{i+1} - f_i}{h} - \frac{f_i - f_{i-1}}{h}\right].$$

7.3 Taylor expansion at $x = x_0$

$$= \frac{1}{2h}\left[-3f + 4\left(f + hf' + \frac{h^2}{2}f'' + \frac{h^3}{6}f''' + \cdots\right) - \left(f + 2hf' + \frac{4h^2}{2}f'' + \frac{8h^3}{6}f''' + \cdots\right)\right]$$

$$= f' - \frac{h^2}{3}f^{(3)}(\xi),\ x_0 < \xi < x_2.$$

Rule = Derivative − LTE.

7.4 Make exact for $1, x, x^2$
(use $x_0 = 0, x_1 = h, x_2 = 2h$)

$$\begin{aligned} 0 &= a_0 + a_1 + a_2 \\ 1 &= a_0x_0 + a_1x_1 + a_2x_2 \\ 2x_0 &= a_0x_0^2 + a_1x_1^2 + a_2x_2^2 \end{aligned}$$

$$\Rightarrow \begin{bmatrix} 1 & 1 & 1 & : 0 \\ 0 & h & 2h & : 1 \\ 0 & h^2 & 4h^2 & : 0 \end{bmatrix}$$

$a_0 = -\frac{3}{2h}, a_1 = \frac{2}{h}, a_2 = -\frac{1}{2h}$
(seen these coefficients before?)

7.5 $E_{\mathrm{T}} = \frac{h^2}{6} M_3, E_R = \frac{10^{-w}}{2h}$,
$E(h) = E_T + E_R$, set $E'(h) = 0$.
$f(x) = \sin x, M_3 = 1, w = 7$,
$h_{\mathrm{opt}} \approx 0.0053$. The $\mathcal{DERIVE}$ expression

```
f(h):=(sin(<Alt>p/4+h)
-sin(<Alt>p/4-h))/(2h)
vector(cos(<Alt>p/4)
-f(10^(-i)),i,0,4)
```

give the following error estimates to the slope of $\sin x$ at $x = \frac{\pi}{4}$. (See Table A.4 below.)

7.6 $E_{\mathrm{T}} = \frac{h^2}{12} M_4, E_{\mathrm{R}} = \frac{2 \times 10^{-w}}{h^2}$,
$E(h) = E_{\mathrm{T}} + E_{\mathrm{R}}$, set $E'(h) = 0$.

7.7 Exact value: $-\cos(0.5) = -0.8776$.
Estimates: -0.8800 ($h = 0.1$),
-0.8775 ($h = 0.2$), -0.8711 ($h = 0.3$).
Best uses $h = 0.2$. Predicted h_{opt} is 0.22.

Chapter 8

8.1 $w_0 = 1, x_0 = b$.

8.2 F is a primitive of f, $E_{\mathrm{R}} = I - R_{\mathrm{R}}$
$= \int_a^b f(x)\,\mathrm{d}x - (b-a)f(b)$.

$$\begin{aligned} E_{\mathrm{R}} &= F(b) - F(a) - (b-a)F'(b) \\ &= F(a+h) - F(a) - hF'(a+h) \\ &= F(a) + hF'(a) + \frac{h^2}{2}F''(a) \\ &\quad + \cdots - F(a) \\ &\quad - h\{F'(a) + hF''(a) + \cdots\} \\ &= -\frac{(b-a)^2}{2} f'(\xi), a < \xi < b. \end{aligned}$$

Same as eqn (8.8) for E_{L} save for the sign. If $f'(x)$ is one-signed on $[a, b]$, R_{L} and R_{R} bracket I.

8.3 $|E_{\mathrm{L}_N}| \le \frac{h}{2}(b-a)M_1$,
$M_1 = \max |f'(x)|$ $(a < x < b)$.
$a = 0, b = 1, f'(x) = \cos x e^{\sin x}$ (max. at $x = 0.666$ – solve $f''(x) = 0 \Rightarrow \cos^2 x = \sin x$). $M_1 \approx 1.459$.

$$\begin{aligned} h \le \frac{2 \times 6 \times 10^{-8}}{(b-a)M_1} &= \frac{12 \times 10^{-8}}{1.459} \\ &\approx 8.23 \times 10^{-8} \\ &\Rightarrow n \approx 1215833. \end{aligned}$$

8.4 $T_4 = 0.3793, M_2 = \frac{0.0099}{h^2}, |E_{\mathrm{T}_4}| \le 0.0003$.

8.5 Trapezium: $\frac{h^2}{12} M_2 \le \frac{1}{2} \times 10^{-2} \Rightarrow h^2 \le 0.06/M_2$.
Error function:

$$f(x) = \frac{2}{\sqrt{\pi}} \mathrm{e}^{-x^2},$$

$$f''(x) = \frac{8}{\sqrt{\pi}} x \mathrm{e}^{-x^2}(4x^2 - 2).$$

$M_2 = \max |f''(x)| = |f''(0)| = \frac{4}{\sqrt{\pi}}$.
$h \approx 0.16$. Use $h = 0.1$, $T_5 = 0.52032$ (exact: 0.52050).
Debye function:

$$f(x) = \frac{x^2}{\mathrm{e}^x - 1},$$

$$f''(x) = \frac{\mathrm{e}^{2x}(x^2 - 4x + 2) + \mathrm{e}^x(x^2 + 4x - 4) + 2}{(\mathrm{e}^x - 1)^3}.$$

Table A.4 See answer 7.5.

h	1	0.1	0.01	0.001	0.0001
$\cos\frac{\pi}{4} - f(h)$	0.112096	0.001178	-6.2×10^{-6}	4.6×10^{-5}	5.4×10^{-4}

$M_2 = \max|f''(x)| = |f''(0)| = 1$.
$h \approx 0.2449$. Use $h = 0.125$, and note that $\mathcal{D} = 0$ at $x = 0$.
$T_4 = 0.10489$ (exact: 0.10547).
Note that the function `tn` is available for $\mathcal{DERIVE}$ users!

8.6 Using one interval, $h = 1$.

$$T_1 = \frac{1}{2}(1 + \sin 1) = 0.9207,$$

$$M_1 = \frac{\sin\frac{1}{2}}{\frac{1}{2}} = 0.9589.$$

Both estimates are reasonable. The 'hidden' danger is that a computer will not generally know about the limiting value of $\frac{\sin x}{x}$ as $x \to 0$, and will baulk at computing $\frac{0}{0}$.

8.7 $S_2 = 0.8690$, $S_4 = 0.8649$, $|E_{S_2}| \leq 0.0111$, $|E_{S_4}| \leq 0.0007$. Extrapolation gives 0.8646. Obs. errors are 0.0043, 0.0002, 0.0001.

8.8 (a) $\frac{\pi}{12} + \frac{\sqrt{3}}{8} \approx 0.4783$, (b) 0.4783, (c) 0.4783.

8.9 Having $\mathcal{DERIVE}$ is the easiest way! Use the function `tn` defined in $\mathcal{DERIVE}$ *Experiment 8.2*. Author and approX

```
vector(tn(sin(29x),0,<Alt>
p/2,<Alt>p/(2n)),n,
    [5,10,20,40,80,160])
```

N	5	10	20	40	80	160
T_N	0.0249	−0.0671	0.0181	0.0307	0.0335	0.0342

No convergence, nowhere near the exact value! *Oscillatory* integrands must be sampled very carefully!

8.10

0.0733	0	0	0
0.4045	0.5149	0	0
0.4182	0.4227	0.4166	0
0.4216	0.4227	0.4227	**0.4228**

8.11 Referring to Section 8.6

$$I = M_1 + \frac{(2h)^2}{24} f''(x_0) + G(2h)^4 + \cdots$$

$$I = M_2 + \frac{h^2}{24} f''(x_0) + Gh^4 + \cdots$$

4 times the second eqn minus the first gives $I = \frac{1}{3}[4M_2 - M_1] - 4Gh^4 + O(h^6)$ – same as T_N-based rule.

0.7358	0	0	0
0.4318	0.3305	0	0
0.4250	0.4227	0.4288	0
0.4233	0.4227	0.4227	**0.4226**

Chapter 9

9.1 One flop per equation. Total: n flops.

9.2 Back substitution:

$$x_3 = -1, x_2 = -2x_3 = 2,$$
$$x_1 = -2x_3 + 8x_2 = 18.$$

Forward substitution:

$$x_1 = 0, x_2 = -2x_1 = 0,$$
$$x_3 = -1 - 2x_1 + 8x_2 = -1.$$

9.3 Forward elimination:

$\boldsymbol{U} =$

$$\begin{bmatrix} 2 & 2 & 3 \\ 4 & 7 & 7 \\ -2 & 4 & 5 \end{bmatrix} \to \begin{bmatrix} 2 & 2 & 3 \\ 0 & 3 & 1 \\ 0 & 6 & 8 \end{bmatrix} \to \begin{bmatrix} 2 & 2 & 3 \\ 0 & 3 & 1 \\ 0 & 0 & 6 \end{bmatrix}$$

$\boldsymbol{L} =$

$$\begin{bmatrix} 1 & 0 & 0 \\ & 1 & 0 \\ & & 1 \end{bmatrix} \to \begin{bmatrix} 1 & 0 & 0 \\ 2 & 1 & 0 \\ -1 & & 1 \end{bmatrix} \to \begin{bmatrix} 1 & 0 & 0 \\ 2 & 1 & 0 \\ -1 & 2 & 1 \end{bmatrix}$$

$m_{21} = 2, m_{31} = -1, m_{32} = 2.$

$$\text{Product } \boldsymbol{LU} = \begin{bmatrix} 2 & 2 & 3 \\ 4 & 7 & 7 \\ -2 & 4 & 5 \end{bmatrix} = \boldsymbol{A}.$$

9.4 (a)

$$\left[\begin{array}{ccc|c} 1 & 3 & 2 & -1 \\ 2 & 2 & -1 & 2 \\ 3 & 5 & 1 & 0 \end{array}\right] \rightarrow \left[\begin{array}{ccc|c} 1 & 3 & 2 & -1 \\ 0 & -4 & -5 & 4 \\ 0 & -4 & -5 & 3 \end{array}\right]$$

$$\rightarrow \left[\begin{array}{ccc|c} 1 & 3 & 2 & -1 \\ 0 & -4 & -5 & 4 \\ 0 & 0 & 0 & -1 \end{array}\right]$$

Last eqn reads $0x + 0y + 0z = 1$ – no soln.

(b)

$$\left[\begin{array}{ccc|c} 1 & 1 & 1 & a^2 \\ 4 & -1 & 1 & a \\ 1 & 1 & 2 & 1 \\ 1 & 6 & 5 & 1 \end{array}\right] \rightarrow \left[\begin{array}{ccc|c} 1 & 1 & 1 & a^2 \\ 0 & -5 & -3 & a - 4a^2 \\ 0 & 0 & 1 & 1 - a^2 \\ 0 & 5 & 4 & 1 \quad a^2 \end{array}\right]$$

$$\rightarrow \left[\begin{array}{ccc|c} 1 & 1 & 1 & a^2 \\ 0 & -5 & -3 & a - 4a^2 \\ 0 & 0 & 1 & 1 - a^2 \\ 0 & 0 & 1 & 1 + a - 5a^2 \end{array}\right]$$

Last 2 eqns require $-a^2 = a - 5a^2 \Rightarrow 4a^2 - a = 0 \Rightarrow a = \in \{0, 1/4\}$.

$$a = 0: \left[\begin{array}{ccc|c} 1 & 1 & 1 & 0 \\ 0 & -5 & 3 & 0 \\ 0 & 0 & -1 & 1 \end{array}\right], \quad z = 1,$$

$y = -3z/5 = -3/5, x = -y - z = -2/5.$

$$a = 1/4: \left[\begin{array}{ccc|c} 1 & 1 & 1 & 1/16 \\ 0 & -5 & -3 & 0 \\ 0 & 0 & 1 & 15/16 \end{array}\right], \quad z = 15/16,$$

$y = -3z/5 = -9/16, x = 1/16$
$-y - z = -5/16.$

(c)

$$\left[\begin{array}{ccc|c} -\lambda & 3 & 0 & 0 \\ 1 & -\lambda & 1 & 0 \\ 0 & 1 & -\lambda & 0 \end{array}\right] \rightarrow \left[\begin{array}{ccc|c} -\lambda & 3 & 0 & 0 \\ 0 & -\lambda^2 + 3 & \lambda & 0 \\ 0 & 1 & -\lambda & 0 \end{array}\right]$$

$$\rightarrow \left[\begin{array}{ccc|c} -\lambda & 3 & 0 & 0 \\ 0 & -\lambda^2 + 3 & \lambda & 0 \\ 0 & 0 & -\lambda^3 + 4\lambda & 0 \end{array}\right]$$

Last eqn. requires $\lambda \in \{-2, 0, 2\}$ – *eigenvalues* of A.
$\lambda = -2$: Infinity of solutions.

$$\left[\begin{array}{ccc|c} 2 & 3 & 0 & 0 \\ 0 & -1 & -2 & 0 \\ 0 & 0 & 0 & 0 \end{array}\right] . \ z = \alpha,$$

$y = -2\alpha, x = 3\alpha.$

$$\boldsymbol{x} = \begin{bmatrix} 3\alpha \\ -2\alpha \\ \alpha \end{bmatrix}.$$

$\lambda = 0$: Infinity of solutions.

$$\left[\begin{array}{ccc|c} 0 & 3 & 0 & 0 \\ 0 & 3 & 0 & 0 \\ 0 & 0 & 0 & 0 \end{array}\right] . \ z = \alpha, y = 0, x = \beta.$$

$$\boldsymbol{x} = \begin{bmatrix} \beta \\ 0 \\ \alpha \end{bmatrix}.$$

$\lambda = 2$: Infinity of solutions.

$$\left[\begin{array}{ccc|c} -2 & 3 & 0 & 0 \\ 0 & -1 & 2 & 0 \\ 0 & 0 & 0 & 0 \end{array}\right] . \ z = \alpha, y = 2\alpha,$$

$$x = 3\alpha. \ \boldsymbol{x} = \begin{bmatrix} 3\alpha \\ 2\alpha \\ \alpha \end{bmatrix}.$$

(d)

$$\left[\begin{array}{ccc|c} 1 & 1 & 3 & 1 \\ 3 & 2 & 1 & 5 \\ 1 & 2 & 11 & -1 \end{array}\right] \rightarrow \left[\begin{array}{ccc|c} 1 & 1 & 3 & 1 \\ 0 & -1 & -8 & 2 \\ 0 & 1 & 8 & -2 \end{array}\right]$$

$$\rightarrow \left[\begin{array}{ccc|c} 1 & 1 & 3 & 1 \\ 0 & -1 & -8 & 2 \\ 0 & 0 & 0 & 0 \end{array}\right]$$

Last eqn reads $0x + 0y + 0z = 0$ – infty. solns. Set $z = \alpha$, $y = 8z + 2 = 8\alpha + 2$, $x = 1 - y - 3z = -11\alpha - 1$.

9.5

$$\left[\begin{array}{cccc|c} 1 & 2 & 1 & 1 & 2 \\ 2 & 3 & 4 & 1 & 6 \\ 1 & 2 & 2 & 2 & 2 \\ 3 & 7 & -1 & -1 & -8 \end{array}\right]$$

$$\rightarrow \left[\begin{array}{cccc|c} 1 & 2 & 1 & 1 & 2 \\ 0 & -1 & 2 & -1 & 2 \\ 0 & 0 & 1 & 1 & 0 \\ 0 & 1 & -4 & -4 & -14 \end{array}\right]$$

$$\rightarrow \left[\begin{array}{cccc|c} 1 & 2 & 1 & 1 & 2 \\ 0 & -1 & 2 & -1 & 2 \\ 0 & 0 & 1 & 1 & 0 \\ 0 & 0 & -2 & -5 & -12 \end{array}\right]$$

$$\rightarrow \left[\begin{array}{cccc|c} 1 & 2 & 1 & 1 & 2 \\ 0 & -1 & 2 & -1 & 2 \\ 0 & 0 & 1 & 1 & 0 \\ 0 & 0 & 0 & -3 & -12 \end{array}\right]$$

Back substitution: $z = 4$, $y = -4$, $x = -14$, $w = 30$.

9.6 Without pivoting (underlined elements rounded to 2 sig. figs.)

$$\left[\begin{array}{ccc|c} 0.5 & 1.1 & 3.1 & 6.0 \\ 2.0 & 4.5 & 0.4 & 0.02 \\ 5.0 & 1.0 & 6.5 & 1.0 \end{array}\right]$$

$$\rightarrow \left[\begin{array}{ccc|c} 0.5 & 1.1 & 3.1 & 6.0 \\ 0 & 0.1 & \underline{-12} & \underline{-24} \\ 0 & -10 & \underline{-25} & \underline{-59.0} \end{array}\right]$$

$$\rightarrow \left[\begin{array}{ccc|c} 0.5 & 1.1 & 3.1 & 6.0 \\ 0 & 0.1 & \underline{-12} & \underline{-24} \\ 0 & 0 & \underline{-1200} & \underline{-2500} \end{array}\right]$$

Back substitution $z = 2.1$, $y = 10$, $x = -24$.
With pivoting, swap rows 1 and 3

$$\left[\begin{array}{ccc|c} 5.0 & 1.0 & 6.5 & 1.0 \\ 2.0 & 4.5 & 0.4 & 0.02 \\ 0.5 & 1.1 & 3.1 & 6.0 \end{array}\right]$$

$$\rightarrow \left[\begin{array}{ccc|c} 5.0 & 1.0 & 6.5 & 1.0 \\ 0 & 4.1 & -2.2 & -0.38 \\ 0 & 1 & 2.5 & 5.9 \end{array}\right]$$

$$\rightarrow \left[\begin{array}{ccc|c} 5.0 & 1.0 & 6.5 & 1.0 \\ 0 & 4.1 & -2.2 & -0.38 \\ 0 & 0 & \underline{3.0} & \underline{6.0} \end{array}\right]$$

Back substitution $z = 2.0$, $y = 0.98$, $x = -2.6$.
Problem: Large pivot $a_{22} = 100$ on eliminating 2nd column.

9.7 (a) Gauss elimination

$$\frac{2n^3}{3} + \frac{3n^2}{2} - \frac{7n}{6} = 80 \times 365 \times 24 \times 60 \times 60 \times 10^6 \Rightarrow n \simeq 155833$$

(b) Cramer's rule

$$(1 + e)(n + 1)! = 80 \times 365 \times 24 \times 60 \times 60 \times 10^6 \Rightarrow n \simeq 16$$

9.8 Forward elimination:

$$\left[\begin{array}{ccc|c} 2 & 2 & 3 & 3 \\ 4 & 7 & 7 & 1 \\ -2 & 4 & 5 & -7 \end{array}\right] \rightarrow \left[\begin{array}{ccc|c} 2 & 2 & 3 & 3 \\ 0 & 3 & 1 & -5 \\ 0 & 6 & 8 & -4 \end{array}\right]$$

$$\rightarrow \left[\begin{array}{ccc|c} 2 & 2 & 3 & 3 \\ 0 & 3 & 1 & -5 \\ 0 & 0 & 6 & 6 \end{array}\right]$$

Back sub.: $z=1$, $y=\frac{1}{3}(-5-z)=-2$, $x=\frac{1}{2}(3-2y-3z)=1$.

$$\boldsymbol{Lc}=\boldsymbol{b}: \left[\begin{array}{ccc|c} 1 & 0 & 0 & 3 \\ 2 & 1 & 0 & 1 \\ -1 & 2 & 1 & -7 \end{array}\right]$$

Forward sub.: $c_1=3$, $c_2=1-2c_1=-5$, $c_3=-7+c_1-2c_2=6$.

$$\boldsymbol{Ux}=\boldsymbol{c}: \left[\begin{array}{ccc|c} 2 & 2 & 3 & 3 \\ 0 & 3 & 1 & -5 \\ 0 & 0 & 6 & 6 \end{array}\right]$$

Back sub.: $z=1$, $y=\frac{1}{3}(-5-z)=-2$, $x=\frac{1}{2}(3-2y-3z)=1$.
Solving two triangular systems gives solution of original square system.
$\boldsymbol{Lc}=\boldsymbol{b} \Rightarrow \boldsymbol{L}(\boldsymbol{Ux})=\boldsymbol{b} \Rightarrow \boldsymbol{Ax}=\boldsymbol{b}$.

9.9 Forward elimination:

$$\left[\begin{array}{cc|c} \varepsilon & 1 & 1 \\ 1 & 1 & 2 \end{array}\right] \rightarrow \left[\begin{array}{cc|c} \varepsilon & 1 & 1 \\ 0 & 1-\frac{1}{\varepsilon} & 2-\frac{1}{\varepsilon} \end{array}\right]$$

Back substitution:

$$x_2 = \frac{2-\frac{1}{\varepsilon}}{1-\frac{1}{\varepsilon}} = \frac{2\varepsilon-1}{\varepsilon-1} \simeq 1,$$

$$x_1 = \frac{1-x_2}{\varepsilon} \simeq 0.$$

Exact solution:

$$x_2 = \frac{2-\frac{1}{\varepsilon}}{1-\frac{1}{\varepsilon}} = \frac{2\varepsilon-1}{\varepsilon-1},$$

$$x_1 = \frac{1-x_2}{\varepsilon} = \frac{1}{1-\varepsilon}.$$

Limit as $\varepsilon \to 0$: $x_1 \to 1$, $x_2 \to 1$.
Partial pivoting: For $\varepsilon \ll 1$, swap rows.
Forward elimination:

$$\left[\begin{array}{cc|c} 1 & 1 & 2 \\ \varepsilon & 1 & 1 \end{array}\right] \rightarrow \left[\begin{array}{cc|c} 1 & 1 & 2 \\ 0 & 1-\varepsilon & 2-\varepsilon \end{array}\right]$$

Back substitution:

$$x_2 = \frac{1-2\varepsilon}{1-\varepsilon} \simeq 1, \; x_1 = 2-x_2 \simeq 1.$$

Pivoting recovers the exact solution to machine accuracy.

9.10

$$-\boldsymbol{D}^{-1}(\boldsymbol{L}+\boldsymbol{U}) = \begin{bmatrix} \frac{1}{5} & 0 & 0 \\ 0 & \frac{1}{10} & 0 \\ 0 & 0 & \frac{1}{4} \end{bmatrix} \begin{bmatrix} 0 & 2 & 1 \\ 1 & 0 & 1 \\ 1 & 1 & 0 \end{bmatrix}$$

$$= \begin{bmatrix} 0 & \frac{2}{5} & \frac{1}{5} \\ \frac{1}{10} & 0 & \frac{1}{10} \\ \frac{1}{4} & \frac{1}{4} & 0 \end{bmatrix}, \boldsymbol{D}^{-1}\boldsymbol{b} = \begin{bmatrix} \frac{3}{5} \\ -\frac{27}{10} \\ -1 \end{bmatrix}.$$

$$\boldsymbol{x}^{(k+1)} = \begin{bmatrix} 0 & \frac{2}{5} & \frac{1}{5} \\ \frac{1}{10} & 0 & \frac{1}{10} \\ \frac{1}{4} & \frac{1}{4} & 0 \end{bmatrix} \boldsymbol{x}^{(k)} + \begin{bmatrix} \frac{3}{5} \\ -\frac{27}{10} \\ -1 \end{bmatrix}.$$

Applying the iterative scheme four times:

k	$\boldsymbol{x}^{(k)}$		
0	0	0	0
1	0.6	−2.7	−1
2	−0.08	−2.74	−1.525
3	−0.801	−2.9205	−1.855
4	−0.9392	−2.9656	−1.9304

Apparent convergence to $[-1\ -3\ -2]^{\mathrm{T}}$.
To check $\|-\boldsymbol{D}^{-1}(\boldsymbol{L}+\boldsymbol{U})\|_\infty$
$= \max\left\{\frac{3}{5}, \frac{2}{10}, \frac{2}{4}\right\} = \frac{1}{2}$.
This is less than 1 (the scheme is convergent).

9.11 Decomposition is

$$\boldsymbol{A} = \begin{bmatrix} 2 & 1 \\ 1 & 2 \end{bmatrix} = \begin{bmatrix} 2 & 0 \\ 0 & 2 \end{bmatrix} + \begin{bmatrix} 0 & 0 \\ 1 & 0 \end{bmatrix} + \begin{bmatrix} 0 & 1 \\ 0 & 0 \end{bmatrix} = \boldsymbol{D}+\boldsymbol{L}+\boldsymbol{U}.$$

Gauss–Seidel:

$$-(\boldsymbol{D}+\boldsymbol{L})^{-1}\boldsymbol{U} = \begin{bmatrix} -\frac{1}{2} & 0 \\ \frac{1}{4} & -\frac{1}{2} \end{bmatrix}\begin{bmatrix} 0 & 1 \\ 0 & 0 \end{bmatrix} = \begin{bmatrix} 0 & -\frac{1}{2} \\ 0 & \frac{1}{4} \end{bmatrix},$$

$$(\boldsymbol{D}+\boldsymbol{L})^{-1}\boldsymbol{b} = \begin{bmatrix} 2 \\ \frac{3}{2} \end{bmatrix}.$$

$$\boldsymbol{x}^{(k+1)} = \begin{bmatrix} 0 & -\frac{1}{2} \\ 0 & \frac{1}{4} \end{bmatrix}\boldsymbol{x}^{(k)} + \begin{bmatrix} 2 \\ \frac{3}{2} \end{bmatrix}.$$

Eqn 2 is $y = \frac{y}{4} + \frac{3}{2} \Rightarrow y = 2$. Eqn 1 is $x = -\frac{y}{2} + 2 \Rightarrow x = 1$.
Error $\boldsymbol{x} - \boldsymbol{x}^{(k+1)} = \boldsymbol{e}^{(k+1)}$
$= -(\boldsymbol{D}+\boldsymbol{L})^{-1}\ \boldsymbol{U}\boldsymbol{e}^{(k)} = \begin{bmatrix} 0 & -\frac{1}{2} \\ 0 & \frac{1}{4} \end{bmatrix}\boldsymbol{e}^{(k)}$. $\boldsymbol{x}^{(0)}$
$= \boldsymbol{0} \Rightarrow \boldsymbol{e}^{(0)} = [1\ \ 2]^{\mathrm{T}}$ and $\boldsymbol{e}^{(1)} = \begin{bmatrix} -1.0 \\ 0.5 \end{bmatrix}$,
$\boldsymbol{e}^{(2)} = \begin{bmatrix} -0.25 \\ 0.125 \end{bmatrix}$, $\boldsymbol{e}^{(3)} = \begin{bmatrix} -0.0625 \\ 0.03125 \end{bmatrix}$.
3 iterations to reduce error by factor of 10.

9.12 For eigenvalue λ, solve

$$|-\boldsymbol{D}^{-1}(\boldsymbol{L}+\boldsymbol{U}) - \lambda\boldsymbol{I}| = 0 \Rightarrow$$
$$|(\boldsymbol{L}+\boldsymbol{U})+\lambda\boldsymbol{D}| = 0 \Rightarrow \begin{vmatrix} -5\lambda & 2 & 1 \\ 1 & -10\lambda & 1 \\ 1 & 1 & -4\lambda \end{vmatrix} = 0,$$

that is $-200\lambda^3 + 23\lambda + 4 = 0$. $\lambda = 0.405$, $-0.203 \pm 0.091\mathrm{i}$. All magnitudes are less than 1, so convergent.

9.13 Jacobi: $|(\boldsymbol{L}+\boldsymbol{U}) + \lambda\boldsymbol{D}| = 0$ gives $-\lambda^3 + 0.25\lambda = 0$, $\lambda \in \{-0.5, 0, 0.5\}$ – convergent (note that norms do not imply convergence in this case!)
Gauss–Seidel: $|\boldsymbol{U} + \lambda(\boldsymbol{D}+\boldsymbol{L})| = 0$ gives $\lambda(\lambda^2 + 1.75\lambda - 2) = 0$. $\lambda \in \{-2.54, 0, 0.79\}$ – not convergent.

Chapter 10

10.1 (a) $f(x,c) \in P_1(x)$, $L = 3$. (b) $f(x,c) \in P_0(x)$, $L = 2$.

10.2 $f(x,c) \in P_0(x)$. $|f(x,y) - f(x,z)| \le |g(y)||y-z|$ (Mean-value thm.), $g(y) = -2y/(\varepsilon + y^2)^2$. $L = \max|g| = 9/(8\varepsilon\sqrt{\varepsilon})$ at $y = \pm\sqrt{\varepsilon/3}$. Require $\varepsilon > 0$.

10.3 $f(x,c) = c\cos x$ – continuous, $L = 1$.

$$y_0(x) = 1,$$
$$y_1(x) = 1 + \int_0^x \cos x\,\mathrm{d}x = 1 + \sin x,$$
$$y_2(x) = 1 + \int_0^x (1+\sin x)\cos x\,\mathrm{d}x = 1 + \sin x - \frac{\cos 2x}{4}.$$

10.4 (a) $y_n = c_1(-k)^n$.
(b) $y_n = \frac{1}{2^n}[c_1(3+\sqrt{5})^n + c_2(3-\sqrt{5})^n]$.
(c) $y_n = (c_1 + nc_2)^n$ (double root $\beta_1 = -1$).
(d) $y_n = c_1(\frac{1}{2})^n - n + 4$.
(e) $y_n = 2^n + 3^n$.

10.5 $A_n = 10 \times 2^n$. Solve $10 \times 2^n = 10^7$ $\Rightarrow n \approx 19.9$, $t = 4n \approx 79.7$ s.

10.6 $E_{n+1} = 0.62E_n$, $E_n = E_0(0.62)^n$ $= 3O_0(0.62)^n$.
$E_n = 0.1E_0 \Rightarrow 0.1 = (0.62)^n \Rightarrow$ $n \approx 4.82$.
$O_{n+1} = 0.87O_n + 0.38E_n = 0.87O_n$ $+1.14O_0(0.62)^n$. Homog. soln. $O_n = c_2(0.87)^n$. Partic. soln. $k = -4.56O_0$. General soln. $O_n = [5.56(0.87)^n - 4.56(0.62)^n]O_0$.

$O_n = 0.1O_0 \Rightarrow 0.1 = 5.56(0.87)^n - 4.56(0.62)^n \Rightarrow n \approx 28.85.$

10.7 (a) $|y_n| \le s\left(\frac{1}{2}\right)^n, s \ge 0$. (b) $|y_n| \le (s-1)3^n + 1, s \ge 1$.

10.8 $f(x,c) \in P_0(x), L = 1. y_{n+1} = y_n + hf_n + \frac{1}{2}h^2 f_n^{(1)}; f_n = \cos y_n, f_n^{(1)} = f_x + ff_y = -\sin y_n \cos y_n, y_{n+1} = y_n + h\cos y_n\left(1 - \frac{h}{2}\sin y_n\right). y_0 = 1, y_1 = 1.0518, y_2 = 1.0992.$

10.9 $y_{n+1} = y_n + hf_n \Rightarrow y_{n+1} = y_n + h(y_n - 1) = (1+h)y_n - h$. Homog. soln. $y_n = c_1(1+h)^n$. Partic. soln. $k = 1$. General soln. $y_n = 1 - 0.75(1+h)^n$.

h	n	y_n
0.1	5	−0.2079
0.05	10	−0.2217
0.025	20	−0.2290

Order $p = \dfrac{\log \frac{-0.2079+0.2217}{-0.2217+0.2290}}{\log 2} \approx 0.919$ (agrees with theory). Note convergence as h is refined (see Figure A.5).

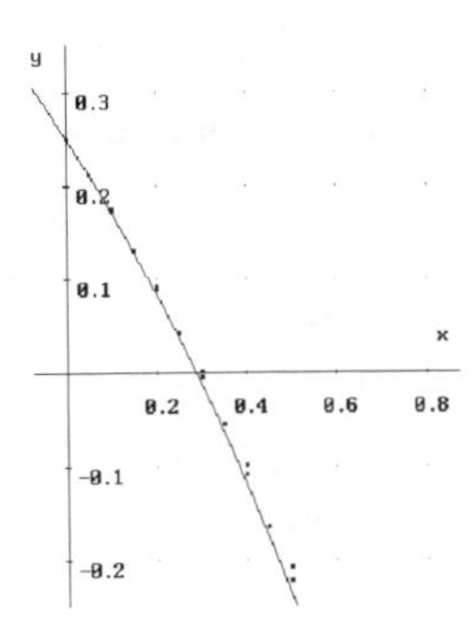

Figure A.5 Euler approximations to $y' = y - 1$.

10.10 Bookwork! Compare $y_{n+1} = y_n + hf_n$ and $y_{n+1} = y_n + \alpha_1 hf_n$.

10.11
$$y_{n+1} = y_n + \frac{h}{2}[f_n + f(x_n + h, y_n + hf_n)]$$
$$= y_n + \frac{h}{2}\left[x_n^2 - y_n + (x_n+h)^2 - y_n - h\left(x_n^2 - y_n\right)\right]$$
$$= \left(1 - h + \frac{h^2}{2}\right)y_n + \frac{h}{2} \times \left[(2-h)x_n^2 + 2hx_n + h^2\right]$$
$$= 0.82y_n + 0.1\left(1.8x_n^2 + 0.4x_n + 0.04\right).$$

n	x_n	y_n
0	0	0.5
1	0.2	0.414
2	0.4	0.35868
3	0.6	0.34292

$y(0.6) = 0.3368$. Use `srk(x^2-y, x,y, 0,.5,h,n)` for refined estimates $y_6 = 0.33827$ ($h = 0.1$) and $y_{12} = 0.33715$ ($h = 0.05$). Order $p = \dfrac{\log \frac{0.3429-0.3383}{0.3383-0.3372}}{\log 2} \approx 2.064$ (agrees with theory).

10.12 $\alpha_1 = \frac{1}{4}, \alpha_2 = \frac{3}{4}, \lambda_2 = \mu_2 = \frac{2}{3}$.
$$y_{n+1} = y_n + \frac{h}{4}\left[f_n + 3f \times \left(x_n + \frac{2h}{3}, y_n + \frac{2hf_n}{3}\right)\right].$$

For Example 10.26, $y' = -y + x + 1$ and
$$y_{n+1} = \left(1 - h + \frac{h^2}{2}\right)y_n + \frac{h}{2}[(2-h)x_n + 2].$$

Same as simple Runge–Kutta and modified Euler's methods – linear ODE in y and x. Suitable *DERIVE* functions are

```
hn2(f,x,y,xn,yn,k1,h):=
   k1+3lim(f,[x,y],[xn +2h/3,
   yn+2hk1/3])
hn1(f,x,y,xn,xy,h):=
   hn2(f,x,y,xn,yn,lim(f,
   [x,y],[xn,yn]),h)
heun(f,x,y,x0,y0,h,n):=
```

```
iterates([xn+h,yn+hhn1(f,
x,y,xn,yn,h)/4],[xn,yn],
[x0,y0],n)
```

10.13 (a)

$$y_{n+1} = y_n + hf\left(x_n + \frac{h}{2}, y_n + \frac{hf_n}{2}\right)$$
$$= y_n + h\left[\left(x_n + \frac{h}{2}\right)^2 - y_n - \frac{h}{2}(x_n^2 - y_n)\right]$$
$$= \left(1 - h + \frac{h^2}{2}\right) y_n + \frac{h}{2}\left[(2-h)x_n^2 + 2hx_n + \frac{h^2}{2}\right]$$
$$= 0.82y_n + 0.1(1.8x_n^2 + 0.4x_n + 0.02).$$

n	x_n	y_n
0	0	0.5
1	0.2	0.412
2	0.4	0.35504
3	0.6	0.33793

Use `meuler(x^2-y,x,y,0,.5,h,n)` to generate refined estimates $y_6 = 0.33708$ ($h = 0.1$) and $y_{12} = 0.33686$ ($h = 0.05$).
Order $p = \frac{\log \frac{0.3379-0.3371}{0.3371-0.3369}}{\log 2} \approx 2.000$ (agrees with theory).

(b)

$$y_{n+1} = y_n + \frac{h}{4}\left[f_n + 3f\left(x_n + \frac{2h}{3}, y_n + \frac{2hf_n}{3}\right)\right]$$
$$= y_n + \frac{h}{4}\left[x_n^2 - y_n + 3\left\{\left(x_n + \frac{2h}{3}\right)^2 - y_n - \frac{2h}{3}(x_n^2 - y_n)\right\}\right]$$
$$= \left(1 - h + \frac{h^2}{2}\right) y_n + \frac{h}{2}\left[(2-h)x_n^2 + 2hx_n + \frac{2h^2}{3}\right]$$
$$= 0.82y_n + 0.1(1.8x_n^2 + 0.4x_n + 0.08/3).$$

n	x_n	y_n
0	0	0.5
1	0.2	0.41267
2	0.4	0.35625
3	0.6	0.33959

Use `heun(x^2-y,x,y,0,.5,h,n)` to generate refined estimates $y_6 = 0.33748$ ($h = 0.1$) and $y_{12} = 0.33695$ ($h = 0.05$).
Order $p = \frac{\log \frac{0.3396-0.3375}{0.3375-0.3370}}{\log 2} \approx 2.070$ (agrees with theory).

10.14 Use *DERIVE* for convenience!

```
euler(x^2+y^2,x,y,0,1,.1,2)
srk(x^2+y^2,x,y,0,1,.1,2)
meuler(x^2+y^2,x,y,0,1,.1,2)
heun(x^2+y^2,x,y,0,1,.1,2)
```

Answers: $y_2 = 1.222$, $y_2 = 1.252$, $y_2 = 1.250$, $y_2 = 1.251$.

10.15

h	n	y_n
0.2	5	1.355748264
0.1	10	1.355751178
0.05	20	1.355751347

Note high accuracy (approx. 6 sig. figs.).
Order $p = \frac{\log \frac{1.355748264-1.355751178}{1.355751178-1.355751347}}{\log 2} \approx 4.108$ (as theory).

10.16 L'Hôpital's rule on eqn (10.51) gives

$$\lim_{L_\phi \to 0} \frac{Nh^p}{L_\phi}\left(e^{(x_n - x_0)L_\phi} - 1\right)$$
$$= \lim_{L_\phi \to 0} Nh^p (x_n - x_0) e^{(x_n - x_0)L_\phi}$$
$$= (x_n - x_0) Nh^p.$$

10.17

$$\text{LTE} = \frac{y(x+h) - y(x)}{h} - f\left(x + \frac{h}{2}, y + \frac{h}{2} f(x, y)\right)$$
$$= \frac{1}{h}\left[y + hy' + \frac{h^2}{2} y'' + \frac{h^3}{6} y''' + \cdots - y\right] - \left[f + \frac{h}{2} f^{(1)} + \frac{h^2}{8} \times (f_{xx} + 2ff_{xy} + f^2 f_{yy}) + \cdots\right]$$
$$= h^2 \left[\frac{y'''}{6} - \frac{(f_{xx} + 2ff_{xy} + f^2 f_{yy})}{8}\right]$$
$$= \frac{h^2}{24}[y''' + 3y'' f_y].$$

$|t(x; h)| \leq Nh^2$ where $N = \frac{1}{24} \max |y''' + 3y'' f_y|$.

10.18

n	x_n	$y(x_n)$	y_n	$\lvert y(x_n) - y_n\rvert$	Add. err.
0	0	1	1	0	
1	0.2	1.00254	1.004	0.00146	0.00146
2	0.4	1.01936	1.02248	0.00312	0.00166
3	0.6	1.06238	1.06723	0.00485	0.00173
4	0.8	1.14134	1.14793	0.00659	0.00174
5	1	1.26424	1.27250	0.00826	0.00167

$N = \frac{1}{24} \max |y''' + 3y'' f_y| = \frac{1}{4} - \frac{1}{3e} \approx 0.127$.
LDE $= |ht| \leq Nh^3 \approx 0.001$.

10.19

n	x_n	y_n
0	0	0
1	0.2	0.0
2	0.4	0.048
3	0.6	0.1504
4	0.8	0.31232
5	1	0.53786

$y(1) = 0.63212$, Obs. error is 0.09426.
Order $p = 1$, $N = \frac{1}{2} \max |y''|$
$= \frac{1}{2} \max \left|1 - \frac{e^{-x}}{2}\right| = 1 - \frac{1}{2e}$, $n = 5$,
$x_n = 1$, $x_0 = 0$, $h = 0.2$ and $L_\phi = 1$.

$$e_5 \leq \frac{Nh^p}{L_\phi}\left(e^{(x_5 - x_0)L_\phi} - 1\right) \approx 0.2804.$$

e_5 certainly bounds the observed error.

10.20 $|\phi(x, y; h) - \phi(x, z; h)|_{\text{SRK}}$

$$= \frac{1}{2}|f(x, y) + f(x+h, y+hf) - f(x, z) - f(x+h, z+hf)|$$
$$\leq \frac{L}{2}(|y - z| + |y + hf(x, y) - z - hf(x, z)|)$$
$$\leq \frac{L}{2}(2|y - z| + hL|y - z|)$$
$$= \boxed{\left(1 + \frac{h}{2} L\right) L |y - z|}$$

$|\phi(x, y; h) - \phi(x, z; h)|_{\text{Heun}}$

$$= \frac{1}{4}\left|f(x, y) + 3f\left(x + \frac{2}{3}h, y + \frac{2}{3}hf\right) - f(x, z) - 3f\left(x + \frac{2}{3}h, z + \frac{2}{3}hf\right)\right|$$
$$\leq \frac{L}{4}\left(|y - z| + 3\left|y + \frac{2}{3}hf(x, y) - z - \frac{2}{3}hf(x, z)\right|\right)$$

$$\leq \frac{L}{4}\left(4|y-z| + \frac{11}{3}h|f(x,y) - f(x,z)|\right)$$

$$\leq L\left(|y-z| + \frac{h}{2}L|y-z|\right)$$

$$= \boxed{\left(1+\frac{h}{2}L\right)L|y-z|}$$

10.21 Min. $E(h) = Nh^p + \frac{\varepsilon}{h}$ w.r.t. h (differentiate and set to zero).

$$E'(h) = pNh^{p-1} - \frac{\varepsilon}{h^2} = 0.$$

$$\Rightarrow h = h_{\text{opt}} = \sqrt[p+1]{\frac{\varepsilon}{pN}}.$$

$y(x) = e^{-x}$, $y''(x) = e^{-x}$, $y'''(x) = -e^{-x}$, $f_y \equiv 0$.

Method	N	p	h_{opt}
Euler	$\frac{1}{2}$	1	0.001
SRK	$\frac{1}{24}$	2	0.018
Mod. Euler	$\frac{1}{12}$	2	0.014

10.22 $y''(x) = 2 - e^{-x}$, $y'''(x) = e^{-x}$, $f_y = -1$.

Method	N	p	h_{opt}
Euler	$1 - \frac{1}{2e^2}$	1	0.00073
SRK	$\frac{1}{4} - \frac{1}{6e^2}$	2	0.00105
Mod. Euler	$\frac{1}{3} + \frac{1}{2e^2}$	2	0.00079
Heun	$\frac{1}{3} - \frac{1}{6e^2}$	2	0.00090

10.23 No. $\lim_{\omega\to-\infty} 1 + \omega \neq 0$.

10.24 Figure A.6 shows the complex A-stable region (outside the closed region) in the $u + iv$ plane. The negative real expression $1 + \omega + \frac{\omega^2}{2}$ is plotted for comparison.

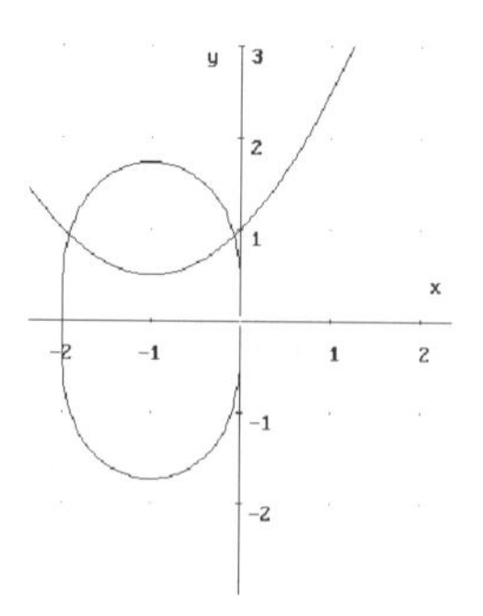

Figure A.6 Complex A-stable region for second-order RK methods.

10.25 Backward Euler method:

$$y_{n+1} = y_n + hf_{n+1}.$$

(a) $y_{n+1} = (y_n + h(x_{n+1} + 1))/(1 + h)$.

```
iterates([x+h,(y+h(x+h +1))/
(1+h)],[x,y],[0,1],h,n).
```

h	n	y_n Euler	y_n Backward Euler
0.1	10	1.34867	1.38554
0.05	20	1.35848	1.37688
0.025	40	1.36323	1.37243

$y(1) = 1.36788$. Backward Euler looks first order!

$p_E = \frac{\log \frac{1.3487-1.3585}{1.3585-1.3632}}{\log 2} \approx 1.060$ (agrees with theory).

$p_B = \frac{\log \frac{1.3855-1.3769}{1.3769-1.3724}}{\log 2} \approx 0.934$ (agrees with conjecture).

(b)
$$\text{LTE} = \frac{y(x+h) - y(x)}{h} - f(x+h, y(x+h))$$

$$= \frac{1}{h}\left[y + hy' + \cdots - y\right] - \left[f + hf_x + \left(hy' + \frac{h^2}{2}y'' + \cdots\right)f_y + O(h^2)\right]$$

$$= -\frac{h}{2}y''(\xi).$$

Consistent of order 1. Note change of sign cf. Euler's method – what might this mean? Take $N = \frac{1}{2}$ $\max|y''|$. Lipschitz $L_\phi = L$.

(c) Test problem gives $y_{n+1} = y_n/(1-\omega) = p(\omega)y_n$.

$$-1 < p(\omega) < 1$$

$$\Rightarrow -1+\omega < 1 < 1-\omega.$$

LHI $\omega < 2$, RHI $\omega < 0$ – both satisfied for $h > 0$ ($\lambda < 0$). Hence A-stable. $\lim_{\omega\to-=\infty} p(\omega) = 0$, so L-stable.

Complex λ: Let $p(\omega) = \mathrm{e}^{\mathrm{i}\theta}$.

$$\mathrm{e}^{\mathrm{i}\theta} = \frac{1}{1-\omega} \Rightarrow 1-\omega = \mathrm{e}^{\mathrm{i}\theta}$$

$$\Rightarrow 1-u-\mathrm{i}v = \cos\theta - \mathrm{i}\sin\theta.$$

$u = 1-\cos\theta$, $v = \sin\theta$ – a circle $(1-u)^2 + v^2 = 1$ of radius 1 and centre $(1,0)$. At $(1,0)$, $p = \frac{1}{0}$ so unstable inside circle. A plot of the A-stable region is shown in Figure A.7(a).

10.26 The $\mathcal{DERIVE}$ plot is shown in Figure A.7(b). For $s < 1$ the solution decays to zero. For $s = 1$ the solution is constant, $y(x) = 1$. For $s > 1$ the solution increases to a vertical asymptote at $x = s$.

10.27 Author `e(h,n):=iterates([x+h, y+hy^2],[x,y],[0,1],n)`. Now Author approX and Plot `e(.1,12)`, and Author and Plot `1/(1-x)` (exact soln.). Euler's method does not 'see' the asymptote and approximates infinite slope at $x = 1$ by finite value – it integrates across the singularity. Hence horrible numerical solution for $x > 1$ (see Figure A.8). This harks back to 'existence and uniqueness'. You can do this 'experiment' with a calculator!

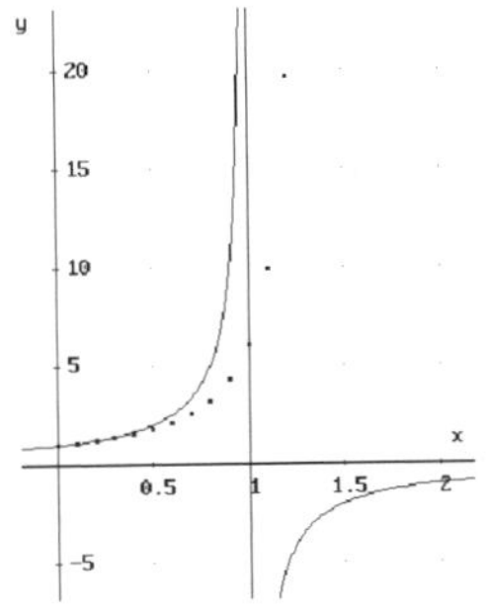

Figure A.8 'Existence and uniqueness' *not* satisfied.

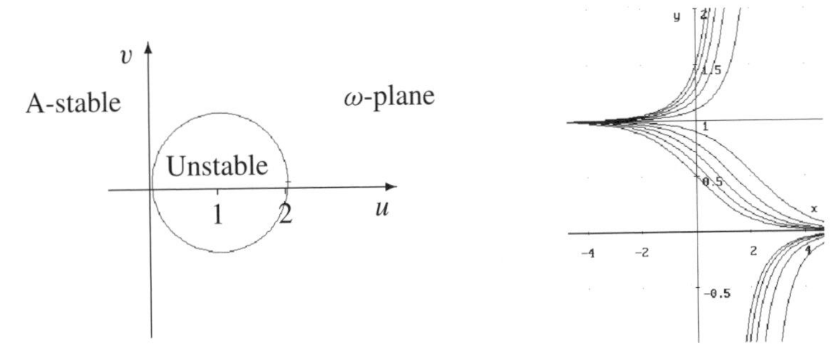

Figure A.7 (a) A-stable region for the backward Euler method. (b) The solution to an ill-conditioned IVP.

B List of Derive Experiments

Index